William Hreha

INTRODUCTION TO
COMMUNICATIONS ENGINEERING

INTRODUCTION TO
COMMUNICATIONS ENGINEERING

ROBERT GAGLIARDI

Department of Electrical Engineering
University of Southern California

A Wiley-Interscience Publication
JOHN WILEY & SONS
New York • Chichester • Brisbane • Toronto • Singapore

Library of Congress Cataloging in Publication Data

Gagliardi, Robert M., 1934–
    Introduction to communications engineering.

    "A Wiley-Interscience publication."
    Includes index.
    1. Telecommunication. I. Title.
TK5101.G24   621.38   77-18531
ISBN 0-471-03099-6

Printed in the United States of America

10 9 8 7 6 5 4 3

To my mother and father
LOUISE and MICHAEL GAGLIARDI

PREFACE

This is a college engineering textbook devoted to the analysis and design of operating communication systems. My objective is to concentrate on the practical system considerations in the end-to-end construction of a typical communication link, allowing a reader to enter the present-day engineering field with some knowledge of the terminology, methodology, and procedures presently being used in the design of modern communication systems. At the same time I wish to introduce to the reader the manner in which more theoretical topics, such as detection theory, information and coding theory, and filter theory, influence and greatly aid design procedures. Although rigorous pursuit of these topics is outside the scope of this book, I hope to motivate and guide further study in these areas by demonstrating their application to communications engineering.

The format of the book is that of a classroom text. Emphasis is placed on deriving design equations of various subsystems, interpreting their implications, and carrying out trade-off studies for system comparisons. In the process the student is introduced to familiar communication systems (telephone, television, teletype, and satellites) through examples and diagrams. The material is arranged to aid an instructor in moving through the topics in logical order, beginning with basic system models, and developing and interfacing the various communication subsystems. Homework problems have been carefully selected to review the text material and give the student practice in design procedures. The early chapters, where material for the novice student is concentrated, have an abundance of homework problems, most of which can be classified as straightforward, ranging in difficulty from easy to average. The chapters in the second half of the book, geared for a more advanced student, cover material at a slightly higher mathematical level with fewer but more challenging problems.

In terms of electrical engineering curricula, the book falls between basic undergraduate signal electronics and the more theoretical graduate topics already stated. The text begins at a basic introductory level suitable for college seniors, yet is wide enough in topic coverage to be usable on a graduate level. Prerequisites for the book are courses in electronics, electromagnetic theory, and linear system theory, all core courses in an electrical engineering curriculum. Chapter 1 is intended to establish the necessary background material for the remainder of the book. A previous course in probability would be advantageous to the student, but is not necessary for one guided by an instructor. The important aspects of noise theory required here are covered as simply as possible in the first chapter. An appendix on probability and random noise is provided for the instructor who wishes to digress further into this area. This appendix can also serve as a convenient review or as brush-up reading for the student having some background in probability theory.

For an undergraduate text Chapters 1 through 5, and parts of 6 and 7, provide a useful, self-contained introduction to the communication engineering field for both the terminating senior and the future graduate student. The classroom instructor must provide the proper match between the rate of topic coverage and the student's background. On a graduate level Chapters 1 through 7 can be easily covered in a semester, and would serve well as an introductory course for a modern graduate communication curriculum. Such a course would give students an overall view of all aspects of communication system design (signal processing, modulation, analog and digital systems, coding, synchronization, etc.). The more theoretical texts can then be used for specialized advanced courses in these areas. A course of this type has been used successfully at USC in a graduate introductory capacity. The remainder of the text can then be used as a second semester or follow-on type of course. In addition, Chapters 6 through 9 alone form a fairly comprehensive one-semester course devoted to modern digital communication system design. The advantage of these chapters is that they concentrate on the entire digital transmission operation (quantization, digitizing, encoding, decoding, and synchronization). The book is also tailored for industrial in-house courses and for one- or two-week short courses covering these aspects of communications engineering.

A book of this type, of course, could not have been compiled without the aid and influence of my friends and colleagues in the communications engineering field. The suggestions (and corrections) of my classroom students helped considerably in structuring the manuscript. I have also benefited immensely from the many technical discussions with graduate students and fellow faculty members at USC, especially Professors W. C. Lindsey, C. Weber, R. Scholtz, I. Reed, S. Golomb, and L. Welch. Most of the serious writing of the book was accomplished while on sabbatical leave at the University of Hawaii, and I wish to acknowledge Professors E. Weldon, Jr., M. Ferguson, and P. Weaver for their cooperation in helping to provide a proper writing atmosphere. Over the years I have also profited from my consulting experiences, which allowed me an opportunity to interact with the staff members of Hughes Aircraft Co., TRW, ITT-Gilfallon, JPL, Lincom, NASA, and other government agencies and institutions. I also received an extensive and beneficial review of my original manuscript from Dr. P. T. McGarty of Comsat Corp., which I greatly appreciated. Lastly, I wish to thank Ms. Corrine Leslie and Ms. Mary Gagliardi for pounding the typewriter keys over the years to keep my manuscript up-to-date.

ROBERT M. GAGLIARDI

Los Angeles, California
January 1978

CONTENTS

INTRODUCTION TO
COMMUNICATIONS ENGINEERING

CHAPTER 1

COMMUNICATION SYSTEM MODELS

The principal objective of a communication system is to transmit *information signals* (waveforms) from one point to another. The information signals may be the result of a voice message, a television picture or a meter reading or may take on a variety of other formats depending on the specific application. The points between which the information is to be sent may be located in close proximity or may be continents, or even planets, apart. *Communications engineering* involves the analysis, design, and fabrication of an operating system that satisfactorily performs the communication objective. In this book we are concerned with the development and application of the basic analytical tools and system alternatives necessary to carry out the communications engineering tasks.

The required transfer of information is most often accomplished by *modulating* (superimposing) the desired information signal onto a *carrier* (sine wave) and converting the carrier to an electromagnetic field. This field is then *transmitted* (propagated) to the desired destination, where it is *received* (intercepted) and the information signal *demodulated* (recovered). The design objective is to construct a suitable system that accomplishes these operations with minimal *distortion* (perturbations) of the information signal itself. To produce such a design, it is necessary to understand the basic functional aspects of each component of the system model, along with their various characteristics and properties that influence the eventual performance.

1.1. Communication Systems

A communication system can be divided into three primary parts, as shown in Figure 1.1. The *transmitter* represents the part of the system where the information signal is generated, modulated, and transmitted. Each of these operations is not necessarily physically located at the same place, so that the transmitter block does not always represent a single, self-contained, unit. The *receiver* portion performs the necessary operation for intercepting the transmitted carrier and recovering the information signal at the desired destination. It also need not represent a single, self-contained unit. The *channel* represents the propagating media or the electromagnetic path interconnecting the transmitter and receiver. It is from this transmission channel that various anomalies and interference effects enter the system operation. Thus, although the design and fabrication of the transmitter and receiver portions are, for the most part, entirely in the hands of the communication engineer, it is often the properties of the channel that ultimately influence and dictate such design procedures. As we shall see, knowledge of these channel properties is mandatory for successful engineering. The combination of transmitter, channel, and receiver is referred to as a communication *link*. Any part of a communication system is called a *subsystem,* and the overall design is accomplished by properly interfacing the individual subsystems.

The transmitter block can be further subdivided into the subsystems shown in Figure 1.2a. The information *source*, whatever its form, generates the electronic signal to be sent to the receiver. Information sources are classified as either *analog* or *digital* sources. An analog source produces a time continuous electronic waveform as its output. A digital source produces sequences of data symbols for its output. In our initial discussion we consider primarily analog sources, deferring detailed discussions of digital sources until Chapter 6. Although Figure 1.2a shows only a single source output, we may in fact have several different outputs from a single source, and perhaps even more than one distinct source. In such cases the communication system must be capable of handling simultaneously the complete set of source signals. When more than a single source signal is involved, the system is referred to as a *multiple source* system.

Figure 1.1.   The basic communication system.

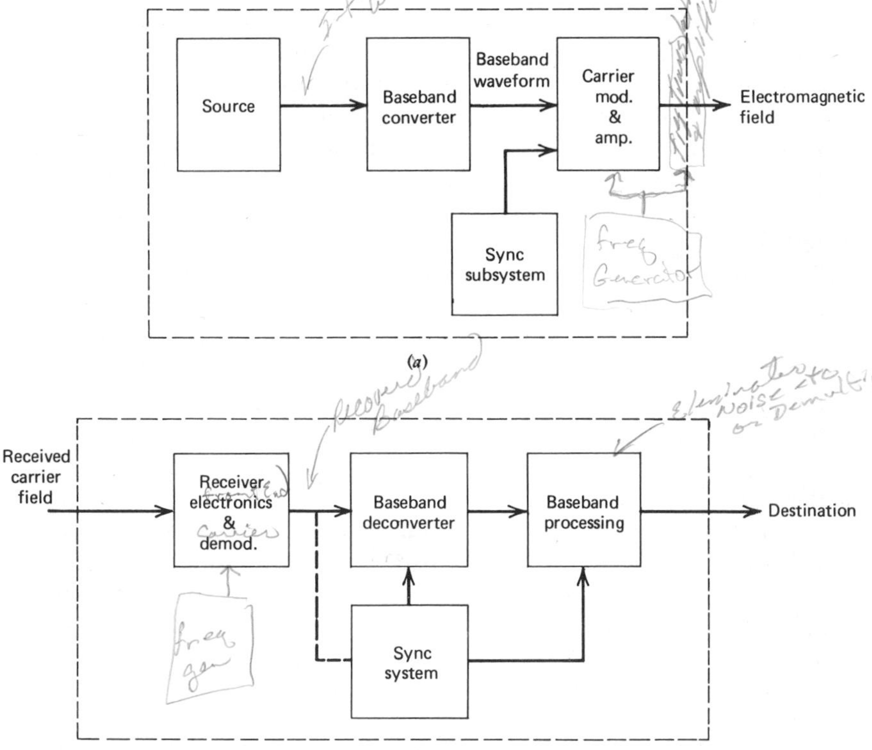

Figure 1.2. The communication system block diagram. (a) The transmitter, (b) the receiver.

The source signals are generally processed in a *baseband converter*. This conversion operation converts the source output into a *baseband* waveform to prepare it for the carrier transmission. The principal objective of this baseband conversion is to insert suitable control and proper formulation of the source output prior to carrier modulation. In some simple systems baseband conversion is not used, and the analog source output is used directly for carrier modulation. In other systems the conversion operation may take on a variety of forms. For example, it may involve only filtering of the source waveform, or it may involve a separate modulation of the source signal onto a secondary carrier (called a *subcarrier*) before modulating onto the main carrier. For digital sources, baseband conversion is used to convert the source symbols into baseband waveforms for transmission, and the conversion operation is called *encoding*. In a multiple source environment the converter involves additionally a combining of all the various source outputs into a single baseband waveform, the latter conversion called *multiplexing*. The selection of the proper baseband conversion operation so as to achieve satisfactory communication becomes an important aspect of transmitter design. In later discussions the various types and alternatives for the conversion operation are examined in detail.

The baseband waveform is modulated onto the carrier to form the carrier waveform. The basic objective of performing a modulation is to generate a transmission waveform that is best suited for the electromagnetic propagation and communication reception. These advantages are made more obvious in subsequent chapters. Here again the modulation may take on various forms, each with its own advantages and disadvantages relative to the channel and receiver. Modulation formats are important in determining the extent of the channel interference and combating the associated distortion. Also, the modulation method must interface properly with the baseband conversion operation and the basic parameters of the receiving system.

In addition to the subsystems devoted to the processing of the source signal, it may also be necessary to provide auxiliary transmitter waveforms that aid in the transmission and recovery of the desired information. These auxiliary waveforms can be classified as transmitter *synchronization* signals whose basic objective is to keep the link operationally aligned and properly interfaced with the transmitter. These synchronization signals can take on a variety of forms and may be superimposed on the source, baseband, or carrier waveforms. In the sophisticated system characterizing modern design, the ability to maintain adequate synchronization and alignment must be a primary consideration and often becomes the ultimate limit to attainable performance.

At the transmitter the modulated carrier waveform generally is amplified to a desired level and used to generate the electromagnetic field for channel propagation. Such fields are characterized by their *polarization* (the orientation of their electric field in space) and their transmitted energy distribution in the various directions of propagation. Electromagnetic transmission channels can be roughly divided into *guided* and *unguided* channels. In an unguided channel the transmitter propagates the field formed from the carrier freely in a medium with no attempt to control its propagation other than by its field propagation pattern. The prime example of an unguided channel is the so-called space channel, in which the medium involved may be free space, the atmosphere, or the ocean (under water). In a guided channel a waveguide is used to confine the field propagation from transmitter to receiver. The prime example of a guided system is the cable, or hard-wire, channel. In Chapter 2 specific properties of these various channels are discussed.

The system receiver can be divided into the subsystems in Figure 1.2*b*. The front end electronics receives and processes the electromagnetic field and attempts to eliminate as much channel interference as possible. The received electromagnetic field is then demodulated to recover the baseband waveform. Channel effects are generally exhibited at this point and cause the recovered baseband no longer to represent the transmitter baseband identically. The recovered baseband is then deconverted to produce the recovered information signal (or signals, in the multiple source case) for the final destination. In general, this deconversion simply attempts to invert the operation performed at the transmitter. It may, for example, involve baseband filtering, subcarrier demodulation, *decoding* (reconverting the baseband back into digital symbols), or possibly *demultiplexing* (splitting the baseband into its various source waveforms). Achievement of better system performance by improved receiver design is a topic that has received considerable attention over the past several decades, in terms of both theoretical analysis and hardware fabrication. Much of this effort is concentrated on more sophisticated receiver baseband processing, and the selection of the appropriate processing alternative becomes a vital part of successful system engineering.

If synchronization signals were provided at the transmitter, they must also be recovered and properly used in the receiver processing. In this case a *synchronization subsystem* must be included within the receiver to separate out the synchronization waveform and to reinsert it into the correct receiver subsystem (Figure 1.2*b*). This synchronization information may be used in the carrier, in the baseband, or in the eventual application of the recovered source signal at the destination. Chapter 9 is devoted to some common synchronization subsystems and their associated design.

The communication system models depicted in Figures 1.1 and 1.2 are simply generalized forms of some of the basic communication systems with which we are personally familiar. Three of the most common examples in our everyday life are the telephone, television, and teletype systems, sketched in Figure 1.3. In the telephone system the speaker's voice is converted to an electrical signal directly in the hand-held phone set, which therefore acts as an analog source for the link. The voice signal is then sent by wire to a central telephone exchange. Here it is carrier modulated and relayed via a complicated switching procedure to the central exchange of the intended listener. The latter exchange is determined by the dialing code sent by the speaker, which therefore acts as synchronization information interconnecting the speaker and listener. If the listener's exchange is in a nearby locality, the modulated carrier may simply be cable connected to it. If it is at a distant location, electromagnetic space propagation may be used. At the listener's exchange, the voice signal

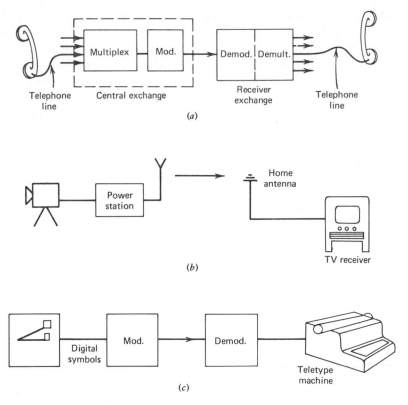

Figure 1.3. Typical communication systems. (*a*) Telephone system, (*b*) television system, (*c*) teletype system.

is demodulated and passed through telephone lines to the listener's phone. Note that in such a system the various subsystems of the transmitter and receiver blocks are separated in their physical location. For long distances a satellite relay may be used. The voice modulated carrier is sent up to a satellite where it is received, amplified, and retransmitted to a receiver exchange at a distant location on the ground.

In commercial television (Figure 1.3*b*) the camera acts as the system analog source, generating the video (television) signal for transmission. Typically, this signal is relayed via cable to a transmitting power station invariably positioned in some electromagnetically strategic point in a high location. Here the carrier is formed and the resulting field is broadcast, either to the immediate locality or perhaps by satellite link to a distant location. For reception the electromagnetic field is intercepted by our rooftop antennas and fed into our receiving subsystem—the home television set. Within the receiver the carrier is demodulated and processed, including voice and picture separation and synchronization alignment. Note that although the source and transmitter are physically separated, in the television system the receiving subsystem is self-contained at a single location.

In teletype systems (Figure 1.3c) customer messages are encoded into digital symbols by a teletype keyboard operator, producing a synchronized digital source for the system. The symbols are then encoded into modulated carrier signals for long range transmission, either over telephone lines or by electromagnetic propagation. At the receiving station the carrier is demodulated, decoded back into symbols, and used to manipulate a synchronized teletype machine for final message printout. In subsequent chapters we examine further the specific details of these systems, while presenting other examples and more general techniques and alternatives for communication operations.

The system model shown in Figure 1.2 implies that a single carrier is involved, which indeed represents the most common situation. In discussing more advanced modern communication systems, however, we find it necessary to extend this basic model to include the case where more than one simultaneous carrier is involved in the transmitting operation. Such a system is called a *multiple carrier* system, and its analysis requires modification and extension of many of the results to be derived for the single carrier case.

1.2. The Electromagnetic Frequency Spectrum

Carrier modulation involves the embedding of baseband or information waveforms onto a carrier sine wave of a fixed frequency. A carrier sine wave has the mathematical form

$$A \cos (2\pi f t + \psi) \tag{1.2.1}$$

where A is the carrier *amplitude*, f is the carrier *frequency* in hertz (Hz), and ψ is its *phase angle* in radians. The modulated carrier is formed from this sine wave and becomes the time variation of the electric field component of the propagated electromagnetic field. The electromagnetic carrier is often labeled by the location of its frequency f within the electromagnetic frequency spectrum (Figure 1.4). The electromagnetic frequency spectrum is divided into groupings in which the frequency bands are denoted as shown. A carrier whose frequency lies in a given band is labeled according to that designation. Hence a carrier selected from the RF (radio frequency) portions of the spectrum is called an RF carrier, one from

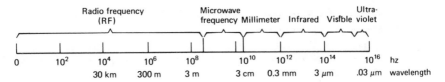

Figure 1.4. The electromagnetic frequency spectrum.

the optical region is an optical carrier, and so on. For high frequencies we find it more convenient to deal with carrier wavelength than with its frequency. The carrier wavelength λ is obtained from the frequency f in hertz by

$$\lambda = \frac{3 \times 10^8}{f} \quad \text{meters} \tag{1.2.2}$$

Thus millimeter wavelength carriers are carriers whose frequencies are higher than 3×10^{11} Hz.

Most modern communication systems use RF carriers, for reasons discussed in Chapter 2. However, increasing use of millimeter and optical carriers is expected as future communication demands increase and the associated technology advances. This extensive use of RF carriers has led to a further subdivision of the RF portion of the spectrum into the designations shown in Table 1.1. In addition, the SHF and EHF bands are further divided as shown in Table 1.2. To control the excessive demands for the S band and X band carrier frequencies, specific assignments for frequency usage within the United States have been made. Table 1.3 lists some of the important frequency band restrictions and their designated uses.

Selection of the proper carrier frequency is one of the initial decisions that must be made in system design. This selection is generally based on three main considerations: (1) the ease with which the baseband waveform can be modulated onto a carrier of a given frequency, (2) the dependence of system hardware (component size and weight) on carrier frequency, and (3) the dependence of channel characteristics on carrier frequency. Each of these factors must be properly taken into account (as is discussed in Chapter 2) before judicious carrier frequency selections can be made. It is precisely these factors that led to the frequency designations of Table 1.3, although further frequency selection must usually be made within a given category.

A basic property of a modulated carrier waveform is the limitation of the amount of information that can be physically modulated onto a carrier of a given frequency. The conditions involved are developed later; it suffices to

Table 1.1.   Frequency Designations

Frequency	Wavelength	Designation
3–30 MHz	100–10 m	HF
30–300 MHz	10–1 m	VHF
300–3000 MHz	1–0.1 m	UHF
(3 GHz)		
3–30 GHz	100–10 mm	SHF
30–300 GHz	10–1 mm	EHF

Table 1.2.   SHF and EHF Bands

S band	1.5–5.2 GHz
X band	5.2–10.9 GHz
K band	10.9–36 GHz
Q band	36–46 GHz
V band	46–56 GHz

Table 1.3.   Assigned Frequency for United States

Application	Frequency Band
Commercial AM radio	0.53–1.6 MHz
Amateur radio and citizen's band	1.6–10
Mobile radio	10–20
Citizen band radio (40 channels)	26.9–27.5
VHF commercial television	54–88
Commercial FM radio	88–108
VHF commercial television	178–216
UHF commercial television	470–890
Military	1.7–1.85 GHz
Studio transmitter link	1.9–2.1
Commercial satellite (downlink)	3.7–4.2
Military	4.4–5.0
Commercial satellite (uplink)	5.9–6.4
Military	7.1–8.4
Community antenna television (CATV)	12.7–12.9
Military	14.4–15.2

point out that the maximum modulating waveform frequency that can be superimposed on a carrier of a fixed frequency is often limited to a certain percent of that carrier frequency. Typical values are 1–10%, depending on the modulaton technique. We immediately see a basic advantage of increasing the carrier frequency being used—higher frequency modulating waveforms can be transmitted. For example, a carrier from the millimeter frequency range can incorporate modulating frequencies 1000 times larger than a carrier in the RF range, and optical carrier can handle frequencies 10^5 times as large. The ability to transmit higher modulating frequencies (i.e., more baseband information) is a prevailing property demanded of modern systems and is the force that continually drives engineers to extend technology into the millimeter and optical carrier frequencies [1]. Design of electronic components and devices for such systems, however, with their accompanying small wavelengths and component limitations, often becomes a major hurdle in this advancement.

1.3. Waveforms and Signals

It is obvious from our basic system model that communication engineering requires us to deal extensively with time waveforms. In general, time waveforms may be classified as either *deterministic* or *random*. A deterministic waveform is one whose value at all time instants is known precisely. Examples of deterministic waveforms are sine waves or square waves of known amplitude, phase, and frequency, exponential functions of known amplitude and decay rate, and pulses of fixed height and width. A random waveform is one whose unobserved values at any instant in time can only be described statistically (i.e., treated as a random variable). Random waveforms are usually referred to as random, or *stochastic*, processes, and model the noise and interference waveforms encountered in system operation. In addition, analog source waveforms are in fact random and noiselike in nature, although their basic properties are often difficult to measure and characterize. For this reason it is sometimes analytically convenient to model source outputs as deterministic waveforms in order to gain some initial insight into design and performance. Care must be used in extending results derived under these conditions to actual operating systems. It is also common to deal with a waveform of known waveshape but containing a random parameter. Such a waveform is actually a random process, requiring statistical description, even though it may not be completely noiselike in appearance.

In dealing with deterministic waveforms it is convenient to introduce the *Fourier transform*. If $x(t)$ is the time waveform involved, its Fourier transform is defined as

$$X(\omega) = \int_{-\infty}^{\infty} x(t) e^{-j\omega t} \, dt \qquad (1.3.1)$$

where ω is the radian frequency variable in radians per second (rps). The radian frequency is related to the frequency f in hertz by

$$\omega = 2\pi f \qquad (1.3.2)$$

Note that the Fourier transform $X(\omega)$ is a complex function (i.e., has a real and imaginary part) in the variable ω. It is well known that $x(t)$ and $X(\omega)$ form a transform pair and each can be uniquely derived from the other. The *inverse* Fourier transform is given by

$$x(t) = \frac{1}{2\pi} \int_{-\infty}^{\infty} X(\omega) e^{j\omega t} \, d\omega \qquad (1.3.3)$$

A tabulation of some useful transform pairs and some basic identities and properties of the Fourier transform are summarized in Appendix A. More extensive tabulations are available [2–4].

The important frequency characteristics of a deterministic waveform are contained in its Fourier transform function. The complex transform function $X(\omega)$ can be written in polar coordinates as

$$X(\omega) = |X(\omega)| \exp [j\angle X(\omega)] \qquad (1.3.4)$$

The function $|X(\omega)|$ is called the transform *amplitude* function, and $\angle X(\omega)$ is the *phase* function. Deterministic waveforms are therefore uniquely described by their amplitude and phase functions. It is easy to show (Problem 1.3) that since $x(t)$ is a real waveform its amplitude function is always even in ω [i.e., $|X(-\omega)| = |X(\omega)|$] and its phase is ways odd [$(\angle X(-\omega) = -\angle X(\omega)$]. Thus these functions need only be described by their positive frequency behavior. Frequency functions defined over the positive frequencies only are called *one-sided* functions.

Transform amplitude functions associated with typical communication waveforms are generally of a form in which the functions are concentrated over a particular frequency range and tend to roll off (decrease in magnitude) at other frequencies (Figure 1.5). If this concentration along the positive frequency axis begins at zero frequency and extends to some higher frequency before roll-off begins, the waveform is said to be a *low pass* waveform. If the concentration is centered along the positive frequency axis, and rolls off toward zero and higher frequencies, the waveform is said to be a *bandpass waveform*. As an example, consider the

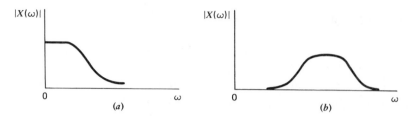

Figure 1.5. Transform amplitude functions. (*a*) Low pass waveform, (*b*) bandpass waveform.

waveform

$$x(t) = A, \qquad o \leq t \leq \tau$$

$$= 0, \qquad \text{elsewhere} \tag{1.3.5}$$

corresponding to a voltage pulse (Figure 1.6*a*). The Fourier transform is given by (Section A.1)

$$X(\omega) = A\left[\frac{1 - e^{-j\omega\tau}}{j\omega}\right] \tag{1.3.6}$$

The amplitude function $|X(\omega)|$ is then

$$|X(\omega)| = A\frac{|1 - \cos \omega\tau - j \sin \omega\tau|}{|j\omega|}$$

$$= A\frac{[(1 - \cos \omega\tau)^2 + \sin^2 \omega\tau]^{1/2}}{\omega}$$

$$= A\tau\left|\frac{\sin (\omega\tau/2)}{(\omega\tau/2)}\right| \tag{1.3.7}$$

and is sketched in Figure 1.6*b*. Thus the pulse waveform has a low pass frequency function extended over the entire frequency axis but primarily concentrated in the range $|f| \leq 1/\tau$. This means that as the pulse is made narrower in time, its frequency function is spread in frequency, and an inverse relation exists between pulse time width and frequency width. This inverse time-frequency behavior is characteristic of transform pairs in general (Problem 1.6).

As a second example, consider the sine wave of (1.2.1), Figure 1.6*c*,

$$x(t) = A \cos (\omega_c t + \psi) \tag{1.3.8}$$

This has the frequency function

$$X(\omega) = \pi A\, e^{j\psi}\delta(\omega - \omega_c) + \pi A\, e^{-j\psi}\delta(\omega + \omega_c) \tag{1.3.9}$$

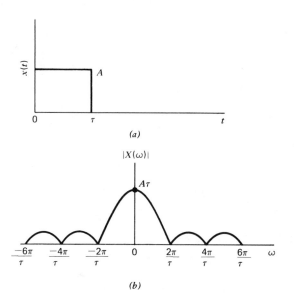

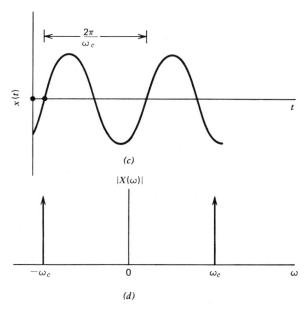

Figure 1.6. Examples of waveform and transforms. (*a*) Pulse waveform, (*b*) pulse transform, (*c*) sine wave, (*d*) sine wave transform.

13

where $\delta(\omega - \omega_c)$ is the *delta*, or *impulse, function* located at ω_c. The amplitude function for this transform is shown in Figure 1.6*d*. Thus a sine wave at frequency ω_c has its frequency function concentrated at the points $\pm\omega_c$ along the frequency axis. Conversely, we can say that delta functions in the frequency domain imply purely sinusoidal components in the corresponding time waveform. Delta functions along the frequency axis are often called frequency *lines*.

In order to discuss noise and random signals in communication systems, we are forced to deal with random processes. Although a rigorous development of the theory of random analysis is well beyond our scope here,* we should point out several basic definitions that continue to arise in a later analysis. Random waveforms can only be described statistically in time. That is, at each point in time t, an unobserved value of the waveform is a random variable described by its probability density at that t. From this density, probabilities, moments, and averages of the process at that t can be determined. For example, if $x(t)$ is a random process, then at $t = t_1$, $x(t_1)$ is a random variable having probability density $p_{x_1}(\xi)$, which statistically describes the process at t_1 (Figure 1.7). The probability that the process $x(t)$ will have a value in the range (a, b) at time t_1 is then given by

$$\text{Prob}\,[a \le x(t_1) \le b] = \int_a^b p_{x_1}(\xi)\,d\xi \qquad (1.3.10)$$

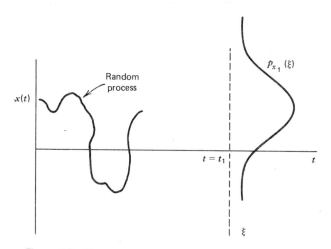

Figure 1.7.   Random process and its probability density.

* A short review of probability, random variables, and random processes is presented in Appendix B. The reader may wish to review this appendix before continuing.

Similarly, the mean and mean squared value (first two moments) of the process $x(t)$ at t_1 are given by

$$\text{mean of } x(t_1) = \mathscr{E}[x(t_1)] = \int_{-\infty}^{\infty} \xi p_{x_1}(\xi)\, d\xi \qquad (1.3.11a)$$

$$\text{mean squared of } x(t_1) = \mathscr{E}[x^2(t_1)] = \int_{-\infty}^{\infty} \xi^2 p_{x_1}(\xi)\, d\xi \qquad (1.3.11b)$$

Here we have introduced the expectation operator $\mathscr{E}[x(t_1)]$, which represents the statistical average of the random variable $x(t_1)$ over its density. Any other average or moment at t_1 can be computed in a similar manner. In general, the probability density of a random process depends on t, and $p_{x_1}(\xi)$ will be different for different points t_1. This means that the probabilities and moments in (1.3.10) and (1.3.11) will be a function of t_1. However, if the process is *stationary*,* then $p_{x_1}(\xi)$ is the same for any point t_1, and the statistical description of the process is the same at all points in time.

Two points in time, t_1 and t_2, define a pair of random variables, $x(t_1)$ and $x(t_2)$, statistically described by their joint second order density $p_{x_1 x_2}(\xi_1, \xi_2)$. The probability that the process $x(t)$ lies in the interval (a, b) at time t_1 and lies in the interval (c, d) at time t_2 is then given by

$$\text{Prob}\,[a \le x(t_1) \le b \quad \text{and} \quad c \le x(t_2) \le d] = \int_a^b \int_c^d p_{x_1 x_2}(\xi_1, \xi_2)\, d\xi_2\, d\xi_1$$

$$(1.3.12)$$

Likewise, a sequence of time points $(t_1, t_2, \ldots, t_n)$ defines a sequence of random variables $x(t_1), x(t_2), \ldots, x(t_n)$, described by their joint nth order probability density. A complete description of a random process requires knowledge of all such joint densities for all possible point locations and orders. The most basic type of noise process encountered in communications is the *Gaussian process*, in which $x(t)$ at any t is a Gaussian random variable. Gaussian processes have the basic property that all joint densities of any order are described by Gaussian probability densities (Section B.5). The modeling of noise and signal processes as Gaussian processes in communication analysis is quite prevalent, and fortunately, in most cases, this assumption can be physically justified as well as being mathematically convenient [5, 6].

---

* There are various degrees of stationarity that can be formally defined for a random process (see Section B.3). For simplicity in discussion we do not distinguish precisely between the specific forms of stationarity in subsequent analysis. Noise processes typically encountered in communication systems are generally considered stationary processes.

Rather than deal with probability densities in analysis, we often choose to work with simpler descriptions of random processes. One simple measure of the degree of randomness of a process $x(t)$ is indicated by the process *autocorrelation* function

$$R_x(t, \tau) = \mathscr{E}[x(t)x(t+\tau)] \qquad (1.3.13)$$

where $\mathscr{E}$ now denotes a statistical average over the joint density of the process $x(t)$ at time t and $t+\tau$. Note that $R_x(t, \tau)$ is simply the correlation moment of the pair of random variables $x(t)$ and $x(t+\tau)$. For stationary processes the autocorrelation in (1.3.13) does not depend on t, but only on the time separation τ between the points. That is, all pairs of values of $x(t)$ at points in time separated by τ sec have the same correlation value. We can therfore denote this stationary correlation function as simply $R_x(\tau)$. It can be easily shown (see Appendix B, Section B.4) that stationary autocorrelation functions $R_x(\tau)$ are always real, even in τ [i.e., $R_x(-\tau) = R_x(\tau)$], and always take on their maximum value at $\tau = 0$. For zero mean processes $R_x(\tau)$ indicates the extent to which the random values of the process separated by τ sec in time are statistically correlated. If $R_x(\tau)$ is highly positive at a particular τ, it implies that values of the process separated by this τ are positively correlated. This means that if the process takes on a positive (negative) value at some point in time, then the value of the process τ sec away will, with a high probability, also be positive (negative). Similarly, if $R_x(\tau)$ is negative at a particular τ, it implies points separated by this τ are negatively correlated, and therefore are most likely of opposite sign. If $R_x(\tau) = 0$ the points involved are uncorrelated, and no statistical inference about one can be inferred from the other. Thus autocorrelation functions present a descriptive indication of the statistical relation between points of the process. A stationary Gaussian process has the further advantage that all its joint densities (and therefore a complete statistical description) are derivable solely from its autocorrelation function.

The frequency characteristics of a stationary random process $x(t)$ are exhibited by its *spectral density*, $S_x(\omega)$, defined as the Fourier transform of the process autocorrelation function

$$S_x(\omega) = \int_{-\infty}^{\infty} R_x(\tau) e^{-j\omega\tau} \, d\tau \qquad (1.3.14)$$

where ω is again the frequency variable in rps. Similarly, by Fourier inversion

$$R_x(\tau) = \frac{1}{2\pi} \int_{-\infty}^{\infty} S_x(\omega) e^{j\omega\tau} \, d\omega \qquad (1.3.15)$$

Thus the correlation function and spectral density of a stationary random process form a transform pair, and each is uniquely derivable from each other. These two functions are two of the more important descriptive functions of noise processes and are used extensively in communication analysis. We point out that since $R_x(\tau)$ is both real and even in τ, the spectral density $S_x(\omega)$ in (1.3.14) must necessarily be real and even in ω (Problem 1.9).

Noise processes are often given special names based on the characteristics of its spectral density. If

$$S_x(\omega) = S_0 \qquad (1.3.16)$$

the process is referred to as a *white noise process*, and the parameter S_0 is called the *spectral level* of the process (Figure 1.8a). A white noise process has an autocorrelation function given by the transform of (1.3.16). From Section A.1 this transform is

$$R_x(\tau) = S_0 \delta(\tau) \qquad (1.3.17)$$

where $\delta(\tau)$ is the delta function, as shown in Figure 1.8b. Note that this correlation function indicates that any point on the process is uncorrelated with any other point of the process. For this reason white noise processes are considered to be the ultimate in randomness. If

$$S_x(\omega) = S_0, \qquad |\omega| \le 2\pi B$$

$$= 0, \qquad \text{elsewhere} \qquad (1.3.18)$$

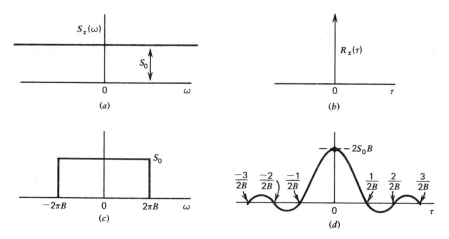

Figure 1.8. Examples of spectral densities and autocorrelation functions. (a) White noise spectrum, (b) white noise autocorrelation, (c) bandlimited white noise spectrum, (d) bandlimited white noise autocorrelation.

the process is called a *bandlimited white noise process* (Figure 1.8c). Such a process has the corresponding autocorrelation function

$$R_x(\tau) = 2BS_0 \frac{\sin(2\pi B\tau)}{2\pi B\tau} \tag{1.3.19}$$

which is shown in Figure 1.8d. Note that the autocorrelation oscillates between positive and negative correlation values, and decays asymptotically to zero as $\tau \to \infty$. As an approximation we can consider $R_x(\tau)$ in (1.3.19) to be approximately zero after several multiples of $1/B$ sec. Hence a bandlimited white noise process, having a spectral extent of B Hz, will have a correlation extent of several times $1/B$ sec. That is, values of the process separated by this correlation extent are approximately uncorrelated. We make use of this property in later studies of receiver noise.

Two different random processes, $x(t)$ and $y(t)$, are described by the joint probability densities between their values at specific points in time. The *cross-correlation* of the two processes can be similarly defined as

$$R_{xy}(t, \tau) = \mathscr{E}[x(t)y(t+\tau)] \tag{1.3.20}$$

The cross-correlation function therefore indicates the statistical correlation between the value of one process at time t and the other at time $t + \tau$. If the processes are jointly stationary then (1.3.20) does not depend on t, and we can simply write $R_{xy}(\tau)$. Note that the ordering of the indexes is now important (i.e., which is written first), and the function $R_{xy}(\tau)$ is not the same as $R_{yx}(\tau)$ (Problem 1.10). If the processes are zero mean and if $R_{xy}(\tau) = 0$ at $\tau = 0$, values of the two processes are uncorrelated at the same t. If $R_{xy}(\tau) = 0$ for all τ, the two processes are everywhere uncorrelated and are referred to as a pair of *uncorrelated random processes*. Two processes generated from two completely independent sources are always uncorrelated, but the converse is not always true. Again $R_{xy}(\tau)$ serves as an indication of the relative randomness between the two processes. The Fourier transform of $R_{xy}(\tau)$ can be defined as a *cross-spectral density*, but it is difficult to associate physical meaning to it as a frequency function.

1.4. Power and Bandwidth

Two of the most significant communication parameters associated with a waveform are its *power* and *bandwidth*. The power is an indication of the strength of the signal, whereas the bandwidth indicates its frequency extent. Both of these parameters play a key role in system analysis. For a deterministic waveform $x(t)$ defined over an interval $(0, T)$ the power is given by

$$P_x = \frac{1}{T} \int_0^T x^2(t)\, dt \tag{1.4.1}$$

If the waveform is defined over all time (i.e., it began an infinite time in the past and will continue an infinite time into the future), P_x must be evaluated as a limiting operation as $T \to \infty$. In this case

$$P_x = \lim_{T \to \infty} \frac{1}{2T} \int_{-T}^T x^2(t)\, dt \tag{1.4.2}$$

For an infinitely long periodic waveform of period t_0, we can let $T = nt_0$ in (1.4.2) and interpret the limit as

$$\begin{aligned}
P_x &= \lim_{n \to \infty} \frac{1}{2nt_0} \int_{-nt_0}^{nt_0} x^2(t)\, dt \\
&= \lim \left(\frac{1}{2nt_0}\right) 2n \int_0^{t_0} x^2(t)\, dt \\
&= \frac{1}{t_0} \int_0^{t_0} x^2(t)\, dt
\end{aligned} \tag{1.4.3}$$

Thus the power of a periodic waveform is that occurring in a single period. Since $x(t)$ is assumed to be deterministic, P_x in either (1.4.1), (1.4.2), or (1.4.3) can be evaluated by straightforward integration, and application of the limit, if required. For example, (1.4.1) shows that the pulse in (1.3.5) has a power of $P_x = A^2 \tau / \tau = A^2$, whereas (1.4.2) yields the sine wave power in (1.3.8) as

$$\begin{aligned}
P_x &= \lim_{T \to \infty} \frac{1}{2T} \int_{-T}^T A^2 \cos^2 (\omega_c t + \psi)\, dt \\
&= \lim_{T \to \infty} \frac{1}{2T} \left[\frac{A^2}{2} \left(2T + \frac{2 \sin (2\omega_c T) \cos (2\psi)}{2\omega_c} \right) \right] \\
&= \lim_{T \to \infty} \left[\frac{A^2}{2} + \left(\frac{A^2}{2}\right) \frac{\sin (2\omega_c T) \cos (2\psi)}{2\omega_c T} \right]
\end{aligned} \tag{1.4.4}$$

Since $\sin (2\omega_c T)$ is bounded by one for any T, the limit of the second term is zero, and

$$P_x = \frac{A^2}{2} \tag{1.4.5}$$

Thus the power in a sine wave is always one-half the amplitude squared. This same result could have been derived by treating the sine wave as a periodic function and using (1.4.3). In the laboratory, power in a waveform is measured by a power meter, which essentially performs the operation in (1.4.2).

Note that in the preceding power definitions P_x has units corresponding to the square of the waveform units (volts^2, amp^2, etc.). These units are converted to watts only if $x(t)$ is properly normalized by impedance units. In analyses we find it convenient to deal with power in terms of *decibels*. The power in decibels (dB) is given by

$$(P_x)_{\mathrm{dB}} = 10 \log_{10} P_x \qquad (1.4.6)$$

If P_x has the units of watts, $(P_x)_{\mathrm{db}}$ has the units of decibel-watts, denoted dBW. If P_x has the units of milliwatts, $(P_x)_{\mathrm{dB}}$ has the units of decibel-milliwatts, or dBM. Thus 100 W is equivalent to 20 dBW or 50 dBM. Similarly, 0.01 W is equivalent to −20 dBW or 10 dBM.

For random processes the power of the process at a point t is defined by

$$P_x = \mathscr{E}[x^2(t)] \qquad (1.4.7)$$

which is simply the mean squared value of $x(t)$ at time t. As such, P_x depends on the choice of the point t. If the process is stationary, $\mathscr{E}[x^2(t)]$ does not depend on t, and the power of the process is simply taken to be its statistical second moment. Note that for a stationary process, it follows from a comparison of (1.4.7) to (1.3.13) that

$$P_x = R_x(0) \qquad (1.4.8)$$

Thus the power is also given by the value of the autocorrelation function at $\tau = 0$. Furthermore, from (1.4.8) and (1.3.15) we can also write

$$P_x = \frac{1}{2\pi} \int_{-\infty}^{\infty} S_x(\omega) e^{j\omega\tau}\, d\omega \bigg|_{\tau=0}$$
$$= \frac{1}{2\pi} \int_{-\infty}^{\infty} S_x(\omega)\, d\omega \qquad (1.4.9)$$

Hence the power in the process can also be determined by integrating over the spectral density of the process. For this reason $S_x(\omega)$ is also referred to as the *power spectrum* of the process and indicates the distribution of power over the frequency axis. Note the integration in (1.4.9) is over the entire frequency axis, and $S_x(\omega)$ is referred to as a *two-sided spectral density*. However, since $S_x(\omega)$ is even in ω, we can also write

$$P_x = \frac{2}{2\pi} \int_{0}^{\infty} S_x(\omega)\, d\omega$$
$$= \frac{1}{2\pi} \int_{0}^{\infty} \hat{S}_x(\omega)\, d\omega \qquad (1.4.10)$$

where we have now defined

$$\hat{S}_x(\omega) = 2S_x(\omega), \qquad \omega \geq 0$$
$$= 0, \qquad \omega < 0 \qquad (1.4.11)$$

Here $\hat{S}_x(\omega)$ is referred to as a *one-sided spectral density* and is generated from $S_x(\omega)$ by folding the negative frequency part onto the positive frequency part, as indicated in (1.4.11). Note that the one-sided density is integrated over positive frequencies only to obtain total power. For this reason we must carefully distinguish in later analysis between one-sided and two-sided spectral densities in power calculations. It is important to recognize that the power of a random process can be determined directly from its autocorrelation function, using (1.4.8), or from its spectrum, using (1.4.9) or (1.4.10) without the necessity of specifying the actual statistics of the process, that is, using (1.4.7).

We see from Figure 1.8a that the white noise process has its power spread evenly over all frequencies. This means it has infinite power and is therefore not truly a physically meaningful process. Nevertheless, the simplicity of its form makes it a mathematically convenient function to deal with in system analysis. The bandlimited white noise process (Figure 1.8c) has a finite power of

$$P_x = \frac{1}{2\pi} \int_{-2\pi B}^{2\pi B} S_0 \, d\omega$$
$$= 2S_0 B \qquad (1.4.12)$$

where S_0 is again the two-sided spectral level. Thus the power is proportional to the spectral level and to twice the positive frequency spectral width.

If two jointly stationary processes $x(t)$ and $y(t)$ are summed to form a new process

$$z(t) = x(t) + y(t) \qquad (1.4.13)$$

the autocorrelation function of the sum process is then

$$R_z(\tau) = \mathcal{E}[z(t)z(t+\tau)]$$
$$= \mathcal{E}[(x(t)+y(t))(x(t+\tau)+y(t+\tau))]$$
$$= R_x(\tau) + R_y(\tau) + R_{xy}(\tau) + R_{yx}(\tau) \qquad (1.4.14)$$

The power of the sum processes $z(t)$ then follows as

$$P_z = R_z(0) = P_x + P_y + 2R_{xy}(0) \qquad (1.4.15)$$

If the processes are uncorrelated at the same t (i.e., $R_{xy}(0) = 0$), then the power of the sum is simply the sum of the powers. On the other hand, if the processes are uncorrelated processes, then $R_{xy}(\tau) = R_{yx}(\tau) = 0$, $P_z = P_x + P_y$, and $R_z(\tau) = R_x(\tau) + R_y(\tau)$. That is, the autocorrelation of $z(t)$ is the sum of the individual autocorrelation functions. This also means that the power spectrum of $z(t)$ is the sum of the individual spectra,

$$S_z(\omega) = S_x(\omega) + S_y(\omega) \tag{1.4.16}$$

In later analyses, when dealing with interference composed of the sum of uncorrelated noise processes, (1.4.16) shows that the spectrum of each can be simply summed to obtain the total interference spectrum.

In analyzing deterministic waveforms it is often necessary to deal with the waveform *energy*, defined as the integral portion of the power definition. That is,

$$E_x = \int_{-\infty}^{\infty} x^2(t)\, dt \tag{1.4.17}$$

Note that a waveform with finite energy may have zero power, according to (1.4.2), whereas a periodic waveform must have infinite energy. Thus a waveform in a system may be constrained in either its power or energy values. From Parceval's theorem (Appendix A, Sec. A.2), (1.4.17) can be equivalently written as

$$E_x = \frac{1}{2\pi} \int_{-\infty}^{\infty} |X(\omega)|^2\, d\omega \tag{1.4.18}$$

where $X(\omega)$ is the Fourier transform of $x(t)$. We see that $|X(\omega)|^2$ plays the role of a waveform *energy spectrum* over which we integrate to get total energy. Thus the energy spectrum of a deterministic waveform is related to waveform energy in the same way that a power spectral density of a random process is related to the process power.

Although the power and energy of a waveform can be defined precisely, the notion of a waveform bandwidth can sometimes be vague and misleading. The basic objective for specifying a waveform bandwidth is to indicate the frequency range over which most of the waveform power is distributed. For deterministic waveforms this bandwidth information is contained within the waveform energy spectral function $|X(\omega)|^2$. For random processes it is contained within the power spectral function, $S_x(\omega)$. However, there are many different ways in which the frequency extent may be specified from these frequency functions (see Figure 1.9). The most common definition of bandwidth is the *3 dB bandwidth*. The 3 dB bandwidth is defined as the frequency width along the positive real frequency axis between the frequencies where the frequency function [either $|X(\omega)|^2$ or $S_x(\omega)$] has dropped to one-half of its maximum value. This definition of spectral bandwidth has the advantage that it can be read directly from the

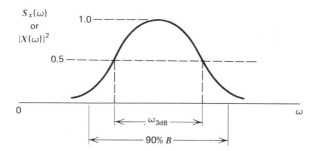

Figure 1.9. Bandwidth definitions.

frequency function, but has the disadvantage that it may be misleading if the function has slowly decreasing tails. Furthermore, it may become ambiguous and nonunique with multiple peak functions. This definition can be extended to a 6 dB bandwidth (width between 25% points) or a 10 dB bandwidth (width between 10% points), and so on.

Another definition of bandwidth is the *90% bandwidth*, defined as the frequency width along the positive axis within which 90% of the one-sided spectrum lies. Note that this bandwidth is always unique, but again may yield misleading values for spectra that have somewhat slowly decaying spectral tails. Although it can sometimes be accurately estimated from the frequency function plot, the 90% bandwidth must be determined by integration for a precise evaluation. Further discussions of the pitfalls that may be encountered with a precise bandwidth definition can be found in Reference 7. Fortunately, most of the waveform spectra encountered in typical communication systems are of a single hump type with relatively fast falloff, much like those in Figure 1.5. For such spectra all the preceding bandwidth definitions yield approximately the same value, and none of these ambiguities occur. Hence it is customary to speak of a general waveform bandwidth without necessarily resorting to a specific definition. We point out that each of the preceding bandwidth definitions are defined along the positive frequency axis only, and always define positive frequency, or *one-sided, bandwidths* only.

The power spectrum of a typical voice (audio) waveform is shown in Figure 1.10a. This would be the spectral density of the voltage waveform observed at the output of a microphone or telephone headset. Voice frequency components may exist out to 20 kHz, but most of the significant voice energy lies in frequency components out to about 4 kHz. This leads to the often quoted 4 kHz audio bandwidth used for voice signals. Figure 1.10b shows the power spectrum of a television (video) waveform (output of a television camera). We see that video waveforms are much wider in

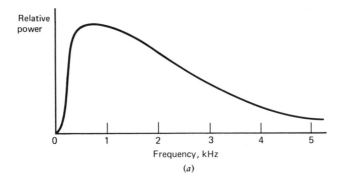

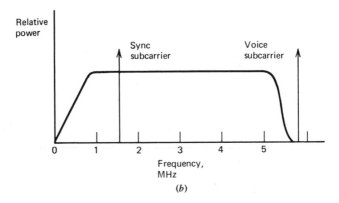

Figure 1.10. Source power spectra. (*a*) Voice, (*b*) television.

frequency extent, occupying frequency components out to about 4.5 MHz. In U.S. commercial television an overall bandwidth of 6 MHz is allowed for transmission of a single television channel, as shown. Note the delta functions within the video spectrum signifying the presence of sinusoidal components used for television synchronization.

1.5. Carrier Waveforms

An important type of waveform encountered in communication systems is the carrier waveform. The *pure*, or unmodulated, carrier was given earlier in (1.3.8) as

$$c(t) = A \cos(\omega_c t + \psi) \qquad (1.5.1)$$

where A, ω_c, and ψ are constants. The pure carrier is therefore a sinusoidal waveform with fixed amplitude A, frequency ω_c in rps selected from the electromagnetic chart in Figure 1.4, and phase angle ψ. Pure carriers are generated by some form of harmonic oscillator, although in practice generation of the ideal carrier of (1.5.1) is hindered by internal oscillator noise and instabilities that cause parameter variations. The assumptions concerning the phase angle ψ become an extremely important part of the carrier waveform characterization. If ψ is considered to be a fixed, known parameter, then the phase of the carrier sinusoid is assumed to be known exactly at any t, and the pure carrier is completely deterministic. Such a waveform has a power of $A^2/2$ and is described in the frequency domain by its Fourier lines at $\pm\omega_c$ (Figure 1.6d). In practice it is necessary to ensure that a known phase is indeed a practical assumption. Knowledge of a carrier phase implies that we know exactly where on a sine wave cycle the carrier will be at the instant it is turned on. For this reason it is often more meaningful to consider ψ in (1.5.1) as a random variable, uniformly distributed over $(0, 2\pi)$. We are then admitting that carrier phase is completely unknown, and subsequent processing must take this into account. Note that a carrier with a random phase ψ becomes a random waveform in time, since at any t its value can only be statistically described. Thus all the averaging operations denoted earlier now involve averages over the probability density of this phase angle. In particular, if ψ is assumed a uniform variable, then $c(t)$ in (1.5.1) has a mean

$$\mathcal{E}[c(t)] = \int_0^{2\pi} A \cos(\omega_c t + \psi)\left(\frac{1}{2\pi}\right) d\psi = 0 \qquad (1.5.2)$$

and an autocorrelation given by

$$R_c(\tau) = \mathcal{E}[c(t)c(t+\tau)]$$

$$= \int_0^{2\pi} A \cos(\omega_c t + \psi) A \cos(\omega_c(t+\tau) + \psi)\left(\frac{1}{2\pi}\right) d\psi \qquad (1.5.3)$$

Expanding the cosine product allows us to write this as

$$R_c(\tau) = \frac{A^2}{2}\int_0^{2\pi} \cos(\omega_c \tau)\left(\frac{1}{2\pi}\right) d\psi + \frac{A^2}{2}\int_0^{2\pi} \cos(2\omega_c t + \omega_c \tau + 2\psi)\left(\frac{1}{2\pi}\right) d\psi$$

$$(1.5.4)$$

The second integral is zero for any t and τ, and (1.5.4) reduces to

$$R_c(\tau) = \frac{A^2}{2}\cos(\omega_c \tau) \qquad (1.5.5)$$

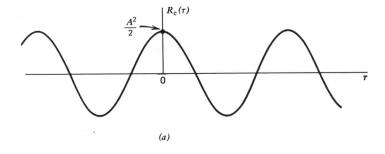

(a)

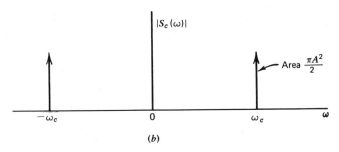

(b)

Figure 1.11. Random phase pure carriers. (*a*) Autocorrelation function, (*b*) power spectrum.

Thus random phased pure carriers have autocorrelation functions that are cosinusoidal in τ (Figure 1.11a) exhibiting periodic positive and negative correlation regions. The power spectral density follows as the transform of $R_c(\tau)$,

$$S_c(\omega) = \frac{\pi A^2}{2} \delta(\omega + \omega_c) + \frac{\pi A^2}{2} \delta(\omega - \omega_c) \qquad (1.5.6)$$

corresponding to spectral lines in the frequency domain* (Figure 1.11b). These lines exhibit the concentration of power at the specific frequency ω_c, a result somewhat obvious from (1.5.1) but shown now to be rigorously true. Conversely, delta functions in the spectral density imply random phased sinusoidal components in the corresponding time process. Note that a pure carrier has a theoretical bandwidth of zero width and is often

---

* The reader should be careful here to distinguish between delta functions in the waveform transform of a deterministic carrier, Figure 1.6d, and delta functions in the power spectrum of a random phased carrier, as in Figure 1.11b.

referred to as a *monochromatic* (single frequency) waveform. The power of the random phased carrier follows from (1.5.5) as $P_c = R_c(0) = A^2/2$, which is identical to (1.4.5) for the deterministic sine wave. Thus the pure carrier has the same power whether the phase angle is assumed random or not.

The modulated carrier is formed by superimposing a baseband waveform onto a pure carrier waveform. The most prevalent type of modulated carriers are the *amplitude modulated* (AM) carrier, the *phase modulated* (PM) carrier, and the *frequency modulated* (FM) carrier. Understanding the properties of these particular waveforms is therefore a prerequisite for later analysis. The general form of a modulated carrier is given by

$$c(t) = a(t) \cos [\omega_c t + \theta(t) + \psi] \qquad (1.5.7)$$

where $a(t)$ is the nonnegative amplitude modulaton of the carrier and $\theta(t)$ is the phase modulation. The AM, PM, and FM carriers are simply special cases of this general carrier waveform. The electronic circuitry and hardware for generating the modulated carriers has been adequately discussed elsewhere [8, 9, 10] and need not be rigorously pursued here. Note that a modulated carrier is no longer a pure carrier, and its frequency function can no longer be represented by a pair of delta functions. This means that the modulated carrier tends to spread its power over a frequency band rather than concentrating it at a single frequency. Frequency functions of the specific modulated carriers are considered in detail in Chapter 2.

It is sometimes convenient to rewrite the modulated carrier of (1.5.7) in the alternate form

$$c(t) = \text{Real} \{a(t) e^{j\theta(t)} e^{j\omega_c t + j\psi}\} \qquad (1.5.8)$$

where Real $\{\cdot\}$ means the real part of $\{\cdot\}$. The time function in braces is now a complex time function, and (1.5.8) is called the *complex representation* of $c(t)$. The term $\exp j(\omega_c t + \psi)$ is called the *complex carrier*, and $a(t) \exp j\theta(t)$ is the *complex envelope function*. By writing the envelope function in terms of its real and imaginary part, $a(t) \cos (\theta(t)) + ja(t) \sin (\theta(t))$, (1.5.8) can be also expanded as

$$c(t) = a(t) \cos (\theta(t)) \cos (\omega_c t + \psi) - a(t) \sin (\theta(t)) \sin (\omega_c t + \psi) \quad (1.5.9)$$

The preceding equation is an expansion of the general modulated carrier into cosine and sine components and is called a *quadrature expansion* of $c(t)$. At certain points in later analysis it is convenient to rewrite the modulated carrier in (1.5.7) in one of the alternate forms in (1.5.8) or (1.5.9).

The *instantaneous frequency* of the carrier waveform $c(t)$ in (1.5.7) is defined as the time derivative of the sinusoidal argument. Hence

$$\omega(t) = \frac{d}{dt}[\omega_c t + \theta(t) + \psi]$$

$$= \omega_c + \frac{d\theta}{dt} \tag{1.5.10}$$

Thus the instantaneous frequency of the modulated carrier does not depend on ψ and varies about the carrier frequency ω_c according to the derivative of the function $\theta(t)$. This phase function $\theta(t)$ may have been intentionally imposed, because of phase or frequency modulation, or may occur unintentionally, because of undesired phase variations of the carrier waveform. If $\theta(t) = 0$, the modulated carrier will have the same instantaneous frequency as the pure carrier. The specific forms of modulated carriers are discussed below.

Amplitude Modulation. The amplitude modulated carrier waveform is produced by linearly multiplying the pure carrier sinusoid by an amplitude modulating waveform $s(t)$. If $m(t)$ represents the baseband waveform to be modulated onto the carrier, the amplitude modulating signal can be written in general form as

$$s(t) = a_0 + b_0 m(t) \tag{1.5.11}$$

where a_0 and b_0 are fixed constants. The AM carrier then has the form

$$c(t) = s(t) \cos(\omega_c t + \psi)$$

$$= (a_0 + b_0 m(t)) \cos(\omega_c t + \psi) \tag{1.5.12}$$

This can be put into the form of (1.5.7) by rewriting it as

$$c(t) = |a_0 + b_0 m(t)| \cos(\omega_c t + \theta(t) + \psi) \tag{1.5.13}$$

where now $\theta(t)$ has the specific form

$$\theta(t) = 0 \qquad \text{if} \quad s(t) \geq 0$$

$$= \pi \qquad \text{if} \quad s(t) < 0 \tag{1.5.14}$$

If the modulating waveform in (1.5.11) is always nonnegative, $\theta(t) = 0$, and the AM carrier reduces to simply

$$c(t) = (a_0 + b_0 m(t)) \cos(\omega_c t + \psi)$$

$$= a_0\left(1 + \frac{b_0}{a_0} m(t)\right) \cos(\omega_c t + \psi) \tag{1.5.15}$$

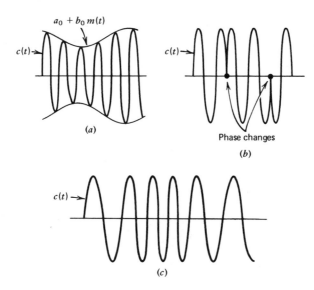

Figure 1.12. Carrier waveforms. (*a*) AM carrier, (*b*) PM carrier, (*c*) FM carrier.

This corresponds to a sinusoid whose amplitude varies about some average value a_0 in accordance with the modulating waveform $m(t)$ (Figure 1.12*a*). Such modulation is referred to as *standard* AM. If $s(t)$ becomes negative for any t, the negative sign can be incorporated into the carrier phase as a shift of π radians. Thus the carrier phase, relative to ψ, shifts π radians every time the amplitude modulating waveform changes polarity. If the latter waveform becomes negative for some t, the carrier is said to be *overmodulated*, and the resulting AM waveform is both amplitude and phase modulated. This unintentional phase modulation of an AM carrier caused by overmodulation may or may not be important, depending on the receiver processing. Proper selection of the constants a_0 and b_0 prevents this overmodulation from occurring. If the most negative value of $m(t)$ is normalized to unity, then the required condition is that the parameter b_0/a_0 in (1.5.15) (referred to as the *amplitude modulation index*) be less than one. If $a_0 = 0$, the carrier is overmodulated whenever the modulating waveform $m(t)$ becomes negative. The resulting AM carrier is called a *carrier suppressed AM waveform*. This nomenclature arises since the a_0 term in (1.5.15) corresponds to a pure sine wave carrier within the modulated waveform.

Phase Modulation.   A phase modulated carrier has the form

$$c(t) = A \cos\left[\omega_c t + \Delta m(t) + \psi\right] \tag{1.5.16}$$

where A is the carrier amplitude, $m(t)$ is again the modulating waveform, and Δ is a *phase deviation coefficient*. This coefficient indicates the amount of phase deviation imparted to the carrier and has units of radians per volt. If the maximum value of $m(t)$ is normalized to one, then Δ represents the maximum phase deviation of the carrier. Note that the modulation directly varies the phase of the carrier, and the carrier phase ψ adds directly to the desired modulation. Phase modulated carriers are generated by voltage control of phase shifting networks into which the carrier is inserted. PM carriers have fixed amplitudes and exhibit the modulation only through their phase behavior (Figure 1.12b). PM is often called *angle modulation*, as distinguished from the linear modulation associated with AM.

Frequency Modulation.   A frequency modulated carrier is a carrier whose frequency is varied according to the modulating waveform and has the form

$$c(t) = A \cos \left[\omega_c t + \Delta_\omega \int m(t) \, dt + \psi \right] \qquad (1.5.17)$$

Hence an FM carrier is simply a phase modulated carrier with the phase modulation proportional to the integral of the modulating waveform. The preceding carrier has an instantaneous frequency defined in (1.5.10) of

$$\omega(t) = \omega_c + \frac{d}{dt} \left[\Delta_\omega \int m(t) \, dt + \psi \right]$$

$$= \omega_c + \Delta_\omega m(t) \qquad (1.5.18)$$

which therefore varies in proportion to the modulating signal $m(t)$ as desired. The proportionality constant Δ_ω is the *frequency deviation coefficient* with units of rps/volt, and determines the amount of frequency variation given to the carrier. Again, if $m(t)$ is normalized to a unit peak value, then Δ_ω is the maximum frequency deviation in rps that the carrier will undergo.

FM carriers are produced by varying the tuning frequencies of a resonant oscillator. Such a device is called a *voltage controlled oscillator* (VCO) and is characterized by the fact that its output frequency can be made to vary in accordance with an applied control (modulating) waveform. Unfortunately, however, there is a physical limit to the maximum amount by which an oscillator sine wave of a given frequency can have its frequency altered. This means that in (1.5.17) a condition is imposed of the form

$$\Delta_\omega \leq \eta \omega_c \qquad (1.5.19)$$

where η is a constant, typically in the range 0.05–0.1. This limit to the amount of frequency modulation that can be inserted on a carrier could become a fundamental constraint in FM carrier systems. Note that the FM carrier also has a fixed amplitude, with the modulation information appearing only in the varying frequency (Figure 1.12c) and therefore represents a form of angle modulation.

In summary, we have reviewed specific forms of the general modulated carrier of (1.5.7). The AM, PM, and FM carriers are generated by superimposing the modulating signal onto the parameters (amplitude, phase, or frequency) of the basic sine wave carrier. These particular carrier formats are the most prevalent in communication systems and are examined in more detail in Chapters 2–4. Some modifications of these basic canonic carrier forms can be made so as to achieve some advantage in specific applications. We later comment on some of these alterations.

1.6. Filtering of Waveforms

A common operation in communication signal processing is to filter a waveform, that is, pass the waveform through a linear filter (Figure 1.13). Linear filters have their output and input related through the filter *impulse response* $h(t)$ or the filter *transfer function* $H(\omega)$, the latter the Fourier transform of the former. [If the filter transfer function is stated in terms of the Laplace variable, then $H(\omega)$ is obtained by replacing this variable by $j\omega$.] The filter output time waveform $y(t)$ is related to the input waveform $x(t)$ by either of the forms

$$y(t) = \int_{-\infty}^{\infty} h(t-\rho)x(\rho)\,d\rho = \int_{-\infty}^{\infty} h(\rho)x(t-\rho)\,d\rho \qquad (1.6.1)$$

These integrals are called *convolutions* of $x(t)$ and $h(t)$. The function $h(t)$ is called the impulse response since $y(t) = h(t)$ if an impulse function is applied at $t = 0$ at the input. For a filter to be physically realizable it is necessary that $h(t) = 0$ for $t < 0$. Similarly, the output waveform transform $Y(\omega)$ is related to the transform $X(\omega)$ of the input by

$$Y(\omega) = H(\omega)X(\omega) \qquad (1.6.2)$$

Hence filter output waveforms can be determined by convolving with the impulse response, or by multiplying with the filter transfer function and

Figure 1.13. Linear filtering.

inverse transforming. By using polar notation we can write

$$Y(\omega) = |H(\omega)| \exp[j\angle H(\omega)]|X(\omega)| \exp[j\angle X(\omega)]$$
$$= |H(\omega)||X(\omega)| \exp[j(\angle H(\omega) + \angle X(\omega))] \qquad (1.6.3)$$

The output transform amplitude function follows as

$$|Y(\omega)| = |H(\omega)||X(\omega)| \qquad (1.6.4)$$

and the output phase function, as defined in (1.3.4), is

$$\angle Y(\omega) = \angle X(\omega) + \angle H(\omega) \qquad (1.6.5)$$

The filter therefore modifies the input transform amplitude by multiplying by $|H(\omega)|$ and adds its phase function to that of the input. The function $|H(\omega)|$ is called the *gain function* of the filter. Thus the filter amplifies the frequencies of the input where its gain function is high and attenuates those frequencies where the gain is low. Undesirable modification of the input amplitude function by the filter is called filter *amplitude distortion*. Addition of a nonlinear phase function by the filter produces *phase distortion* (Problem 1.19). To prevent significant amplitude distortion by the filter it is necessary that $|H(\omega)|$ be constant over all ω within the bandwidth of the input waveform. To prevent significant phase distortion by the filter it is necessary that $\angle H(\omega)$ be linear in ω over the bandwidth of the input.

Typical filters have gain functions that are constant over a specific band of frequencies (called the filter *passband*) and fall off fairly rapidly outside the band, as shown in Figure 1.14. This means filters have bandwidths that can be defined just as with waveform frequency functions. To reduce amplitude distortion it is, therefore, necessary that the filter passband equal or exceed the bandwidth of the input waveform. The usual filter bandwidth definition is the 3 dB bandwidth, defined from the squared gain function $|H(\omega)|^2$. Filter phase functions are generally linear over the pass-

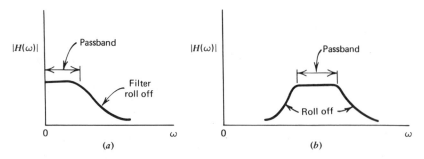

Figure 1.14. Filter gain functions. (*a*) Low pass filter, (*b*) bandpass filter.

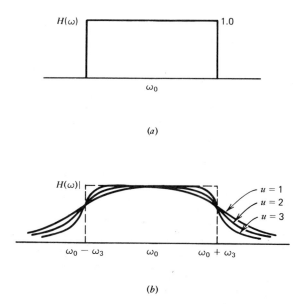

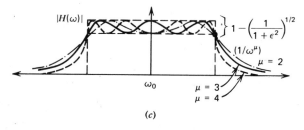

Figure 1.15. The ideal filter function. (a) The exact form, (b) approximation by Butterworth filters, (c) approximation by Chebychev filters.

band of the filter gain function and depart from linearity during filter roll-off. Hence maximum phase distortion occurs for values of ω at band edges. If the passband of the filter begins at zero frequency and extends to a higher frequency, the filter is said to be a *low pass filter* (Figure 1.14a). If the passband extends over a range of frequencies away from zero, the filter is said to be a *bandpass filter* (Figure 1.14b).

An *ideal*, or *rectangular*, filter is one having a constant passband gain function with perfect rejection, as shown in Figure 1.15. Although truly ideal filter functions are unrealizable (Problem 1.25), they can be approximated arbitrarily closely by realizable frequency functions. The two most common are the Butterworth and Chebychev filter functions. The Butterworth function (Figure 1.15b) is given by

$$|H(\omega)|^2 = \frac{1}{1+[(\omega-\omega_0)/\omega_3]^{2\mu}} \qquad (1.6.6)$$

where ω_3 is the 3 dB bandwidth, ω_0 is the filter center frequency, and μ is the order of the filter, the latter determining its falloff rate. For low pass filters, $\omega_0 = 0$. As the order of the filter increases, the Butterworth function approaches the ideal bandpass (low pass, if $\omega_c = 0$) filter function. Chebychev filter functions (Figure 1.15c) have the form,

$$|H(\omega)|^2 = \frac{1}{1+\varepsilon^2 C_\mu^2[(\omega-\omega_0)/\omega_3]} \qquad (1.6.7)$$

where $C_\mu(\omega)$ is the Chebychev polynomial [11] of order μ. Chebychev filter functions exhibit a passband ripple of height $\varepsilon^2/2$ and a fall off rate that increases with order for improved out-of-band rejection. For a specific value of μ and the same 3 dB bandwidth, a Butterworth filter has flatter inband gain, whereas a Chebychev filter has slightly better out-of-band rejection. Both Butterworth and Chebychev filter functions can be constructed by passive networks with properly located filter poles [12, 13].

If the input to the filter is a random process, then the output in (1.6.1) will also evolve as a random process. The statistics of the output process, however, are related in a rather complicated way to those of the input process (see, for example, [14], Chapter 7). Two important exceptions are when the input is a Gaussian process, in which case the output is also a Gaussian process, and when the bandwidth of the input process is much larger than the filter bandwidth. In the latter case the output has been shown to be approximately Gaussian no matter what the input statistics [15]. There are, however, relatively straightforward expressions relating input and output correlation and spectral densities for any process. The autocorrelation function of the output of a linear filter can be derived by direct substitution from (1.6.1)

$$R_y(\tau) = \mathscr{E}[y(t)y(t+\tau)]$$

$$= \mathscr{E}\left[\int_{-\infty}^{\infty} x(t-\rho)h(\rho)\,d\rho \int_{-\infty}^{\infty} x(t+\tau-\alpha)h(\alpha)\,d\alpha\right] \qquad (1.6.8)$$

where $\mathscr{E}$ is the expectation operator over the statistics of the integrals. However, we see that the integrals involve the random variables $x(t-\rho)$ and $x(t+\tau-\alpha)$, and we need only average over their joint density. Interchanging averaging and integration, and moving the expectation operation inside the integrals, allows us to rewrite (1.6.8) as

$$R_y(\tau) = \int_{-\infty}^{\infty} \int_{-\infty}^{\infty} \mathscr{E}[x(t-\rho)x(t+\tau-\alpha)]h(\rho)h(\alpha)\,d\alpha\,d\rho \qquad (1.6.9)$$

We now note that the expectation in the integrand is precisely the definition of the autocorrelation function of the input process $x(t)$ at points $(t-\rho)$ and $(t+\tau-\alpha)$. If $x(t)$ is a stationary input process, this expectation will depend only on the time difference of the points, and we can substitute

$$\mathscr{E}[x(t-\rho)x(t+\tau-\alpha)] = R_x(\tau+\rho-\alpha) \qquad (1.6.10)$$

This simplifies (1.6.9) to

$$R_y(\tau) = \int_{-\infty}^{\infty} \int_{-\infty}^{\infty} R_x(\tau+\rho-\alpha)h(\rho)h(\alpha)\,d\alpha\,d\rho \qquad (1.6.11)$$

Equation (1.6.11) therefore indicates how the output autocorrelation is related to the input autocorrelation when a stationary process is filtered by a linear filter having impulse response $h(t)$. The output power spectrum is obtained by Fourier transforming (1.6.11). Thus

$$S_y(\omega) = \int_{-\infty}^{\infty} R_y(\tau)e^{-j\omega\tau}\,d\tau$$

$$= \int\int\int_{-\infty}^{\infty} R_x(\tau+\rho-\alpha)h(\rho)h(\alpha)e^{-j\omega\tau}\,d\alpha\,d\rho\,d\tau \qquad (1.6.12)$$

Interchanging the order of integration, and integrating out the τ variable first, changes this to

$$S_y(\omega) = \int_{-\infty}^{\infty} h(\rho)\,d\rho \int_{-\infty}^{\infty} h(\alpha)\,d\alpha \left[\int_{-\infty}^{\infty} R_x(\tau+\rho-\alpha)e^{-j\omega\tau}\,d\tau \right]$$

$$(1.6.13)$$

The last bracket can be recognized as $S_x(\omega)e^{-j\omega(\alpha-\rho)}$, and (1.6.13) reduces to

$$S_y(\omega) = S_x(\omega)\int_{-\infty}^{\infty} h(\rho)e^{j\omega\rho}\,d\rho \int_{-\infty}^{\infty} h(\alpha)e^{-j\omega\alpha}\,d\alpha$$

$$= S_x(\omega)H^*(\omega)H(\omega)$$

$$= S_x(\omega)|H(\omega)|^2 \qquad (1.6.14)$$

where the asterisk denotes a complex conjugate. Hence the square of the filter gain function filters the power spectral density of the input to produce the spectral density of the output. Note that the resulting power in the output process follows as

$$P_y = \frac{1}{2\pi} \int_{-\infty}^{\infty} S_x(\omega) |H(\omega)|^2 \, d\omega \qquad (1.6.15)$$

If the input process is a white noise process of spectral level S_0 W/Hz, then (1.6.15) becomes

$$P_y = S_0 \left(\frac{1}{2\pi} \int_{-\infty}^{\infty} |H(\omega)|^2 \, d\omega \right) \qquad (1.6.16)$$

It is conventional to define

$$B_n = \frac{1}{2\pi} \int_0^{\infty} |H(\omega)|^2 \, d\omega \qquad (1.6.17)$$

as the *noise bandwidth* of the filter. The output power in (1.6.15) can then be written as simply

$$P_y = 2S_0 B_n \qquad (1.6.18)$$

Thus the filter noise bandwidth is the effective bandwidth determining the output noise power when a white noise process is inserted at the input. Note that if a filter has a constant transfer function, $H(\omega) = \sqrt{G}$, the output power is given by

$$P_y = GP_x \qquad (1.6.19)$$

The parameter G is referred to as the *power gain* of the filter. Power gain is often stated in decibels, with

$$(G)_{dB} = 10 \log_{10} G \qquad (1.6.20)$$

If $G < 1$, representing a power loss or an attenuation, $(G)_{dB}$ will be negative.

If two filters are placed in cascade (Figure 1.16*a*), their overall filter function $H(\omega)$ is given by the product of their individual functions. Thus

$$H(\omega) = H_1(\omega) H_2(\omega) \qquad (1.6.21)$$

where the $H_i(\omega)$ are the individual filter functions. If a filter function has a peak value of $\sqrt{G}$, it is convenient to factor its filter function as $\sqrt{G} H(\omega)$, where $H(\omega)$ is now a normalized filter function of unit peak value. Using (1.6.20) we can then interpret the filter as the cascade of a device with constant power gain G and a unit gain filter function $H(\omega)$. This separation allows us always to deal with unit level frequency functions in our subsequent analysis. If the filters are placed in parallel (Figure 1.16*b*),

$$H(\omega) = H_1(\omega) \pm H_2(\omega) \qquad (1.6.22)$$

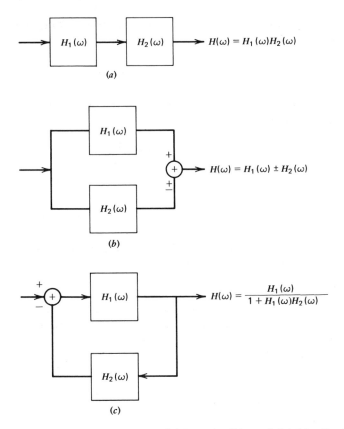

Figure 1.16. Filter combinations. (*a*) Cascade, (*b*) parallel, (*c*) feedback.

whereas if the filters are placed in a negative feedback arrangement (Figure 1.16c),

$$H(\omega) = \frac{H_1(\omega)}{1 + H_1(\omega)H_2(\omega)} \tag{1.6.23}$$

These results can be easily derived by basic linear system theory (Problem 1.30). Their basic advantage is they allow us to build up complicated filter functions from relatively simple filters.

1.7. Performance Criterion; Signal to Noise Ratio

In the actual operation of a communication system the recovered waveforms at the receiver do not generally correspond exactly to the desired waveforms produced in the transmitter. This is due to the anomalies occurring during transmission and reception that cause signal distortion and the insertion of interference and noise waveforms. These effects cause a basic deterioration of the desired waveform and a degradation in the overall communication operation. To assess the capability of the link it is therefore necessary to account for these effects in system analysis and design. Typically, a specific performance criterion relating desired and actual operation is first decided on. Subsequent design or system comparison is then based on satisfying the criterion. When more than one system design is being considered, a comparison can be made with respect to the decided criterion and the most favorable systems can be determined.

One of the most convenient and widely used measures of performance in communication analysis is the concept of *signal to noise ratio* (SNR). When dealing with a situation in which a desired waveform, whether it be an information signal, baseband waveform, or modulated carrier, is contaminated with an additive, interfering waveform, the SNR of the combined waveform at that point in the system is defined as

$$\text{SNR} = \frac{\text{power in the desired waveform}}{\text{power of the interfering waveform}} \qquad (1.7.1)$$

Thus SNR indicates how much stronger the desired waveform is relative to the interference at the point in question. If SNR is greater than one, there is more power in the desired signal than in the interference, and the opposite is true if SNR is less than one. Adequate system design generally requires the desired signal power to be many times larger than that of the interference. Thus $\text{SNR} \gg 1$ is generally required in design. Note that the SNR parameter depends only on power levels of the signals involved, which is the basic reason for its wide usage. It is, therefore, easily measured, and applicable whether dealing with deterministic waveforms or random processes. Furthermore, its computation does not require knowledge of the properties and statistics of the random waveforms involved.

It is common to express SNR in decibels. Hence if P_d is the power of the desired signal and P_n is the power of the interference, then

$$\begin{aligned}
(\text{SNR})_{\text{dB}} &= 10 \log_{10} \frac{P_d}{P_n} \\
&= 10 \log_{10} P_d - 10 \log_{10} P_n \\
&= (P_d)_{\text{dB}} - (P_n)_{\text{dB}} \qquad (1.7.2)
\end{aligned}$$

Thus $(\text{SNR})_{\text{dB}}$ is the difference of the two powers expressed in decibels. When the interference is composed of the sum of two separate uncorrelated, interfering waveforms, say $n_1(t)$ and $n_2(t)$, the SNR in (1.7.1) is then

$$\text{SNR} = \frac{\text{power of the desired signal}}{\text{power of } n_1(t) + \text{power of } n_2(t)}$$

$$= \frac{1}{1/(\text{SNR})_1 + 1/(\text{SNR})_2} \qquad (1.7.3)$$

where $(\text{SNR})_i$ is the SNR due to $n_i(t)$ alone. We see that the overall SNR is not simply the sum of the individual SNR_i, but is related by the expression in (1.7.3). The result can be easily extended to more than two interfering waveforms as well. We emphasize that the SNR is associated with a particular point or location within the system. Different points often generate different values for SNR since the interference may vary throughout the link. When the desired signal involved is a carrier waveform, we often speak instead of the *carrier to noise ratio* (CNR) instead of SNR.

In comparing several systems it is quite common to select the SNR value at some specified point (usually the final reconstructed information waveform) for comparison. The system producing the highest SNR at that point is considered the most favorable. Conversely, the design of a particular system may be based on guaranteeing that at particular points the resulting SNR is above a specified level, the latter often called a *threshold* SNR. The ratio of the actual SNR to the threshold SNR at a particular point is called the *margin* at that point. Thus

$$\text{design margin} = \frac{\text{SNR}}{\text{threshold SNR}} \qquad (1.7.4)$$

Margin values are extremely important since they indicate the amount of excess that is available for allowing for unexpected events and possible subsystem failure.

Care must be used in dealing with SNR when signal distortion has occurred. The recovered signal waveform may no longer correspond to the desired form, even though its power remains the same. To include distortion effects, we define a desired signal error waveform

$$e(t) = s_d(t) - s(t) \qquad (1.7.5)$$

between the recovered signal $s(t)$ and the desired signal $s_d(t)$. By rewriting this as

$$s(t) = s_d(t) - e(t) \qquad (1.7.6)$$

it appears as if the recovered signal has been formed by adding an error "noise" to the desired waveform. Neglecting any additive interference, we can treat the error waveform as an equivalent noise and define a *signal to distortion ratio* SDR as

$$\text{SDR} = \frac{\text{power in desired signal}}{\text{power in the error waveform}} \tag{1.7.7}$$

Hence SDR is comparable in interpretation to SNR, and is generally treated as such. However, it must be remembered that the "noise" is actually a distortion error, and in fact may be itself dependent on the desired signal. When the distorting system corresponds to a linear filter and $s(t)$ and $s_d(t)$ are the output and input signals, respectively, the error waveform in (1.7.5) has the transform

$$
\begin{aligned}
E(\omega) &= F_{s_d}(\omega) - F_{s_d}(\omega)H(\omega) \\
&= [1 - H(\omega)]F_{s_d}(\omega)
\end{aligned} \tag{1.7.8}
$$

where $H(\omega)$ is the frequency transfer function of the linear system and $F_{s_d}(\omega)$ is the transform of the desired waveform $s_d(t)$. When written as in (1.7.8), it appears as if the error waveform is generated by an effective filtering of $S_d(t)$ with the function $[1 - H(\omega)]$. If $s_d(t)$ is a random process with power spectral density $S_d(\omega)$, then the error process has the equivalent spectrum $|1 - H(\omega)|^2 S_d(\omega)$. The corresponding SDR in (1.7.7) then becomes

$$\text{SDR} = \frac{P_d}{(1/2\pi)\int_{-\infty}^{\infty} |1 - H(\omega)|^2 S_d(\omega)\,d\omega} \tag{1.7.9}$$

If there is additional interference to the distorted waveform, the overall SNR can be evaluated as in (1.7.3), using SDR to replace one of the individual SNR_i terms. In effect, this combines the waveform error and the waveform interference into a total interference superimposed on the desired waveform.

In summary, this first chapter has presented an overview of a typical communication system, from information source to final destination. Several examples of familiar systems were discussed with respect to this model. Our prime objective was to delineate the overall system structure and indicate the various subsystems and functional operations comprising the typical link. A review of waveform analysis, noise theory, and linear system theory was presented, along with some key system definitions that will be used extensively. We now begin to examine somewhat more rigorously the component subsystems and to indicate their physical and mathematical characterization. It is often convenient to examine these subsystems in groups so as to portray more clearly their interconnections.

We begin in the next chapter by considering the carrier portion of the link (i.e., the subsystems involved with generating and propagating the carrier waveform).

References

1. Gagliardi, R. and Karp, S. *Optical Communications*, Wiley, New York, 1976.
2. Bracewell, R. *The Fourier Transform*, McGraw-Hill, New York, 1965.
3. Abramowitz, M. and Stegun, I. *Handbook of Mathematical Functions*, National Bureau of Standards, Washington, D.C., 1965, Chap. 29.
4. Sneddon, I. *Fourier Transforms*, McGraw-Hill, New York, 1958.
5. Wax, N. *Noise and Stochastic Processes*, Dover, New York, 1954.
6. Davenport, W. and Root, W. *Random Signals and Noise*, McGraw-Hill, New York, 1958.
7. Scholtz, R. "How Do You Define Bandwidth," *Proc. ITC*, Los Angeles, October 1972, pp. 281–288.
8. Terman, F. *Electronic and Radio Engineering*, McGraw-Hill, New York, 1955.
9. Black, H. S., *Modulation Theory*, Van Nostrand, New York, 1953.
10. Panter, P. *Modulation, Noise, and Spectral Analysis*, McGraw-Hill, New York, 1965.
11. Abramowitz and Stegun, loc. cit., Chap. 22.
12. Van Valkenberg. *Modern Network Synthesis*, Wiley, New York, 1960, Chap. 13.
13. Balabanian, N. *Network Synthesis*, Prentice-Hall, Englewood Cliffs, N.J., 1958.
14. Papoulis, A. *Probability, Random Variables, and Stochastic Processes*, McGraw-Hill, New York, 1965.
15. Middleton, D. *Statistical Communication Theory*, McGraw-Hill, New York, 1960.

Problems*

1. (1.2) Engineers deal with frequency separation measured in hertz. Physicists use equivalent separation measured in wavelength λ. Determine the relation between a frequency difference df about a center frequency f_c and its equivalent wavelength difference $d\lambda$ about the corresponding λ_c.

2. (1.3) Show that if a deterministic waveform is periodic with period T[i.e., $x(t+T)=x(t)$], then the Fourier transform $X(\omega)$ reduces to a Fourier series. Identify the coefficients.

3. (1.3) Show that if $x(t)$ is real, then $|X(\omega)|$ is even in $\omega[|X(-\omega)| = |X(\omega)|]$ and $\angle X(\omega)$ is odd $[\angle X(-\omega) = -\angle X(\omega)]$.

* Problem numbers in parentheses refer to the section required for their solution.

4. (1.3)   Determine the amplitude function $|X(\omega)|$ of the waveform

$$x(t) = A \cos(\omega_0 t + \psi), \qquad 0 \le t \le T$$
$$= 0, \qquad\qquad\qquad \text{elsewhere}$$

[*Hint:* Treat $x(t)$ as the product of a sinusoid and a pulse and use frequency convolution (Section A.1).]

5. (1.3)   Using transforms, verify the frequency spreading that occurs as pulses are made narrower, using pulses that are time waveforms composed of (a) Gaussian-shaped time pulses (b) pulses shaped by the half period of a cosine wave (*raised cosine* pulse) between $(-\pi/2, \pi/2)$.

6. (1.3)   Show that for any transform pair with bounded time and frequency functions

$$\int_{-\infty}^{\infty} |X(\omega)| \, d\omega \int_{-\infty}^{\infty} |x(t)| \, dt = \text{finite constant}$$

Comment on what this means in terms of the relation between the time extent and frequency extent of the waveform.

7. (1.3)   A random voltage $x(t)$ has a first order density

$$p_{x_t}(x) = \tfrac{1}{20}, \qquad -10 \le x \le 10$$

at any t. What is the probability that $x(t)$ will have a voltage value between 5 and 7 V at time $t = 5$ sec?

8. (1.3)   What is the probability that a stationary Gaussian random process with zero mean and correlation $R_x(\tau) = 20e^{-10|\tau|}$ will be in the interval $(4, 6)$ V at $t = 5$ sec and in the interval $(5, 8)$ V at $t = 10$ sec?

9. (1.3)   Show that since $R_x(\tau)$ is real and even in τ, the spectral density $S_x(\omega)$ must necessarily be real and even in ω (i.e., $S_x(-\omega) = S_x(\omega)$).

10. (1.3)   Let $x(t)$ and $y(t)$ be jointly stationary random processes. Show that (a) $R_{xy}(-\tau) = R_{yx}(\tau)$. (b) $R_{xy}(\tau) + R_{yx}(\tau)$ is even in τ.

11. (1.3)   Using (B.2.16) of Appendix B show that the second order characteristic function of a zero mean stationary Gaussian process at times t and $t + \tau$ is given by

$$\Psi(\omega_1, \omega_2, \tau) = \exp \tfrac{1}{2}[R_x(0)\omega_1 + R_x(0)\omega_2 + 2R_x(\tau)\omega_1\omega_2]$$

12. (1.4)   Convert the following power values to decibels: (a) 3 W, (b) 0.25 W, (c) 25 mW, (d) 230 W. Convert the following to watts: (e) 3 dBW, (f) 70 dBM, (g) -15 dBW.

13. (1.4) Determine the 3 dB, 6 dB, and 90% bandwidths of the frequency function

$$|X(\omega)|^2 = \frac{2}{1+0.25\omega^2}$$

14. (1.4) The *rms bandwidth* of a low pass power spectral density $S_x(\omega)$ is defined by

$$B_{rms} = \left[\frac{1}{2\pi P_x} \int_{-\infty}^{\infty} \omega^2 S_x(\omega)\, d\omega\right]^{1/2}$$

where P_x is the power in $S_x(\omega)$. Show that the power in a bandwidth of kB_{rms} is bounded by $[1-(1/k^2)]P_x$.

15. (1.4) Show that the variable transformation

$$j\omega \rightarrow \frac{\omega_c}{B}\left[\frac{j\omega}{\omega_c} + \frac{\omega_c}{j\omega}\right]$$

converts a low pass function to a bandpass function.

16. (1.5) Let $x(t) = A\cos(\omega_c t + \psi)$, where ψ is a random phase variable with probability density $p_\psi(\xi)$. (a) Find the mean of $x(t)$ if ψ has a Gaussian distribution with mean zero and variance σ_ψ^2. (b) Find the condition on $p_\psi(\xi)$ needed for the mean of $x(t)$ to be zero.

17. (1.5) Let $q(t)$ be a $(0, 1)$ square wave. Show that phase modulating with $\pi q(t)$ is identical to amplitude modulating with $(2q(t)-1)$.

18. (1.5) Using the complex envelope representation of a modulated carrier, show how its imaginary part is related to the instantaneous frequency of the carrier.

19. (1.6) Show that a filter with a unit gain and linear phase function $H(\omega) = e^{-j\alpha\omega}$, does not distort, but only delays, any input waveform.

20. (1.6) Given the RC network in Figure P1.20 (a) Find $|H(\omega)|$ and $\angle H(\omega)$. (b) Repeat for N cascaded stages of this type (neglect network loading).

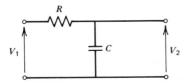

Figure P1.20.

21. (1.6) Consider the network in Figure P1.20, with R = 10,000 ohms and C = 10^{-6} F. The sine wave $10 \sin(2\pi 10t + \pi/9)$ is applied. (a) State the output time function. (b) If the output has an amplitude of 5 when the input sine wave amplitude is 10, what is the frequency of the input sine wave?

22. (1.6) Given the linear system described by the differential equation

$$\frac{dy}{dt} + 2y(t) = 3x(t)$$

where $x(t)$ is the input and $y(t)$ the output. Determine $h(t)$ and $H(\omega)$ for this system.

23. (1.6) A square pulse of width T sec and height A is applied to a first order low pass filter with impulse response $h(t) = 2\pi B\, e^{-2\pi Bt}$, $t \geq 0$, where B is the filter 3 dB bandwidth. (a) Compute and sketch the output response. (b) Assume $T \gg 1/B$ and find the time in terms of B for the output to rise to 90% of its peak amplitude.

24. (1.6) The impulse response of network is $h(t)$. Show that its response to a unit pulse of width τ is given by

$$\int_{-\infty}^{t} h(\rho)d\rho - \int_{\tau}^{t} h(\rho - \tau)\, d\rho$$

(*Hint:* Treat the pulse as two step functions τ sec apart.)

25. (1.6) A realizable filter must have an impulse response that is zero for negative t. Show that the ideal, rectangular low pass filter is unrealizable.

26. (1.6) What must the order of a low pass Butterworth filter be in order that $H(\omega)$ is down by at least 30 dB at $\omega = 10\omega_3$?

27. (1.6) Determine the noise bandwidth of a low pass Butterworth filter function with a 3 dB bandwidth of ω_3 and order μ.

28. (1.6) Given a linear system with impulse response $h(t)$, show that the input–output cross-correlation is related to the input autocorrelation function $R_x(\tau)$ of a stationary random process by

$$R_{yx}(\tau) = \int_{-\infty}^{\infty} h(\rho)R_x(\tau + \rho)\, d\rho$$

29. (1.6) A differentiator has the filter transform function $H(\omega) = j\omega$. If the input is a random process $x(t)$, show that the output autocorrelation is the negative second derivative of the input autocorrelation, that is,

$$R_y(\tau) = R_{dx/dt}(\tau) = -\frac{d^2}{d\tau^2}R_x(\tau)$$

(*Hint:* First determine the output spectral density.)

30. (1.6) Prove the overall transfer function result for the cascade, parallel, and feedback networks in (1.6.21)–(1.6.23). (*Hint:* Apply input inpulse functions, trace the response through the system, and write the transform of the output.)

31. (1.7) (a) A dc signal of 5 V is desired at a point in a system. Instead the waveform observed is $5+0.2 \sin(2\pi 10t + \pi/3)$. Determine the SDR of the waveform at that point. (b) Repeat with the observed waveform at $5+n(t)$ V, where $n(t)$ is a stationary, Gaussian random process with correlation function $R_n(\tau) = 0.02e^{-\tau/10}$ V^2.

32. (1.7) The input to a linear system $H(\omega)$ is a random signal with spectral density $S_d(\omega)$ and additive independent noise with spectral density $S_n(\omega)$. Show that the output SDR is

$$SDR = \frac{P_d}{(1/2\pi)\int_{-\infty}^{\infty}|1 - H(\omega)|^2 S_d(\omega)\,d\omega + (1/2\pi)\int_{-\infty}^{\infty}|H(\omega)|^2 S_n(\omega)\,d\omega}$$

CHAPTER 2

CARRIER TRANSMISSION

In the basic communication system information is sent from the transmitter by a subsystem similar in form to that shown earlier in Figure 1.1a. The baseband waveform, generated from the source information signal by the baseband conversion operation, is modulated onto the carrier to form the modulated carrier waveform. This carrier waveform is amplified to a desired power level by the transmitter power amplifier and is then either coupled into a transmitting antenna for space wave propagation or coupled directly into a transmission line for guided propagation, depending on the particular application. In order to analyze and design the transmitting subsystem properly, it is first necessary to understand the specific properties of the modulated carrier waveforms involved and the effect of the channel on them during propagation. In this chapter we present a more detailed modulated carrier description, in terms of power levels and spectral properties, and investigate the transmission propagation channels. In the following several sections the specific modulated carrier waveforms introduced in Section 1.5 are reconsidered. We assume an arbitrary modulating baseband signal and temporarily ignore its exact waveform structure, other than its specified power and bandwidth values. Later, in Chapter 5, specific forms for the modulating baseband signal are considered.

2.1. The Amplitude Modulated Carrier

Let us first consider the amplitude modulated carrier. The standard AM carrier was given in (1.5.15) as

$$c(t) = a(t) \cos (\omega_c t + \psi) \qquad (2.1.1)$$

where $a(t)$ is the amplitude function, ω_c is the carrier frequency in rps, and ψ is an arbitrary carrier phase. The amplitude function $a(t)$ is related to the baseband modulating function $m(t)$ by

$$a(t) = a_0 + b_0 m(t) \qquad (2.1.2)$$

where we assume the waveform $m(t)$ has a power of P_m and an upper frequency of B_m Hz, and the parameters a_0 and b_0 are selected to prevent overmodulation. To determine the power in a standard AM carrier we apply the results of Section 1.4. For deterministic carrier waveforms we must evaluate (1.4.2). Hence

$$P_c = \lim_{T \to \infty} \frac{1}{2T} \int_{-T}^{T} a^2(t) \cos^2 [\omega_c t + \psi] \, dt \qquad (2.1.3)$$

By expanding the squared cosine term we can rewrite this as

$$P_c = \lim_{T \to \infty} \left[\frac{1}{2T} \int_{-T}^{T} \frac{a^2(t)}{2} \, dt + \frac{1}{2T} \int_{-T}^{T} \frac{a^2(t)}{2} \cos (2\omega_c t + 2\psi) \, dt \right] \qquad (2.1.4)$$

The second integral requires specification of $a(t)$ in order to be evaluated exactly. However, we can upper bound the magnitude of this integral by

$$\left| \frac{1}{2T} \int_{-T}^{T} \frac{a^2(t)}{2} \cos (2\omega_c t + 2\psi) \, dt \right| \le \frac{\max a^2(t)}{2} \left| \frac{1}{2T} \int_{-T}^{T} \cos (2\omega_c t + 2\psi) \, dt \right|$$

$$= \frac{\max a^2(t)}{2} \left| \frac{\sin (2\omega_c T) \cos (2\psi)}{2\omega_c T} \right| \qquad (2.1.5)$$

where $\max a^2(t)$ is the maximum value of $a^2(t)$. Since the sine function is bounded in magnitude by one, the right-hand side is bounded by $(\max a^2(t))/4\omega_c T$. In the limit as $T \to \infty$ this magnitude bound approaches zero for all $\omega_c > 0$, so long as $a(t)$ is finite for all t. Hence the second integral (2.1.4) must necessarily approach zero in the limit. The power is therefore given by the first term,

$$P_c = \lim_{T \to \infty} \frac{1}{2T} \int_{-T}^{T} \frac{a^2(t)}{2} \, dt$$

$$= \tfrac{1}{2} P_a \qquad (2.1.6)$$

where P_a is the power in the amplitude function $a(t)$. From (2.1.2)

$$P_a = \lim_{T \to \infty} \frac{1}{2T} \int_{-T}^{T} [a_0 + b_0 m(t)]^2 \, dt$$

$$= a_0^2 + b_0^2 P_m \qquad (2.1.7)$$

where P_m is the power in the modulation $m(t)$, and we have assumed the limiting time average of $m(t)$ is zero. Hence for the deterministic AM carrier, the power is given by

$$P_c = \tfrac{1}{2}[a_0^2 + b_0^2 P_m]$$ (2.1.8)

Note that the total carrier power is composed of a contribution from the modulation power P_m and from the carrier component a_0. Conversely, we can state that of the total available carrier power P_c only a portion is available as modulation power. Of course, it must be remembered that the overmodulation condition requires $b_0^2 P_m \leq a_0^2$.

To determine the frequency function of a deterministic AM carrier we compute the frequency transform of (2.1.1). For a deterministic waveform, $a(t)$ will have the Fourier transform

$$A(\omega) = 2\pi a_0 \delta(\omega) + b_0 M(\omega)$$ (2.1.9)

where $M(\omega)$ is the transform of $m(t)$. The carrier $c(t)$ in (2.1.1) therefore has the transform

$$C(\omega) = A(\omega) \oplus F_c(\omega)$$ (2.1.10)

where $\oplus$ denotes frequency convolution and $F_c(\omega)$ is the Fourier transform of $\cos(\omega_c t + \psi)$, given in (1.3.9). Integrating out the delta functions in the frequency convolution integral (Appendix A, A.1) yields

$$C(\omega) = \frac{e^{j\psi}}{2} A(\omega - \omega_c) + \frac{e^{-j\psi}}{2} A(\omega + \omega_c)$$ (2.1.11)

The Fourier transform of the AM carrier is therefore simply a shift of the amplitude transform in (2.1.9) to the carrier frequency $\pm \omega_c$, accompanied by a phase shift and a division by 2. We see that AM carrier modulation causes a shift of the amplitude frequency function up to the carrier frequency. Since $M(\omega)$ is generally low pass in nature (or at least confined to low frequencies), $C(\omega)$ will have its spectrum concentrated in the vicinity of ω_c, as shown in Figure 2.1. Thus AM waveforms are typically bandpass waveforms with a center frequency at the carrrier frequency. Note that since $|A(\omega)|$ is necessarily symmetric about $\omega = 0$, $|C(\omega)|$ will be symmetric about ω_c. The frequency components of $C(\omega)$ above ω_c are called *upper sideband* frequencies, and those below are *lower sideband* frequencies. We also see that if $m(t)$ has a one-sided bandwidth of B_m Hz, then $C(\omega)$ has a bandwidth of $2B_m$ Hz around its center frequency. The spectral line (delta function) at $\omega = \pm \omega_c$ indicates a sinusoidal component at the carrier frequency within the AM spectrum. This is called the *carrier component* of the modulated signal, and is often important in receiver

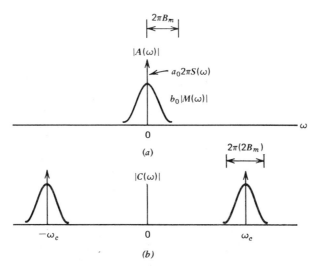

Figure 2.1. The AM frequency function. (*a*) Modulating transform, (*b*) carrier transform.

processing. In suppressed carrier AM signals $a_0 = 0$ and this carrier component is not present.

The ability to shift the baseband function to higher frequencies via amplitude modulation begins to illustrate an advantage of the carrier modulation operation, and why we use it at all in communication systems. By modulating we can arbitrarily select (to within certain constraints) the set of frequencies to be transmitted, while still sending the baseband information. For example, rather than having to transmit a frequency ω of $M(\omega)$ we can, by use of AM, instead be concerned with transmitting frequency $\omega_c \pm \omega$ of $C(\omega)$. As we shall see, it is generally easier to transmit higher frequencies over long distances than lower frequencies. Thus modulation allows us to match together properly two basically conflicting conditions—that baseband information is generally low frequency, whereas communication transmission is best accomplished at high frequencies. It is precisely this match-up that governs most of our transmitter and receiver design.

Consider now the case where the carrier amplitude function $a(t)$ is a stationary random waveform, with the carrier phase angle ψ a random phase angle uniformly distributed over $(0, 2\pi)$ and statistically independent of the $a(t)$ process. The standard AM carrier $c(t)$ becomes a random carrier process, dependent on the randomness in $a(t)$ and ψ. Its autocorrelation function can be determined from (1.3.13) as

$$R_c(\tau) = \mathscr{E}[c(t)c(t+\tau)]$$

$$= \mathscr{E}[a(t)a(t+\tau)\cos(\omega_c t + \psi)\cos(\omega_c(t+\tau) + \psi)] \quad (2.1.12)$$

where $\mathscr{E}$ represents the statistical average over the statistics of $a(t)$ and ψ. Expanding the product of the cosine terms into sum and difference angle terms, and making use of the independence of $a(t)$ and ψ, yields

$$R_c(\tau) = \tfrac{1}{2}\mathscr{E}[a(t)a(t+\tau)]\cos\omega_c\tau$$

$$+ \tfrac{1}{2}\mathscr{E}[a(t)a(t+\tau)]\mathscr{E}[\cos(2\omega_c t + \omega_c\tau + 2\psi)] \quad (2.1.13)$$

The last average term is zero because of the uniform assumption on ψ, just as in (1.5.4). The first average is precisely the definition of the autocorrelation function $R_a(\tau)$ of the $a(t)$ process and

$$R_c(\tau) = \tfrac{1}{2}R_a(\tau)\cos\omega_c\tau \quad (2.1.14)$$

Thus the autocorrelation of the random AM carrier is one-half the product of the autocorrelation of the amplitude and pure carrier processes. The power in $c(t)$ follows as

$$P_c = R_c(0) = \tfrac{1}{2}P_a \quad (2.1.15)$$

identical to (2.1.6). Hence the power in the standard AM carrier is always given by one-half the power in the amplitude waveform, whether it is considered as a deterministic or random waveform.

The power spectral density of the random AM carrier is obtained by Fourier transforming $R_c(\tau)$. We again see that this involves a frequency convolution of the spectrum $S_a(\omega)$ of $a(t)$ and the delta function spectrum in (1.5.6). Thus

$$S_c(\omega) = \tfrac{1}{2}\{S_a(\omega) \oplus [\pi\delta(\omega - \omega_c) + \pi\delta(\omega + \omega_c)]\}$$

$$= \tfrac{1}{4}S_a(\omega - \omega_c) + \tfrac{1}{4}S_a(\omega + \omega_c) \quad (2.1.16)$$

The power spectrum of $c(t)$ is therefore again obtained by shifting the power spectrum of the process $a(t)$ to $\pm\omega_c$, just as in (2.2.11). For $a(t)$ in (2.1.2)

$$S_a(\omega) = 2\pi a_0^2\delta(\omega) + b_0^2 S_m(\omega) \quad (2.1.17)$$

where $S_m(\omega)$ is now the power spectrum of $m(t)$. When (2.1.17) is used in (2.1.16) we see again that $S_c(\omega)$ occupies a one-sided spectral bandwidth equal to twice the bandwidth of the power spectrum of $m(t)$. Whether dealing with Fourier transforms or power spectra, the AM carrier frequency spectrum always appears as a shifted version of the baseband spectrum to the carrier frequency. This is all, of course, simply a consequence of the fact that amplitude modulation produces a carrier waveform that is linearly proportional to the modulating signal.

The fact that the upper and lower sidebands of the AM carrier are related suggests the possibility of tansmitting only one such part without loss of baseband information. This allows a reduction of the carrier bandwidth by one-half. This notion is the basis of *single sideband AM* (SSB-AM) modulation and is quite commonly used in modern voice communications. The objective of SSB-AM is to generate a carrier waveform in which only an upper (or lower) sideband appears. Simple filtering off of the undesired sideband is not satisfactory, since practical filters will not truly eliminate one sideband and will invariably distort the desired sideband. Instead, SSB-AM is implemented by forming at the modulator the

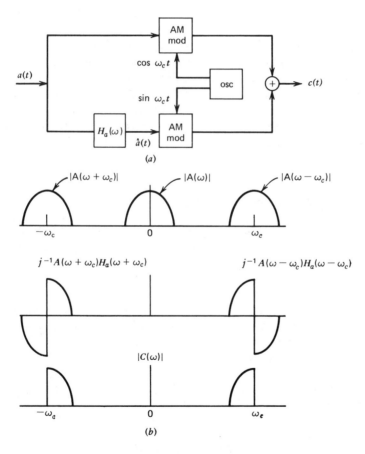

Figure 2.2. Single sideband AM. (*a*) Modulator, (*b*) spectrum.

quadrature carrier signal

$$c(t) = a(t) \cos(\omega_c t + \psi) + \hat{a}(t) \sin(\omega_c t + \psi) \qquad (2.1.18)$$

This is simply a modified AM signal in which each quadrature component is formed separately as an AM signal, with $\hat{a}(t)$ obtained by passing $a(t)$ through a linear filter $H_a(\omega)$ with transfer function (Figure 2.2a)

$$\begin{aligned} H_a(\omega) &= -j, & \omega \geq 0 \\ &= j, & \omega < 0 \end{aligned} \qquad (2.1.19)$$

The transform of $c(t)$ is then

$$C(\omega) = \tfrac{1}{2}[A(\omega - \omega_c) + A(\omega + \omega_c)]$$

$$+ \frac{1}{2j}[A(\omega - \omega_c)H_a(\omega - \omega_c) - A(\omega + \omega_c)H_a(\omega + \omega_c)] \quad (2.1.20)$$

Now since $H_a(\omega - \omega_c) = -j$ for $\omega > \omega_c$ and j for $\omega < \omega_c$, whereas $H_a(\omega + \omega_c) = -j$ for $\omega > -\omega_c$ and j for $\omega < -\omega_c$, the terms in (2.1.20) sum to eliminate one sideband, as shown in Figure 2.2b. Thus the quadrature modulation of (2.1.18) generates a SSB-AM carrier. The prime difficulty is to construct the filter $H_a(\omega)$ in (2.1.19), which requires basically a constant phase network over the bandwidth of $a(t)$. Fortunately, such networks can be adequately approximated over relatively small bandwidths, as in the case of voice systems.

2.2. The Frequency Modulated Carrier

In frequency modulation the baseband waveform $m(t)$ is used to modulate the frequency of the RF carrier. The general FM carrier has the form

$$c(t) = A \cos\left[\omega_c t + \Delta_\omega \int m(t)\, dt + \psi\right] \qquad (2.2.1)$$

where A is the carrier amplitude, Δ_ω is the frequency deviation coefficient in rps/V, and ψ is again the carrier phase. The power in a deterministic FM carrier can be determined by a procedure similar to (2.1.3)–(2.1.6). By substitution into (2.1.3) we have

$$P_c = \lim_{T \to \infty}\left[\frac{1}{2T}\int_{-T}^{T} \frac{A^2}{2}\, dt + \frac{1}{2T}\int_{-T}^{T} \frac{A^2}{2} \cos(2\omega_c t + 2\theta(t) + 2\psi)\, dt\right] \quad (2.2.2)$$

where

$$\theta(t) = \Delta_\omega \int m(t)\, dt \qquad (2.2.3)$$

To evaluate the second term, we expand and bound the magnitude of the second integral as

$$\left| \frac{A^2}{4T} \int_{-T}^{T} \cos 2\theta \cos 2(\omega_c t + \psi) \, dt + \frac{A^2}{4T} \int_{-T}^{T} \sin 2\theta \sin 2(\omega_c t + \psi) \, dt \right|$$

$$\leq \frac{A^2}{4} \left| \max \cos 2\theta(t) \right| \left| \frac{\sin (2\omega_c T) \cos 2\psi}{\omega_c T} \right|$$

$$+ \frac{A^2}{4} \left| \max \sin 2\theta(t) \right| \left| \frac{\cos (2\omega_c T) \sin 2\psi)}{\omega_c T} \right|$$

$$\leq \frac{A^2}{2\omega_c T} \tag{2.2.4}$$

In the limit as $T \to \infty$, the magnitude bound approaches zero, so that in (2.2.2),

$$P_c = \frac{A^2}{2} \tag{2.2.5}$$

for any deterministic frequncy modulating waveform $m(t)$. Thus the power in the FM carrier depends only on the amplitude A in (2.2.1) and is not a function of the modulation power P_m, as was the case in AM.

Computation of the FM frequency function is, unfortunately, more difficult, because of the nonlinearity of this type of modulation, and no simple relation exists for relating the frequency spectrum of $c(t)$ to that of $m(t)$ as in the AM case. However, we can investigate several special cases and use these results to approximate more general situations.

Sine Wave Modulation. Consider the case where the baseband modulating signal is a pure sinusoid at frequency ω_m and phase θ_m. That is, let

$$m(t) = \cos (\omega_m t + \theta_m) \tag{2.2.6}$$

We then have

$$c(t) = A \cos [\omega_c t + \beta \sin (\omega_m t + \theta_m) + \psi] \tag{2.2.7}$$

where

$$\beta = \frac{\Delta_\omega}{\omega_m} \tag{2.2.8}$$

is referred to as the *FM modulation index*. Since the peak of $m(t)$ is one, we see that Δ_ω is also the peak frequency deviation from ω_c imported to the carrier by the modulation. We wish to determine the frequency transform of $c(t)$. However, rather than formally transforming, we can instead write

$$c(t) = A \text{ Real } \{\exp (j\omega_c t + j\psi) \exp [j\beta \sin (\omega_m t + \theta_m)]\} \tag{2.2.9}$$

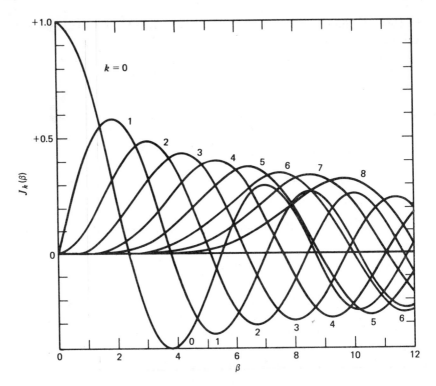

Figure 2.3.   Bessel functions.

We now use the identity

$$\exp\left[j\beta\sin\left(\omega_m t + \theta_m\right)\right] = \sum_{k=-\infty}^{\infty} J_k(\beta)\exp\left[jk\left(\omega_m t + \theta_m\right)\right] \quad (2.2.10)$$

where $J_{-k}(\beta) = (-1)^k J_k(\beta)$ and $J_k(\beta)$ is the kth order Bessel function in the argument β [1, 2]. Several of these Bessel functions are shown in Figure 2.3. Substituting into (2.2.9) and taking the real part allows us to write

$$c(t) = A\sum_{k=-\infty}^{\infty} J_k(\beta)\cos\left[\left(\omega_c + k\omega_m\right)t + k\theta_m + \psi\right] \quad (2.2.11)$$

Note that (2.2.11) appears as an infinite sum of sine waves, with amplitudes $AJ_k(\beta)$ and having frequency components above and below the RF carrier frequency ω_c. These sine waves transform to delta functions in frequency, as shown in Figure 2.4. Again, the frequency lines about the carrier frequency form upper and lower sideband frequencies of the modulated carrier. The actual shape of the spectrum (strength of the delta functions)

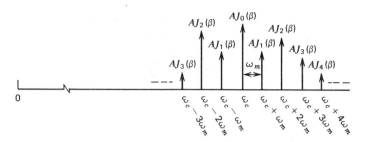

Figure 2.4. Spectrum of FM carrier with sinusoidal modulation.

varies with β, as is evident from (2.2.11). As β is changed, the various carrier and sideband components vary in their relative strengths, some increasing while others are decreasing. Contrast this result with the AM case, where a single modulating frequency produces only a single upper and lower sideband. Thus FM, being a nonlinear type of modulation, produces a carrier spectrum significantly different from a simple shift of the baseband spectrum. Note that the carrier phase angle ψ is added to every sideband, whereas multiples of the phase of the modulating sinusoids are added to each corresponding sideband.

Although (2.2.11) theoretically has an infinite number of sine terms, we note from Figure 2.3 that many higher order Bessel terms produce relatively insignificant amplitude values. In particular we see that $J_k(\beta) \ll 1$ for $\beta \ll k$. Hence only significant terms exist up to values of k slightly greater than the value of β. Roughly speaking, $c(t)$ can be considered to have $\beta + 1$ significant harmonics on each side of the carrier, corresponding to frequency components out to $[\omega_c \pm (\beta + 1)\omega_m]$. This means $c(t)$ can be considered to occupy a significant frequency bandwidth of approximately

$$B_c \cong 2(\beta + 1)f_m \text{ Hz} \qquad (2.2.12)$$

where $f_m = \omega_m/2\pi$. The bandwidth B_c is called the *Carson rule* bandwidth for $c(t)$ and is used as a rough rule of thumb for assessing the spectral width of a sinusoidally modulated FM carrier. This bandwidth approximation is convenient because of its simplicity and has been found to be fairly satisfactory in FM analysis. Note that if $\beta \gg 1$, then B_c occupies a bandwidth $\beta + 1$ times larger than that which would be produced by AM. Thus FM systems with large values of β produce wideband carriers and are often referred to as *wideband FM* systems. On the other hand, *narrowband FM* occurs if $\beta \leq 1$, in which case the FM bandwidth is comparable to that of an AM system.

To assess the meaning of the Carson rule bandwidth we might consider the effect on total power of using only M of the sidebands on each side of

ω_c. The harmonics excluded therefore constitute lost carrier power. We can therefore define a normalized power factor as

$$\mathscr{P} = \frac{\text{power in } M \text{ sidebands of } c(t)}{\text{total power of } c(t)}$$

$$= \frac{(A_2/2)\sum_{k=-M}^{M} J_k^2(\beta)}{A^2/2}$$

$$= J_0^2(\beta) + 2 \sum_{k=1}^{M} J_k^2(\beta) \qquad (2.2.13)$$

A plot of $\mathscr{P}$ in (2.2.13) is shown in Figure 2.5 for several β as a function of $2M$, the latter being the total number of sideband components used. We note a continual increase in $\mathscr{P}$ as M is increased, and appropriate bandwidths can be selected for any desired value of $\mathscr{P}$. Superimposed are the values obtained if the filter has the Carson rule bandwidth in (2.2.1), that is, if $M = \beta + 1$. We see that the Carson rule bandwidth produces power values of more than 99% of the total power. If larger values of $\mathscr{P}$ are desired, more sidebands and therefore larger bandwidths must be selected. Plots as in Figure 2.5 help us to specify the bandwidths needed.

Sum of Sine Waves. Consider now the case where $m(t)$ is the sum of K separate sine waves. That is, let

$$m(t) = \sum_{i=1}^{K} C_i \cos(\omega_i t + \theta_i) \qquad (2.2.14)$$

where C_i, ω_i, and θ_i are the corresponding deviations, frequency, and phase angles. We then have

$$c(t) = A \cos\left[\omega_c t + \sum_{i=1}^{K} \beta_i \sin(\omega_i t + \theta_i) + \psi\right] \qquad (2.2.15)$$

where $\beta_i = C_i/\omega_i$ is the modulation index of the ith sine wave. When written as in (2.2.9), and expanded using (2.2.10), we derive

$$c(t) = A \sum_{k_K=-\infty}^{\infty} \cdots \sum_{k_1=-\infty}^{\infty} \left[\prod_{i=1}^{K} J_{k_i}(\beta_i)\right] \cos\left[\omega_c t + \sum_{i=1}^{K} k_i(\omega_i t + \theta_i) + \psi\right] \qquad (2.2.16)$$

The preceding equation represents the general expression for the FM carrier modulated by K sinusoids. Note that it corresponds to a collection of harmonic frequencies at all the sidebands $\sum_{i=1}^{K} k_i\omega_i$, where all combinations of integers for the $\{k_i\}$ must be considered. Each such combination $\{k_1, k_2, \ldots, k_K\}$ yields a different sinusoid, each with its own

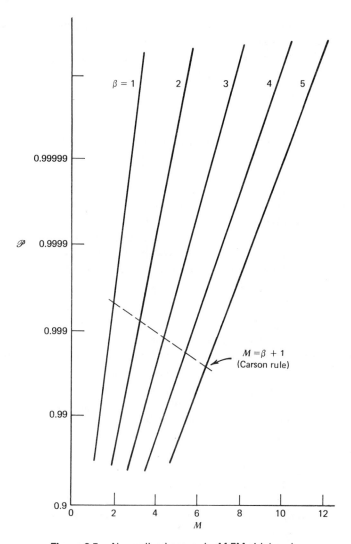

Figure 2.5. Normalized power in M FM sidebands.

phase, $\sum_{i=1}^{K} (k_i \theta_i + \psi)$ and its own amplitude $\{A \prod_i^K J_{k_i}(\beta_i)\}$. In particular, we note that the carrier component at ω_c corresponds to $k_1 = k_2 \ldots = k_K = 0$ and has amplitude $\{A \prod_i^K J_0(\beta_i)\}$, whereas the frequency component at frequency $(\omega_c + \omega_j)$ corresponds to $k_1 = 0$, $k_2 = 0, \ldots, k_j = 1$, $k_{j+1} = 0, \ldots, k_K = 0$ and has amplitude $\{A J_1(\beta_j) \prod_{i=1, i \neq j}^{K} J_0(\beta_i)\}$. We also note that the component at $(\omega + \omega_j)$ contains the exact phase angle of the jth sine wave in (2.2.14) added to that of the carrier.

The bandwidth of the FM carrier in (2.2.16) can be estimated in several ways. One way is simply to use the Carson rule bandwidth with ω_m taken to be ω_K, the highest frequency of $m(t)$, and using its index β_K. A more accurate estimate is to use an rms deivation and define an effective β as

$$\beta = \frac{[(1/K)\sum_{i=1}^{K} C_i^2/2]^{1/2}}{\omega_K} \tag{2.2.17}$$

to use the Carson rule bandwidth. Altough these procedures indicate roughly the frequency extent of (2.2.16), they do not give the true spectral shape of the carrier spectrum.

Arbitrary Modulation.   When the modulating signal $m(t)$ is deterministic, but more complicated than simply sine waves, an exact expression for the spectrum of $c(t)$ is indeed difficult to generate. Bandwidth extent can be estimated by using Carson rule bandwidths with ω_m taken as the largest significant frequency of $m(t)$, and by using an effective index of

$$\beta = \frac{\Delta_\omega \sqrt{P_m}}{\omega_m} \tag{2.2.18}$$

where Δ_ω is the peak frequency deviation and P_m is the power of $m(t)$. A more accurate method is to approximate the modulation frequency function with a sum of sine waves at appropriately selected frequencies within the band, each with proper amplitude so as to represent the spectrum in its vicinity, and using (2.2.17). This allows the analysis to account for portions of the modulation frequency spectrum within the band that may have significantly more power, and therefore more effect on carrier spectrum, than the band edge frequencies.

To account for random modulating signals we must reconsider (2.2.1) with $m(t)$ modeled as a random process of known statistics. In this case it is more convenient mathematically to treat $c(t)$ as a carrier that is phase modulated with the integral of the baseband process and to derive spectral information in this way. This we do in the next section.

2.3. The Phase Modulated Carrier

If the baseband signal $m(t)$ phase modulates the RF carrier, the carrier waveform is

$$c(t) = A \cos[\omega_c t + \theta(t) + \psi]$$
$$\theta(t) = \Delta m(t) \tag{2.3.1}$$

where $\theta(t)$ is the resulting baseband phase modulation. In this case the phase of the carrier varies directly with the desired modulation, and ψ represents an added phase shift that may appear with the carrier. In later work we also are interested in the situations where ψ itself is a function of t, representing extraneous or unintentional phase modulations superimposed on the carrier waveform. The deterministic PM carrier has a power of

$$P_c = \frac{A^2}{2} \tag{2.3.2}$$

obtained by the same development as in (2.2.2)–(2.2.5). To determine the frequency transform we again examine specific types of modulating waveforms.

Sine Wave. For the case of a single sine wave modulation at ω_m,

$$\theta(t) = \Delta \sin(\omega_m t + \theta_m) \tag{2.3.3}$$

where Δ is now the phase modulation index in radians per volt. When (2.3.3) is substituted into (2.3.1), the carrier waveform is identical to (2.2.7) with β replaced by Δ. The resulting carrier expansion in (2.2.11), and Carson rule bandwidth in (2.2.12), can therefore be used directly with this substitution. We point out, however, that (2.2.11) is also valid for PM even if θ_m depended on t [i.e., if θ_m represented phase or frequency modulation of the modulating sine wave]. In this case the sidebands in (2.2.11) now become phase modulated sidebands located at the same sideband frequencies. This was not true for the FM case, because the integral of a frequency modulating sine wave containing phase modulation is not equivalent to phase modulating with a sine wave having the same modulation. Often this problem is circumvented in FM, however, by arguing that the time modulation of the sine wave is so much slower than that of the sine wave frequency itself that it can be considered a constant. This assumption allows us to replace the frequency modulating sine wave by a phase modulating sine wave with the same angle modulation, and by using our earlier results in Section 2.2.

Sum of Sine Waves. The results for the case of phase modulating with a sum of sine waves also carry over from the FM case. Let the phase modulation be

$$\theta(t) = \sum_{i=1}^{K} \Delta_i \sin(\omega_i t + \theta_i) \tag{2.3.4}$$

where Δ_i is the phase modulation index of the ith sine wave. The carrier $c(t)$ in (2.3.1) therefore has the same form as (2.2.15) with β_i replaced by

Δ_i. The expansion of (2.2.16) is again valid with this substitution. The resulting carrier spectrum therefore appears as a collection of frequency components at all the beat frequencies of the modulation in (2.3.4). Again we point out that the result is valid even if the individual sine waves were themselves modulated [i.e., each θ_i in (2.3.4) depended on t]. The spectral lines in (2.2.16) now broaden into individual PM spectra at the same center frequencies with a bandwidth determined by the modulation on the phase of each component. Bandwidth extent can again be estimated by the Carson rule bandwidth with Δ replacing β and Δ approximated as $[(\sum_{i=1}^{K} \Delta_i^2/2K)]^{1/2}$.

Binary Waveforms. In digital systems we often deal with pulsed waveforms that can only take on one of two possible values at any t. Such waveforms are called *binary waveforms*. Consider phase modulating with a binary waveform of the form

$$\theta(t) = \Delta q(t) \tag{2.3.5}$$

where $q(t)$ takes on only the value ± 1 at any t and Δ is the phase index. The waveform $q(t)$ can correspond to a unit square wave or can jump randomly between $+1$ and -1, as in some type of coded binary signal. The corresponding PM waveform is then

$$c(t) = A \cos [\omega_c t + \Delta q(t) + \psi]$$
$$= A \cos (\Delta q(t)) \cos (\omega_c t + \psi) - A \sin (\Delta q(t)) \sin (\omega_c t + \psi) \tag{2.3.6}$$

Now using the fact that

$$\sin (\Delta q(t)) = q(t) \sin \Delta$$
$$\cos (\Delta q(t)) = \cos \Delta \tag{2.3.7}$$

we can rewrite (2.3.6) as

$$c(t) = (A \cos \Delta) \cos (\omega_c t + \psi) - (A \sin \Delta) q(t) \sin (\omega_c t + \psi) \tag{2.3.8}$$

The first term represents a carrier component with amplitude $A \cos \Delta$, whereas the second represents a carrier signal amplitude modulated by the waveform $q(t)$. The results in Section 3.1 can therefore be applied, and the second term will have a frequency spectrum given by the shift of the frequency function of spectrum of $q(t)$ to the frequency ω_c. In later work, where the waveform $q(t)$ is more specifically described, (2.3.8) is useful for estimating the resulting carrier PM spectrum.

Random Modulation. We now want to examine the spectral density of a PM wave when phase modulated by a stationary random process. Consider again the PM carrier in (2.3.1) where now $\theta(t)$ is a random modulation waveform having autocorrelation $R_\theta(\tau)$ and ψ is a random phase angle, uniformly distributed over $(0, 2\pi)$ and independent of $\theta(t)$. The carrier $c(t)$ has autocorrelation

$$R_c(\tau) = \frac{A^2}{2} \mathscr{E}[\cos[\omega_c\tau - \theta(t) + \theta(t+\tau)]]$$

$$+ \frac{A^2}{2} \mathscr{E}[\cos[2\omega_c t + \omega_c\tau + \theta(t) + \theta(t+\tau) + 2\psi]] \quad (2.3.9)$$

The average over ψ in the second term is zero, because of the uniform assumption on ψ, which means

$$R_c(\tau) = \frac{A^2}{2} \mathscr{E}[\cos[\omega_c\tau - \theta(t) + \theta(t+\tau)]] \quad (2.3.10)$$

Since the average of a real part of a complex random variable is equal to the real part of its complex average, we can instead write

$$R_c(\tau) = \frac{A^2}{2} \operatorname{Real}\{e^{j\omega_c\tau}\mathscr{E}[e^{-j[\theta(t)-\theta(t+\tau)]}]\}$$

$$= \frac{A^2}{2} \operatorname{Real}\{e^{j\omega_c\tau}\Psi_\theta(1, -1, \tau)\} \quad (2.3.11)$$

where now we have defined

$$\Psi_\theta(\omega_1, \omega_2, \tau) = \mathscr{E}[e^{j\omega_1\theta(t+\tau)+j\omega_2\theta(t)}] \quad (2.3.12)$$

and the average is over the random phase $\theta(t)$. We recognize (2.3.12) as the second order *characteristic function* of the process $\theta(t)$ [see Appendix B (B.2.5)]. Since the modulating process $\theta(t)$ is stationary, its characteristic function depends only on τ. Equation (2.3.12) represents a general expression for the autocorrelation function of any randomly modulated PM carrier with uniform phase variable and requires only the knowledge of the characteristic function of the modulation. Nevertheless, the carrier power follows as in (2.3.2) since $P_c = R_c(0) = A^2/2$ for any random modulation. Thus the PM carrier power is always $A^2/2$ whether the modulation is deterministic or random, just as in the FM case.

The result in (2.3.11) can be expanded further by noting that if we define this new random variable at t and $t+\tau$ by

$$z_1 = e^{j\theta(t+\tau)}$$

$$z_2 = e^{-j\theta(t)} \quad (2.3.13)$$

then

$$\Psi_\theta(1, -1, \tau) = \mathscr{E}[z_1 z_2]$$
$$= \mathscr{E}[(z_1 - \bar{z}_1)(z_2 - \bar{z}_2)] + \bar{z}_1 \bar{z}_2 \qquad (2.3.14)$$

where

$$\bar{z}_1 = \mathscr{E}[z_1] = \Psi_\theta(1, 0, 0)$$
$$\bar{z}_2 = \mathscr{E}[z_2] = \Psi_\theta(0, -1, 0) = \Psi_\theta^*(1, 0, 0) \qquad (2.3.15)$$

and * denotes the complex conjugate. The last identity follows from the stationarity of the process. Note that the first term in (2.3.14) depends on τ and the second term does not. If we denote the first term as $\tilde{\psi}_\theta(\tau)$, then (2.3.13) can be rewritten as

$$R_c(\tau) = \frac{A^2}{2} \text{Real} \{e^{j\omega_c\tau}[|\Psi_\theta(1, 0, 0)|^2 + \tilde{\Psi}_\theta(\tau)]\}$$
$$= \frac{A^2}{2} |\Psi_\theta(1, 0, 0)|^2 \cos \omega_c\tau + \frac{A^2}{2} \text{Real} \{\tilde{\Psi}_\theta(\tau) e^{j\omega_c\tau}\} \qquad (2.3.16)$$

In this way we have separated out the constant portion of the function $\Psi_\theta(1, -1, \tau)$. The first term will transform to a delta function (spectral line) at $\omega = \pm\omega_c$ in the power spectrum. This therefore represents the carrier component of the spectrum, and it is interesting that it depends only on the factor $|\Psi_\theta(1, 0, 0)|^2$ in (2.3.15). Since $A^2/2$ is the total carrier power and since $|\Psi(1, 0, 0)|^2 \le 1$, the strength of the carrier component is a computable fraction of the total carrier power. For this reason the factor $|\Psi(1, 0, 0)|^2$ is often called a carrier component *suppression factor* and can be determined directly from the modulation statistics. The second term in (2.3.16) transforms to a continuous spectrum cnetered at $\pm\omega_c$ the actual shape dependent on the form of the function $\tilde{\Psi}_\theta(\tau)$.

When the phase modulation process $\theta(t)$ is assumed to be a Gaussian random process, the general autocorrelation in (2.3.16) can be expanded further. In this case it is known that the second order characteristic function of a Gaussian process is

$$\Psi_\theta(\omega_1, \omega_2, \tau) = \exp -\left[\frac{\Delta^2\omega_1^2}{2} + \frac{\Delta^2\omega_2^2}{2} + R_\theta(\tau)\omega_1\omega_2\right] \qquad (2.3.17)$$

where $\Delta^2 = R_\theta(0)$. We therefore have

$$|\Psi_\theta(1, 0, 0)|^2 = (e^{-\Delta^2/2})^2 = e^{-\Delta^2} \qquad (2.3.18a)$$
$$\tilde{\Psi}_\theta(\tau) = e^{-\Delta^2 + R_\theta(\tau)} - e^{-\Delta^2} = e^{-\Delta^2}(e^{R_\theta(\tau)} - 1) \qquad (2.3.18b)$$

and

$$R_c(\tau) = \frac{A^2}{2} e^{-\Delta^2} \cos \omega_c \tau + \frac{A^2}{2} e^{-\Delta^2} (e^{R_\theta(\tau)} - 1) \cos \omega_c \tau \quad (2.3.19)$$

By expanding

$$e^{R_\theta(\tau)} = 1 + \sum_{i=1}^{\infty} \frac{R_\theta^i(\tau)}{i!} \quad (2.3.20)$$

we can write (2.3.19)

$$R_c(\tau) = \left(\frac{A^2}{2} e^{-\Delta^2}\right) \left[\cos \omega_c \tau + R_\theta(\tau) \cos \omega_c \tau + \frac{R_\theta^2(\tau)}{2!} \cos \omega_c \tau \cdots \right]$$

$$(2.3.21)$$

The transform of (2.3.21) yields the power spectrum. Hence

$$S_c(\omega) = \frac{A^2 e^{-\Delta^2}}{2} \left[\pi \delta(\omega \pm \omega_c) + \tfrac{1}{2} S_\theta(\omega \pm \omega_c) + \tfrac{1}{4} [S_\theta(\omega) \oplus S_\theta(\omega)]_{\omega = \omega \pm \omega_c} \cdots \right]$$

$$(2.3.22)$$

where $S_\theta(\omega)$ is the power spectrum of $\theta(t)$ and $\oplus$ denotes frequency convolution. The first term is the carrier component, the second is a shift of the modulating spectrum to $\pm \omega_c$, the third is a shift of the convolved modulating spectrum, and so on. The remaining terms involve higher order convolutions of the baseband spectrum, each shifted to $(\pm \omega_c)$. (Recall that convolving two functions spreads the resulting function while reducing its amplitude.) The resulting PM spectrum appears as sketched in Figure 2.6. We see that the spectrum is actually built up from an infinite sequence of overlapping convolved spectra. Hence the phase modulated spectrum is much more complicated than that of the modulation itself and does not simply correspond to a shift of the baseband up to the carrier frequency. The bandwidth extent can usually be estimated by superimposing the first few terms of (2.3.22). More quantitative estimates can often be derived (Problem 2.9).

In summary, these first sections have attempted to outline the intrinsic properties of the modulated carrier waveform generated at the transmitter.

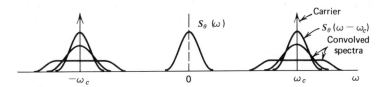

Figure 2.6. PM spectrum due to random Gaussian phase modulation.

We have concentrated particularly on the power and on the modulated carrier frequency functions and the manner in which they are derived from the baseband waveform. We have shown that in all cases the modulated carrier has a bandpass frequency spectrum that theoretically can be located anywhere along the frequency axis by proper choice of carrier center frequency. As stated earlier, this basic property of frequency spectral control represents the primary reason for utilizing a transmitter modulation operation in communication systems. In subsequent discussions these advantages become more apparent.

2.4. Carrier Amplification

At the transmitter modulator the carrier waveform is formed from the baseband modulating waveform, as in Figure 2.7a. The resulting carrier $c(t)$ has the waveform, power, and frequency characteristics described in Sections 2.1–2.3, depending on the format selected. The transmitter power amplifier then must amplify the carrier signal to the desired power value for transmission, over either the antenna or cable propagation system. (That is, the amplifier multiplies the waveform by a gain constant such that the resulting carrier power P_c has the desired value.) We shall find that system performance invariably depends on the amount of carrier amplification that can be generated at the transmitter. Therefore there always exists a requirement for extremely high power amplification at carrier frequencies. Furthermore this amplification must extend over the complete bandwidth of the modulated carrier, so as not to cause carrier amplitude or phase distortion. In some transmitters there may be multiple carriers to be amplified by a single power amplifier. The power amplifier must have sufficient gain and bandwidth to properly amplify all carriers simultaneously.

Conventional electronic amplifiers are limited by parasitic capacitance and electron transit times, both tending to reduce gain at the higher carrier frequencies. For this reason, cavity amplifiers are necessary for producing higher power levels over wide, high frequency bandpass bandwidths. Such devices are physically large and usually have relatively poor amplifier *efficiency*. Efficiency is the ratio of the carrier power output to the prime power (batteries, generators, solar cells, etc.) that must be provided to operate the amplifier. The lower the efficiency, the less the carrier power for a given amount of prime power, or, conversely, the more prime power to achieve a desired carrier power. Also important is the fact that the unused prime power (one minus the efficiency times the prime power) represents power that must be dissipated as heat in some way at the

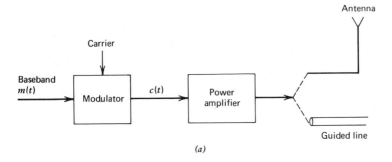

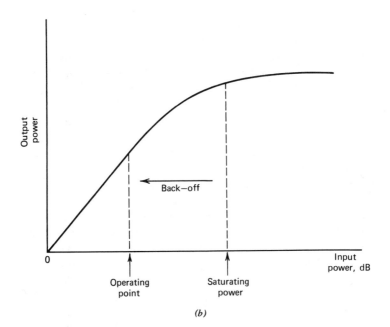

Figure 2.7. The transmitter carrier subsystem. (*a*) Block diagram, (*b*) typical power amplifier characteristic.

transmitter. In systems where prime power is constrained by weight or cost limitations, the efficiency of the power amplifier becomes a critical design parameter, often dictating the ultimate hardware components.

The most common types of high gain cavity amplifiers are the *klystron* and the *traveling wave tube* (TWT) [3, 4]. Both devices can produce amplification factors on the order of 50–60 dB with efficiencies of about 40–60%. However, the klystron is generally quite large in physical size

(3–6 ft long), which limits its usefulness to Earth-based operation. The TWT, because it is more compact, has emerged as the basic spaceborne transmitter amplifier for satellites and space vehicles. However, TWT amplifiers exhibit saturation effects when operated at high input (*drive*) power levels. Figure 2.7*b* shows a typical TWT curve for available output power versus input drive power, and indicates the pronounced saturation of the amplification. For low values of input power the output power is linearly related to the input, and the device operates as a conventional linear amplifier. As the input waveform amplitude is increased, the amplifier becomes nonlinear, with the output amplitude eventually becoming saturated as the attainable power from the amplifier levels off. Achievement of the maximum power in the amplifier is therefore accompanied by a nonlinear amplification of the input signal, as exhibited by the saturation effect. When only a single RF carrier is being amplified, this nonlinearity poses no serious problem, since the amplifier harmonics produced will be well outside the RF bandwidth of the carrier. Thus single carrier power amplifiers are usually operated with as large an input amplitude as possible so as to achieve the maximum available output power. The amplifier is said to be *saturated*, and the drive power at which the saturation occurs is called its *saturating power*. When multiple carriers are involved, the nonlinearity of the amplifier causes carrier cross-products, or *intermodulation* interference, to occur. By reducing the total input drive power from its saturating power, a more linear amplification is produced (with less intermodulation) but with less total output power to be divided among the carriers. The ratio of saturating drive power to desired operating drive power is called the amplifier *backoff*. Increasing backoff produces less output power but less intermodulation interference, and a tradeoff study must be made to assess the effect of each in selecting power amplifier operating points.

Solid state amplifiers are also used for power amplification. The two most common solid state devices are the *tunnel diode* [3] amplifier and the *maser* [5, 6]. Tunnel diodes are compact, have fairly good efficiencies and low noise levels, and can provide 20–40 dB of amplification, but they are usually limited in their peak power output capability. Thus tunnel diodes are mostly used in low level transmitters or as drive amplifiers in cascade with more powerful cavity amplifiers. Masers provide much larger gain values but usually must be cooled to operate well below room temperatures for best performance. This requires additional cooling apparatus that must also be incorporated at the transmitter.

In any transmitter the parameter of particular importance is the actual amount of amplified carrier power coupled into the antenna or guide system from the amplifier. This depends not only on the amplifier size but

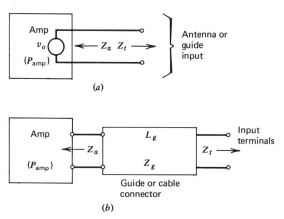

Figure 2.8. Amplifier coupling circuitry. (*a*) Direct amplifier coupling, (*b*) cable connected amplifier·coupling.

also on the connecting circuitry from the amplifier output to the antenna or cable input terminals. Power levels quoted for carrier amplifiers are usually for matched impedance loading (i.e., the input impedance of the cable or antenna is equal to the output impedance of the amplifier). When the impedances are not matched, a power loss occurs in the coupling. This can be accounted for as shown in Figure 2.8*a*. If v_0 is the open circuit output voltage of the amplifier, Z_a its output resistance, and Z_t the input resistance of the transmission antenna or cable terminals, the transmitter power (voltage squared divided by resistance) at the antenna or cable input is then

$$P_t = v_0^2 \left[\frac{Z_t}{Z_t + Z_a} \right]^2 \left(\frac{1}{Z_t} \right) \qquad (2.4.1)$$

Under matched load conditions $Z_a = Z_t$ and $P_t = v_0^2/4Z_a \triangleq P_{amp}$, which is the stated amplifier output power. For the mismatched case, however, (2.4.1) becomes

$$P_t = P_{amp} \left[\frac{4Z_a Z_t}{(Z_a + Z_t)^2} \right] \qquad (2.4.2)$$

The bracket accounts for the power coupling loss into the transmission terminals. If the antenna is remote from the amplifier, connecting lines or waveguides are necessary, which inject additional loss. Now coupling losses may occur at both the waveguide and antenna input, in addition to losses in the guide itself (Figure 2.8*b*). If L_g is the ratio of the guide output power to its input power when fed and terminated by a matched impedance, (2.4.2) becomes instead

$$P_t = P_{amp}L_g\left[\frac{4Z_tZ_g}{(Z_t+Z_g)^2}\right]\left[\frac{4Z_aZ_g}{(Z_a+Z_g)^2}\right] \qquad (2.4.3)$$

where Z_g is the ohmic impedance of the guide. When properly matched $Z_t = Z_g = Z_a$ and $P_t = L_gP_{amp}$. Thus the brackets in (2.4.3) account for the additional coupling losses. Waveguide loss parameters L_g are usually specified in decibel loss per length, and therefore the power loss in the guide is directly dependent on its length. This can become a rather serious effect when the antenna is remotely located from the power amplifier. It should also be pointed out that waveguide impedance values Z_g are functions of frequency, and the impedance matching condition is frequency dependent. Ideally, Z_g is designed to be matched at the carrier band center, that is, the carrier center frequency.

2.5. Transmitting Antennas

In communication systems using unguided space transmission, the carrier waveform is propagated from the transmitter by use of a transmitting antenna. An antenna is simply a transducer that converts electronic signals into electromagnetic fields, or vice versa. A transmitting antenna converts the amplified carrier signal into a propagating electromagnetic field. This field is transmitted by the antenna as a propagating plane wave with a prescribed polarization and spatial distribution of its field power density. The spatial power distribution of a transmitting antenna is described by its *antenna pattern*, the latter defined as

$$w(\phi_l, \phi_z) = \frac{\text{power transmitted per unit solid}}{\text{angle in the direction } (\phi_l, \phi_z)} \qquad (2.5.1)$$

where (ϕ_l, ϕ_z) are the elevation and azimuth angles defining a ray line direction from the geometrical center of the antenna, as shown in Figure 2.9. Thus the antenna pattern indicates the amount of field power that will pass through a unit solid angle* in a given direction out from the antenna center. The antenna pattern is often normalized to form the *antenna gain function*

$$g(\phi_l, \phi_z) = \frac{w(\phi_l, \phi_z)}{\left[\begin{array}{c}\text{power per unit solid angle of an isotropic} \\ \text{radiator with same total transmitted power}\end{array}\right]} \qquad (2.5.2)$$

---

* Recall that the solid angle of a cone enscribed on a sphere of radius D is given by the surface area of its spherical cap, divided by D^2.

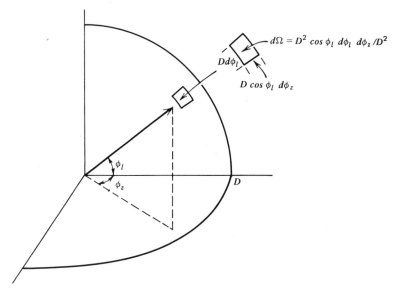

$$d\Omega = D^2 \cos \phi_l \, d\phi_l \, d\phi_z / D^2$$

$$Dd\phi_l$$

$$D \cos \phi_l \, d\phi_z$$

$$D$$

Figure 2.9. Antenna geometry.

An *isotropic* radiator is an antenna that radiates equally in all directions. The total power transmitted by an antenna P_T is the integral part of $w(\phi_l, \phi_z)$ over the unit sphere,

$$P_T = \int_{\text{unit sphere}} w(\phi_l, \phi_z) \, d\Omega = \int_{-\pi}^{\pi} \int_{-\pi/2}^{\pi/2} w(\phi_l, \phi_z) \cos \phi_l \, d\phi_l \, d\phi_z \tag{2.5.3}$$

where $d\Omega = (\cos \phi_l) \, d\phi_l \, d\phi_z$ is the differential solid angle (Figure 2.9). The power per unit solid angle of an isotropic antenna with the same total power is then $P_T / 4\pi$. Therefore

$$g(\phi_l, \phi_z) = \frac{w(\phi_l, \phi_z)}{P_T / 4\pi} \tag{2.5.4}$$

Thus the gain function of a transmitting antenna is simply a normalized form of the antenna pattern. Note that the area under the gain function is always 4π. This means that if the gain function is increased in one direction, it must be suitably decreased in other directions so as to maintain constant area. In general, a transmitting antenna is designed to transmit its radiated field in one basic direction, and therefore the gain function is usually peaked along one specific ray line, referred to as the antenna *main lobe*. Transmissions in other directions are called *sidelobes* of the antenna

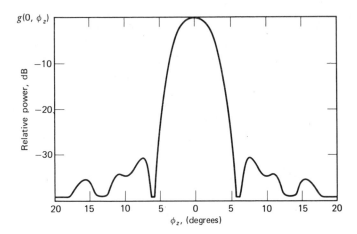

Figure 2.10. Antenna gain function in azimuth ($\phi_l = 0$).

and represent power transmission in unwanted directions. Figure 2.10 shows a typical parabolic antenna pattern plotted as a function of azimuth angle in degrees, for the elevation plane $\phi_l = 0$, showing clearly the main lobe in the forward ($\phi_z = 0$) direction and the associated sidelobe directions. A similar (but sometimes different) pattern exists for the elevation angle in the plane $\phi_z = 0$.

Rather than deal with the actual antenna gain function, we often simply specify its *gain* and *field of view*. The gain g of a transmitting antenna is the maximum value of its gain function:

$$g \triangleq \max_{\phi_l, \phi_z} g(\phi_l, \phi_z) \qquad (2.5.5)$$

and is often stated in decibels using (1.6.20). The antenna solid angle field of view is a measure of the solid angle into which most of the transmitted field power is concentrated. That is, the field of view is a measure of the directional properties of the antenna. Although field of view can be defined in several ways from $g(\phi_l, \phi_z)$ (3 dB angle, main lobe angle, etc.), the simplest is to define the field of view as the solid angle Ω_{fv} through which all the radiated power would pass if the antenna pattern in this angle were constant and equal to its maximum value. Thus $\Omega_{fv}[\max w(\phi_l, \phi_z)] = P_T$, or

$$\Omega_{fv} \triangleq \frac{P_T}{\max w(\phi_l, \phi_z)}$$

$$= \frac{4\pi}{g} \text{ steradians} \qquad (2.5.6)$$

Field of view is therefore inversely related to gain, and high gain antennas are commensurate with narrow fields of view. Instead of using the solid angle field of view, we often instead deal with the planar angle *beamwidth* in radians or degrees, defined separately in the azimuth and elevation planes. For example, the antenna pattern in Figure 2.10 has a main lobe azimuth beamwidth of about 12°. (This is just enough to cover the Earth when transmitting from a satellite at an altitude of 22,000 miles.) For a symmetric gain pattern the planar beamwidth Φ_b, in radians in any plane, is related to the solid angle field of view in steradians by (Problem 2.20)

$$\Omega_{fv} = 2\pi[1 - \cos(\Phi_b/2)]$$

$$\approx \frac{\pi}{4}\Phi_b^2, \qquad \Phi_b \ll 1 \tag{2.5.7}$$

Since transmitting antennas inherently have ohmic losses, the power P_T radiated is less than the power fed into the input terminals. It is therefore convenient to define an *effective antenna transmitting gain*, relative to the power actually coupled into the antenna input terminals [P_t in (2.4.2) or (2.4.3)]. Hence we define the effective gain function

$$\tilde{g}(\phi_l, \phi_z) \triangleq \frac{w(\phi_l, \phi_z)}{P_t/4\pi} \tag{2.5.8}$$

This means

$$\tilde{g}(\phi_l, \phi_z) = \rho_r g(\phi_l, \phi_z) \tag{2.5.9}$$

where $\rho_r \triangleq P_T/P_t$ is the antenna *radiation efficiency* factor. In essence, ρ_r is a measure of the radiation losses of an antenna. Note that $\rho_r \leq 1$, so that the effective antenna gain $\tilde{g}$ is always less than the gain g of the antenna pattern. Typical radiation efficiencies are on the order of $0.90 \leq \rho_r \leq 0.99$, corresponding to losses of about 0.5 dB in gain.

The value of an antenna gain function in communication analysis is that it allows immediate calculation of the amount of transmitted field power that will impinge in a normal receiving surface area of size $\mathscr{A}$ located a distance D in direction (ϕ_{l0}, ϕ_{z0}) from the transmitting antenna (Figure 2.11). The power over $\mathscr{A}$, P_a, is from (2.5.1)

$$P_{\mathscr{A}} = w(\phi_{l0}, \phi_{z0})\Omega_a \tag{2.5.10}$$

where Ω_a is the solid angle subtended by $\mathscr{A}$ when viewed from the antenna. If $D \gg \sqrt{\mathscr{A}}$, then

$$\Omega_a \cong \frac{\mathscr{A}}{D^2} \tag{2.5.11}$$

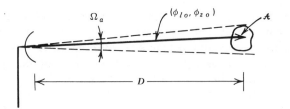

Figure 2.11. Antenna propagation diagram.

and from (2.5.8), (2.5.9), and (2.5.10) we have

$$P_{\mathscr{A}} = w(\phi_{l0}, \phi_{z0})\frac{\mathscr{A}}{D^2}$$

$$= \left[\frac{P_t \rho_r g(\phi_{l0}, \phi_{z0})}{4\pi D^2}\right]\mathscr{A} \qquad (2.5.12)$$

The quantity in brackets has units of power/area, and therefore represents the power density (called the *field intensity*, or *flux density*) of the propagating electromagnetic field at the distance D. The term in the numerator is often referred to as the *effective isotropic radiated power* (EIRP) in the direction ϕ_{l0}, ϕ_{z0}). That is,

$$\text{EIRP} = P_t \rho_r g(\phi_{l0}, \phi_{z0}) \qquad (2.5.13)$$

and represents the equivalent power that an isotropic antenna must transmit in order to provide the same field power density at $\mathscr{A}$. Note that electromagnetic field intensity decreases as distance squared when propagating in an undistorting (free space) medium.

An intrinsic property of any antenna is that its transmitting field of view at wavelength λ is related to the physical area of the antenna, $\mathscr{A}_a$, by [3, 7, 8]

$$\Omega_{fv} = \frac{\lambda^2}{\rho_{ap}\mathscr{A}_a} \qquad (2.5.14)$$

where ρ_{ap} is the antenna *aperture loss factor*. This aperture loss factor accounts for antenna diffraction losses and typically has a value in the range 0.5–0.75. This means that from (2.5.6) the antenna gain is given by

$$g = \left(\frac{4\pi}{\lambda^2}\right)\rho_{ap}\mathscr{A}_a \qquad (2.5.15)$$

Equations (2.5.14) and (2.5.15) are the key transmitting antenna equations that are of most use to communication engineers, since they relate gain and field of view to antenna size and carrier frequency. Note that for a fixed antenna size, higher frequencies have larger gains and narrower field of views. Alternatively, at a specific carrier frequency, increasing gain and narrowing field patterns require larger antennas. For a parabolic antenna having a diameter of d meters and $\rho_{\text{ap}} = 1$ the gain and planar beamwidth equations [the latter obtained by using (2.5.7)] at frequency f_c reduce to

$$g = \left(\frac{\pi}{3}\right)^2 \left(\frac{df_c}{10^8}\right)^2 \qquad (2.5.16a)$$

$$\Phi_b = (1.9)^2 \left(\frac{10^8}{f_c d}\right) \qquad (2.5.16b)$$

For $d = 1$ m, we need $f_c = 10^9 = 1$ GHz for a 20 dB ($g = 100$) gain, whereas a frequency of 100 MHz requires $d = 10$ m for the same gain. Thus, for practical antenna sizes, we must transmit fairly high waveform frequencies in order to obtain sufficient antenna gain. This now demonstrates a basic reason why modulation is used prior to transmission. It allows low baseband frequencies to be shifted up to high carrier frequency ranges where sufficient antenna gains and reasonable antenna sizes are available. If modulation was not used we would have to deal with the antenna parameters based on baseband frequencies instead of carrier frequencies. We obtain the same conclusion if we instead designed an antenna using (2.5.16b) instead of specifying a desired gain. This would occur, for example, if our prime objective was to achieve a specified beamwidth in order to confine power transmission properly. This equation also dictates large antennas and high carrier frequencies. For example, a 2 m antenna would have to be operated at 1 GHz to have the beamwidth in Figure 2.10.

A nomogram relating gain and beamwidth to parabolic antenna diameter d and carrier frequency f_c in (2.5.16) is shown in Figure 2.12. (A straight line connecting two of the parameters will read off the corresponding value of the third parameter.) It should be remembered that the listed value of d is for a unit value of aperture loss, $\rho_{\text{ap}} = 1$. The true physical antenna diameter needed is related to the diameter value listed by $d_{\text{phy}} = d/\sqrt{\rho_{\text{ap}}}$.

In transmitting from satellites to Earth several antennas may be available on the satellite for carrier transmissions. These antennas are generally separated into *global* antennas and *spot beam* antennas. The global antenna has a pattern that provides coverage of the entire Earth surface, as viewed from the satellite. From Problem 2.10 such a pattern has a beamwidth of approximately 18°. A spot beam has a much narrower beamwidth, approximately 4°, and is designed to illuminate only a specific portion of the Earth's surface. Spot beam antennas produce more gain but require careful control of pointing direction. When several spot beams are used on

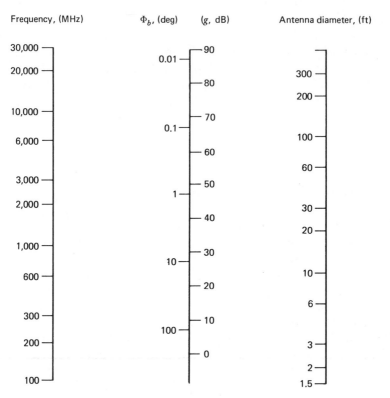

Frequency, (MHz) Φ_b, (deg)   (g, dB)   Antenna diameter, (ft)

Figure 2.12. Nomogram for parabolic antenna gain and beamwidth (g = gain, Φ_β = beamwidth in degrees).

board a satellite for downlink transmission, each can be directed to a separate region of the Earth. An immediate advantage of this is that the same frequency band can be used to transmit separate downlink carriers on each spot beam. This is referred to as *frequency reuse.*

2.6. Carrier Propagation Channels

When designing a communication system, the properties of the propagation path from transmitter to receiver must be taken into account. Proper characterization of the path is equivalent to defining the communication channel of Figure 1.1. Electromagnetic propagating channels can be roughly separated into guided and unguided transmissions. Again, a thorough discussion of electromagnetic propagation in all types of media is beyond our scope here. In this section, however, we review the salient features of each of the channels and attempt to summarize much of the extensive work dealing with channel measurements and models [9–20]. Our basic objective is to indicate the principal characteristics of the channel and to make the system engineer aware of the manner in which they affect the overall communication operation.

Guided Channels. In a guided channel the electromagnetic field is confined to a closed path or "pipe" from transmitter to receiver. Such channels occur whenever hardwire, cabling, or waveguides are used, such as for telephone lines, cable transmissions, or short range internal communications. The guided channel is used primarily for convenience. It has the advantage of being completely shielded from external interference and is therefore extremely suitable for establishing a link in areas dense with electromagnetic fields. Also, the guided channel can be inserted to avoid propagation obstacles and to establish privacy links, as in home cable television. It also has the advantage of allowing complete control of the electromagnetic field during propagation by the insertion of internal amplifiers (*repeaters*) spaced along the line. The field can be amplified and regenerated at these points to restore it to proper power levels, as is discussed in Section 3.4. This becomes significant when other forms of field amplification are not possible, such as with transoceanic telephone cabling.

The basic disadvantages of guided channels are the cost of implementing such channels and the excessive attenuation of the field in the guide. Guide attenuation values are typically stated in terms of decibel loss per unit length, so that the total attenuation applied to the field from trnasmitter to receiver depends on the guide length. Although such attenuation factors depend on the type, size, and material of the cable, in general, lower attenuation is achieved only with larger cable sizes and therefore increased cost. Low cost, lossy lines are limited in their usable length and confine the available link to relatively short distances (several tens of miles). As stated, cable attenuation can be overcome either by amplifier insertion or by resorting to larger cables at increased cost. One must therefore trade off the increased cost of less lossy cables against the additive cost of amplifier insertions and their accompanying amplifier noise.

An added complexity is that attenuation factors also depend on propagating frequency. Figure 2.13 shows a typical plot of coaxial cable attenuation versus propagating frequency for two representative coaxial sizes. The line attenuation increases with frequency and decreases with cable size. This rapid increase with frequency prevents use of carrier frequencies above UHF and generally confines operation to the 10–20 MHz range. The variation with frequency also means that different frequencies of the transmitted waveform may undergo different attenua-

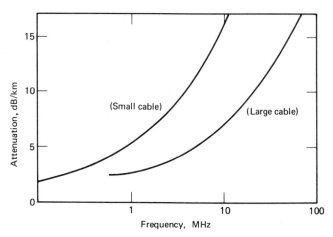

Figure 2.13. Typical cable attenuation versus frequency (small cable, 0.17 in., large cable, 0.37 in.).

tion and therefore produce an effective filtering (amplitude distortion) of the signal. This is important when low frequency baseband waveforms are transmitted directly, such as with telephone lines. This effect requires line *equalization* to even out the frequency response over the waveform bandwidth, adding to the cost of the line. By modulating onto a carrier and operating at higher frequencies, where the bandwidth is a smaller percentage of the propagating frequencies, this effect becomes negligible, although the overall attenuation is greater. In addition, guides must be properly terminated to avoid reflections and allow most efficient coupling out of the propagating energy. This termination requires impedance matching to the receiver input circuitry. Unfortunately, cable impedances are quite low (50–1000 ohms) while the input impedance of receiver amplifiers is much larger. This means impedance matching transformers must be inserted, again adding to the system cost.

Unguided Channels. An unguided channel occurs when an antenna is used and the transmitted electromagnetic field is allowed to propagate freely from the antenna. As stated earlier, when an electromagnetic wave propagates in free space it undergoes an intrinsic power loss that increases with the square of the distance propagated. When the wave propagates in the Earth's atmosphere, however, additional losses may be superimposed on the free space loss because of the particulate nature of the gases and vapors constituting the atmosphere. In addition, further losses may be

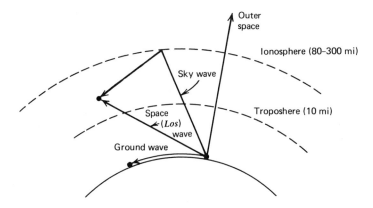

Figure 2.14. Channels for unguided carrier propagation.

imposed by the Earth itself. The magnitude of these additional losses will depend on the manner in which the wave is transmitted and on the particular properties of the atmosphere itself.

There are four basic propagation channels that can be defined for the unguided carrier wave (Figure 2.14). These are (1) the ground wave channel; (2) the space wave channel; (3) the sky wave channel; and (4) the outer space channel. Each is appropriate in a specific application, and each defines a slightly different propagation channel model. A communication engineer must carefully distinguish the most accurate channel model before proper assessment of propagation losses can be made. In the following discussion, we summarize the important characteristics of each of these channels.

Ground Wave Channel. When both transmitter and receiver are located within a few meters of the Earth, radio waves propagate as guided waves with the Earth itself serving as a waveguide wall for the propagating field. Such models are applicable in land based mobile and hand-held communication systems. Ground wave channels are characterized by strong attenuation with distance, and attenuation values that increase with frequency. These effects are generated identically to waveguide attenuation in guided channels. For this reason ground wave transmission is restricted to frequencies HF and below, and over relatively short (10–50 km) distances even when no obstructions (mountains, buildings, trees, etc.) are present. Figure 2.15 shows a typical attenuation characteristic associated with ground wave propagation, showing the rapid deterioration of field strength over that predicted purely from free space losses.

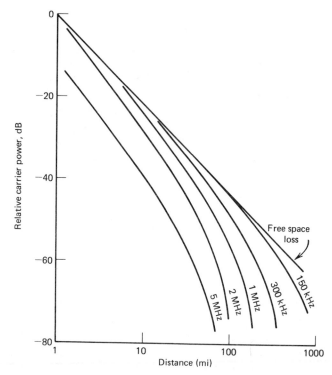

Figure 2.15. Ground wave carrier power loss with distance and frequency.

Space Wave Line of Sight (LOS) Channel. In LOS space wave channels
the wave propagates directly from transmitter to receiver through space
without influence from the Earth. Such systems characterize Earth-air-
plane and airplane-airplane links, as shown in Figure 2.14. For land based
systems the LOS space model is appropriate only if the antennas are raised
high enough (using towers, rooftops, or mountain tops) to avoid the ground
attenuation. This raising of the antennas to achieve a space channel
generally restricts their size, and higher frequencies (UHF, VHF) must be
used to achieve suitable gains. Space wave propagation however is susceptible to extraneous attenuation due to particles in the atmosphere. This
effect is particularly severe when the wavelength of the field becomes
commensurate with the particle sizes making up the atmosphere. Figure
2.16*a* shows a typical plot of atmospheric attenuation as a function of
frequency when a space wave is being transmitted. Note the rapid increase
as the carrier frequency increases beyond 10 GHz, and also at the
resonances of particular gases and vapors. Atmospheric attenuation is

generally stated as an attenuation factor per unit length, and therefore the total effective loss will depend on the transmission distance.

Atmospheric losses also depend on the elevation angle of the transmission and on altitude, being greatest when a wave is propagating horizontally near sea level. This is due to the fact that the atmospheric particle structure becomes less dense at the higher altitudes. Thus vertical transmissions experience a transmission path with a decreasing density,

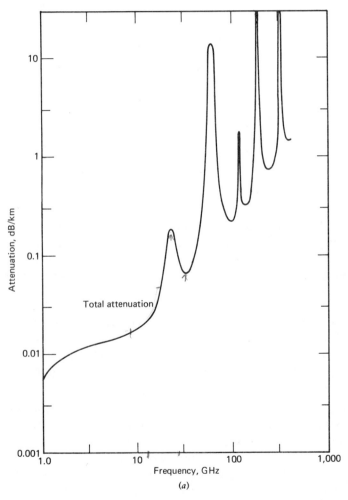

Figure 2.16. Atmospheric absorption (a) [*above*] at sea level, (b) [*page 80*] as a function of elevation angle.

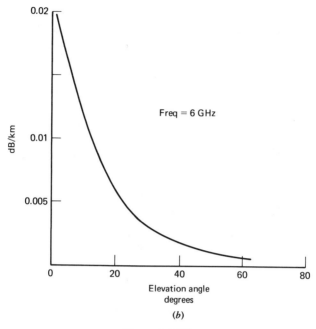

Figure 2.16(b).

producing less loss. Figure 2.16*b* shows a typical plot of wave attenuation as a function of transmission angle measured from the vertical.

Rainfall is also particularly significant in space wave transmission, since water drop sizes may approach the length of carrier wavelength at the higher microwave frequencies. For this reason rainfall effects become more severe as the carrier frequency is increased. Figure 2.17 shows a typical plot of rainfall attenuation versus carrier frequencies, for several values of the severity of the ranfall (i.e., the rainfall rate). Since exact knowledge of rainfall rate and rainfall extent during a particular storm is not always available, rainfall attenuation is generally determined using an average rainfall value characteristic of the location.

Sky Wave Channels. Sky wave channels are obtained by reflecting electromagnetic fields at the ionosphere or troposphere boundaries (Figure 2.14). The ionosphere is a dense collection of free electrons trapped in a belt around the Earth. To the longer wavelength carriers (HF and below) this ionosphere appears as a conducting surface that reflects carrier energy. This reflected energy may be used as a primary path in a long range non-LOS link. Such reflections often allow communications over distances where LOS space wave propagation is no longer possible. Unfortunately,

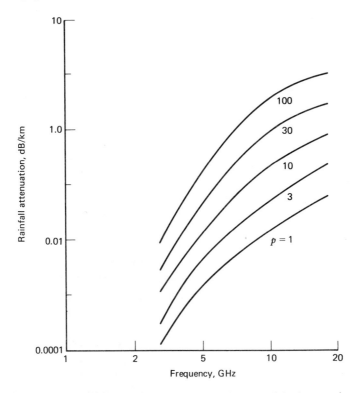

Figure 2.17. Rainfall attenuation versus frequency. p = precipitation rate in mm/hr.

the attenuation of the field caused by boundary reflection is generally quite high. In addition, the movement of the belt from hour to hour and its dependence on time of the year and galactic activity (such as solar flares), produces a variable channel that often varies from a strong to a weak link in short time periods. For this reason a reliable sky wave channel can only be defined for limited time periods. Nevertheless, sky wave links have been successfully implemented and used for long range, over-the-horizon systems. In addition, sky waves are often responsible for commercial radio stations being heard far outside their transmission range on certain clear evenings.

Using sky wave channels at frequencies below HF is basically disadvantageous because of required antenna sizes. On the other hand, as frequency is increased into the VHF and UHF range, these smaller wavelengths penetrate the ionosphere and are absorbed or scattered rather than reflected. Hence sky wave channels cannot be maintained at the higher frequency bands. The *critical frequency* of a sky wave channel is that frequency below which a reflected sky wave can be satisfactorily maintained. This critical frequency depends on the electron density of the ionosphere, which in turn depends on the time of day, season of the year, and so on. Published data, obtained from years of experimental measurements, are available for specifying this frequency. Critical frequencies generally fall in the range 5–20 MHz. This is why long range commercial radio is possible via sky waves but commercial television is not.

Outer Space Channel.   An outer space channel involves transmission out through the Earth's atmosphere to satellites and space stations. In this channel energy must purposely pass through the ionosphere belt rather than be reflected. Hence outer space channels must use frequencies well above the critical frequency. As the carrier frequency is increased into the VHF and SHF bands a larger portion of the field energy passes through the ionosphere, rather than being scattered or reflected. Thus deep space channels are almost always operated in the SHF band or above (see Table 1.3). The deep space link is still, of course, susceptible to the atmospheric absorption, but the effect is minimized somewhat when vertical paths are used. In addition, typical deep space links to stationary satellites ($\approx$20,000 mi), the moon ($\approx$232,000 mi), or nearby planets ($\approx 10^8$ mi) clearly involve distances only a small part of which includes the Earth's atmosphere. Thus the added decibel loss introduced by the atmospheric path is a relatively small part of the total electromagnetic loss caused by the propagation over such long distances. Of course, transmission outside the Earth's atmosphere (e.g., between two space satellites) would have no atmospheric losses.

In summary, we have qualitatively investigated the carrier propagation channel and its effect on carrier transmission. We have presented some representative curves and figures to aid in understanding these effects. Unfortunately, most of these propagation phenomena are functions of many external and immeasurable parameters, specialized to each specific situation, such as weather conditions, geographical location, season of the year, and so on. For this reason a single curve is inadequate to describe all cases. In addition, most documented results report average data, and in any given channel, variations from the average may be quite significant. We point out that other types of channels can be defined that have not been considered here but that arise in special applications. Underwater communications (sonar systems) and seismic propagation channels represent two such special examples. Such systems require more complicated electromagnetic analyses for proper modeling.

Throughout our channel discussion we have repeatedly emphasized the dependence of the channel effect on frequency, sometimes favoring higher carrier frequencies and sometimes lower. This again illustrates the importance of transmitter modulation to select the transmission frequencies that best interface with channel. This then serves as another important consideration in the carrier frequency designation.

2.7. Other Channel Effects

The discussion in the previous section was primarily concerned with the power losses introduced by the channel during field propagation. However, in addition to these power losses the channel may introduce other effects on the carrier waveform. These additional effects may take the form of channel filtering, Doppler shifting, fading, or multipaths. Although these effects are usually considered secondary, and are often ignored in first order analyses, the system designer should be aware of their possibility, since in some situations they may become the limiting factor on performance. In this section we review their causes and characteristics.

Filtering. When they propagate in a cable or in the atmosphere, different frequencies may actually function at different velocities. Thus different frequencies within the modulating waveform may experience slightly different propagation times, referred to as *dispersion*. The overall effect is to produce inherent nonuniform frequency delays, which causes a phase distortion of the carrier waveform. This distortion can be interpreted as being caused by an effective channel filtering on the carrier waveform. This filtering tends to reduce the available carrier modulation bandwidth. For relatively quiet atmospheric conditions, the space channel dispersion bandwidth is generally quite large, on the order of hundreds of gigahertz, and is usually not of concern in the RF carrier range, where bandwidths are never this large. Turbulent atmospheric conditions, however, may reduce this channel bandwidth to several gigahertz, and therefore may be a significant factor in microwave and millimeter carrier systems. Channel filtering is usually measured by observing pulse spreading. An RF carrier is amplitude modulated by a pulse and sent over the space channel. The pulses are then made continually narrower, and the received pulse is monitored until a fixed percentage of pulse spreading has been observed. Channel bandwidth is then related to the reciprocal of the pulse width at which the spreading was noted.

Doppler. Whenever relative motion exists between the transmitter and receiver, the received carrier frequency is not the same as the transmitted carrier frequency. This frequency shifting during transmission is due to the

Doppler effect of propagating waves between nonstationary points. The effect is to cause an overall shift of the carrier frequency. A system designer must allow for, or compensate for, this Doppler shift whenever it becomes significant. In certain cases the Doppler may be completely predictable, as with satellites with fixed orbits. Doppler effects due to unintentional transmitter or receiver motion may have to be estimated and properly compensated for in system design.

If the relative velocity of the transmitter toward (away from) the receiver is $+v$ $(-v)$ m/sec, then the Doppler frequency shift of the RF carrier at f_c Hz is given by*

$$f_{\text{Dop}} = \left(\frac{\pm v}{c}\right) f_c \qquad (2.7.1)$$

where c is the speed of light. Note that v in (2.7.1) is the velocity component along the line-of-sight vector between transmitters and receiver. For velocity directions at other angles, only the velocity component along this line will produce the Doppler effect. It should also be pointed out that if there is a constant acceleration—linearly changing velocity—between transmitter and receiver, the Doppler frequency will also appear to be linearly changing when observed at the receiver.

Fading. The channel losses discussed in Section 2.6 represented an average type of power loss. However, over a short period of time a communication signal may have severely reduced power values corresponding to deep "fades" of the transmitted signal. This fading can be considered as an instantaneous attenuation of the carrier amplitude. Analysis or design of a system in the presence of these fades requires one to describe these attenuation factors statistically.

In general, fading is divided into *slow* and *fast* fading. In the former case the rate of fading is much slower than the reciprocal of the highest frequency of the modulated carrier. Here the carrier waveform can be considered to be undistorted but effectively multiplied by a random amplitude factor. Probability statistics of this random amplitude therefore describe the properties of the fade. With fast fading, the rate of fade is considered to be significant relative to the carrier amplitude time variation. The overall result is an effective multiplication of the carrier by a rapidly varying amplitude function that represents the fading phenomena. In this case the carrier is in fact "modulated" by the fading amplitude, which

---

*To be more rigorous, if f_c is the transmitted frequency, the received frequency is given by $(1+v/c)f_c/[1-(v/c)^2]^{1/2}$. Since $v/c \ll 1$, (2.7.1) is an accurate approximation to the difference frequency.

appears as a carrier distortion at the time of reception. The rapidly varying fading effect is generally random and described by a suitable stochastic process model. Fast fading can generally only be corrected by "dividing out" the fading effect by some type of automatic gain control or fading estimation, the latter via channel probing.

Multipath. Often the receiver receives not only the direct transmission from the transmitter, but secondary transmissions caused by reflections or echoes of the direct transmission. These secondary receptions generally arrive later than the primary transmission (since they propagate over longer paths) and therefore appear as interfering electromagnetic waves. These secondary paths, called *multipaths*, are predominant when reflecting surfaces appear near the receiver. Such mutlipath effects are particularly noticeable in Earth based mobile systems, where reflections from buildings and mountains are unavoidable. It is precisely this multipath that produces "ghosts" in home television reception.

Although multipaths may be well reduced in amplitude, a large number can accumulate to cause an effective interfering noise. Often one component of the multipath, called the *specular* component, tends to dominate. The remaining multipaths tend to accumulate into an effective noise, called the *diffuse* multipath components. Thus the typical multipath carrier signal is generally modeled as

$$c_m(t) = \alpha_c c(t) + \alpha_s c(t - \tau_s) + [\text{diffuse noise}] \qquad (2.7.2)$$

where α's account for power losses and τ_s is the transmission delay of the specular component. Specific statistics for the diffuse noise are difficult to measure, and the latter noise is often modeled as an additive noise process. Elimination of the effect of the specular component is often of fundamental importance in multipath environments. Of particular significance is the fact that the specular interference is itself the transmitted carrier signal, and the desired carrier term cannot be made to dominate by simply increasing the transmitter carrier power. In fact, all three terms in (2.7.2) depend on transmitter power. In certain cases the effect of the specular part can be reduced by proper signal selection if the delay τ_s is known or can be estimated (Problem 2.24). Sometimes specular multipath can be eliminated by accurate antenna pointing and antenna pattern control to avoid reception of the secondary path.

Doppler, fading, and multipath are, of course, only of concern in unguided space field transmission systems and are not characteristic of guided links. Cables, though having attenuation and often dispersive filtering effects, do not exhibit these other properties, which is a primary advantage in their use.

References

1. Abromowitz, M. and Stegun, I. *Handbook of Mathematical Functions*, National Bureau of Standards, Washington, D.C., 1965, Chap. 9.

2. Watson, G. *A Treatise on Bessel Functions*, Cambridge University Press, Cambridge, England, 1958.

3. Augelakos, D. and Everhart, T. *Microwave Communications*, McGraw-Hill, 1968.

4. Reich, H. *Microwave Theory and Tehcniques*, Van Nostrand, Princeton, N. J., 1953.

5. Singer, J. *Masers*, Wiley, New York, 1959.

6. Meyer, J. "Systems Applications of Solid State Masers," *Electronics*, vol. 33, no. 45, November 1960, pp. 58–63.

7. Kraus, J. *Antennas*, McGraw-Hill, New York, 1950.

8. Silver, S. *Microwave Antenna Theory and Design*, MIT Radiation Laboratory Series, vol. 12, McGaw-Hill, New York, 1949.

9. Terman, F. E. *Electronic and Radio Engineering*, 4th ed., McGraw-Hill, New York, 1956.

10. Jordan, E. C. *Electromagnetic Waves and Radiating Systems*, Prentice-Hall, Englewood Cliffs, N.J., 1950.

11. Norton, K. A. The Propagation of Radio Waves over the Surface of the Earth and in the Upper Atmosphere, Part I, *Proc. IRE*, vol. 24, October 1936, p. 1367; Part II, *Proc. IRE*, vol. 25, September 1937, pp. 1203–1236; "The Calculation of Ground-wave Field Intensities over a Finitely Conducting Spherical Earth," *Proc. IRE*, vol. 29, December 1941, pp. 623–629.

12. Booker, H. G. and Gordon, W. E. "A Theory of Radio Scattering in the Troposphere," *Proc. IRE*, vol. 28, April, 1950, pp. 401–417.

13. *Proc. IRE*, vol. 43, October 1955, scatter propagation issue.

14. *Proc. IRE*, vol. 45, June 1957, VLF propagation issue.

15. Bullington, K. "Radio Propagation at Frequencies above 30 Megacycles", *Proc. IRE*, vol. 35, October 1947, p. 1122.

16. Burrows, C. R. and Attwood, S. S. "Radio Wave Propagation," Academic, New York, 1949.

17. Reed, H. R. and Russell, C. M. *Ultra High Frequency Propagation*, Wiley, New York, 1953.

18. Millman, G. H. "Atmospheric Effects on VHF and UHF Propagation," *Proc. IRE*, vol. 46, August 1958, pp. 1492–1501.

19. Van Vleck, J. H. "The Absorption of Microwaves by Oxygen," *Phys. Rev.*, vol. 71, April 1, 1947, pp. 413–424.

20. Van Vleck, J. H. "The Absorption of Microwaves by Uncondensed Water Vapor," *Phys. Rev.*, vol. 71, April 1, 1947, pp. 425–433.

Problems

1. (2.1)   A sine wave of frequency 1 kHz is modulated onto a carrier at 2 GHz. Assuming standard double-sided AM, determine and sketch the amplitude frequency function at the modulated carrier.

2. (2.1) Determine the frequency spectrum of the waveform $a(t)p(t)$, where $a(t)$ is a positive waveform with transform $A(\omega)$ and $p(t)$ is a periodic, unit square wave with period T sec.

3. (2.1) Show that the power in the SSB-AM in (2.1.18) is given by $P_c = P_a$.

4. (2.1) Determine $c(t)$ and its corresponding transform for the special case of an SSB-AM carrier with modulation $a(t) = b \sin \omega_a t$. Show a plot of the transform magnitude. Show that the required $H_a(\omega)$ for SSB-AM has an impulse response

$$h(t) = \frac{1}{\pi t}$$

5. (2.2) A voice tone with frequency of $f_m = 4$ kHz is sent by FM over a system with bandwidth of 24 kHz. What is the maximum permitted value of frequency deviation Δ_ω assuming a Carson rule bandwidth?

6. (2.2) A baseband waveform of 10 kHz is frequency modulated onto a carrier. Find the Carson rule bandwidth when the following frequency deviation coefficients are used: (a) $\Delta_\omega = 1$ kHz/V, (b) $\Delta_\omega = 100$ kHz/V, (c) $\Delta_\omega = 1$ MHz/V.

7. (2.2) An oscillator can be frequency deviated only 10% of its resonant frequency. Assume sine wave modulation and determine the maximum modulating frequency that can be used to frequency modulate the oscillator when operated at a resonant frequency of 2 GHz and with $\beta \geq 5$. What is the Carson rule bandwidth that is required with this FM signal?

8. (2.2) A carrier is frequency modulated by two sine waves

$$C_1 \cos (\omega_1 t + \theta_1) + C_2 \cos (\omega_2 t + \theta_2)$$

Derive the result in (2.2.16) for the case $K = 2$.

9. (2.2) We wish to determine the rms bandwidth (Problem 1.14) of a PM carrier phase modulated with a Gaussian noise process with zero mean and autocorrelation $R_\theta(\tau)$.

(a) Show that the spectral spread is due only to the term $\exp[R_\theta(\tau)]$.

(b) Write the rms bandwidth B_{rms} in terms of $S_{ex}(\omega)$, the transform of $\exp[R_\theta(\tau)]$.

(c) Using transform theory, show that (b) is

$$B_{rms} = \frac{2[-e^{R_\theta(0)} R_\theta''(0)]^{1/2}}{e^{R_\theta(0)/2}}$$

$$= 2[-R_\theta''(0)]^{1/2}$$

where the primes denote derivatives [recall $R_\theta'(0) = 0$].

(d) Use Problem 1.29 to show

$$B_{\text{rms}} = 2[R_{\theta'}(0)]^{1/2}$$

Explain what this means in terms of carrier frequency.

10. (2.2) An FM carrier is frequency modulated with a random process $m(t)$. The frequency variation at any t is given by $\Delta_\omega m(t)$. The probability that its frequency lies in the interval $(f_1, f_1 + df)$ is therefore the probability that $m(t)$ lies in the interval $(f_1/\Delta_\omega, f_1/\Delta_\omega + df/\Delta_\omega)$. The power spectral value at f_1 is therefore given by that of a sine wave at f_1 times the probability that $f = f_1$. Use these facts to show that the FM spectrum must approximate in shape the probability density of $m(t)$ at any t.

11. (2.3) A carrier at ω_c is phase modulated by the baseband waveform in (2.3.4). How much power is in the frequency components at (a) $\omega_c + 2\omega_1$, (b) $\omega_c - 3\omega_1 + 2\omega_2$, (c) $\omega_c - 5\omega_4$, (d) $\omega_c + \omega_1 - 3\omega_2 + 4\omega_4$?

12. (2.3) An RF carrier is phase modulated with the sum of K sine waves $\sum_{i=1}^{K} \Delta_i \sin(\omega_i t + \theta_i)$ plus a binary waveform $\Delta_r r(t)$, where $r(t) = \pm 1$ and has transform $F_r(\omega)$. (a) Derive the resulting PM spectrum and roughly sketch its shape. (b) Write the amplitude of the carrier component.

13. (2.3) Determine the condition on the carrier phase angle ψ in (2.3.9) for the autocorrelation function to be stationary (not depend on t).

14. (2.3) A Gaussian white noise process of spectral level N_0 one-sided, is filtered in an ideal filter of bandwidth B_1 Hz, and is used to phase modulate a pure carrier at frequency ω_c and amplitude A. What is the power in the carrier component of the modulated carrier?

15. (2.4) Consider the transmitter amplifier system in Figure 2.8b. The amplifier has a 50 W output under matched loading and an output impedance of 500 ohms. The cable is 20 ft long and has an impedance of 100 ohms from either end and a power loss of 0.1 dB/ft. The antenna has an input impedance of 1000 ohms. Determine the amplifier power coupled into the antenna.

16. (2.4) A 100 W amplifier with 1000 ohms output impedance feeds into a 1 m antenna. The antenna has its input impedance approximated by $Z_g \cong 800 \, \mathcal{A}/\lambda^2$. Determine the power coupled into the cable at each of the frequencies: (a) 30 MHz, (b) 300 MHz, (c) 3 GHz.

17. (2.5) (a) A rectangular antenna of width a and length b produces at wavelength λ the gain function

$$g(\phi_l, \phi_z) = K \left[\frac{\sin\left[(\pi a \sin \phi_l)/\lambda\right] \sin\left[(\pi b \sin \phi_z)/\lambda\right]}{\left[(\pi a \sin \phi_l)/\lambda\right]\left[(\pi b \sin \phi_z)/\lambda\right]} \right]^2$$

where K is a proportionality constant. Determine the half-power beam-width in degrees for the pattern in the $\phi_l = 0$ and in the $\phi_z = 0$ plane.

(b) Repeat for

$$g(\phi_l, \phi_z) = K\left[\frac{2J_1[(\pi a \sin \phi_l)/\lambda]J_1[(\pi b \sin \phi_z)/\lambda]}{[(\pi a \sin \phi_l)/\lambda][(\pi b \sin \phi_z)/\lambda]}\right]^2$$

where $J_1(x)$ is the Bessel function.

18. (2.5) A stationary satellite, placed in orbit at 20,000 mi above the Earth, is to transmit to the Earth at the frequency of 1 GHz. A parabolic antenna is to be used having a radiation efficiency of 90% and an aperture efficiency of 50%. (a) Determine the size of the antenna that will transmit a field of view essentially covering the Earth. (b) What is the resulting effective transmitting gain. (The Earth has a diameter of approximately 6876 mi.)

19. (2.5) A communication station on Earth is to transmit to the moon at 2 GHz. Design an antenna pair (transmitting plus receiving) so that the transmission from Earth essentially covers the moon, and the total effective antenna gain (product of each gain) is 70 dB. Use the same efficiencies as in Problem 2.18. Assume the distance to the moon is 200,000 mi and its diameter is 0.27 of the Earth's.

20. (2.5) Derive the relation between antenna solid angle Ω_{fv} and planar angle beamwidth Φ_b for a symmetric gain function. [*Hint*: relate the area of the spherical cap of a cone to its planar angle.] Determine the approximate behavior as $\Phi_b^2 \ll 1$.

21. (2.5) A 2 m circular antenna at 1 GHz has a circular gain pattern shown in Figure P2.21 as a function of azimuth angle. It is fed by 30 dBW

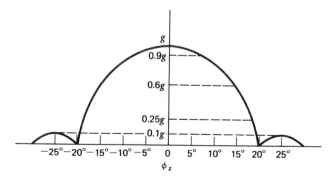

Figure P2.21.

of power and has a radiation efficiency of 0.95. What is the transmitted field power density at a distance of 1000 m located at 15° from the main antenna lobe direction?

22. (2.7) An Earth station communicates at S band with a satellite at 22,000 mi. Determine the maximum Doppler shift that can be expected because of the rotation of the Earth. Assume the satellite is stationary while the Earth rotates under it. (The diameter of Earth is 6876 mi.)

23. (2.7) A pulse 10 nsec wide is spread to a 50 nsec width during propagation through a medium. Using Problem 1.6, estimate the filtering bandwidth of the medium.

24. (2.7) A T sec burst of a sine wave of amplitude A and frequency ω_c is transmitted to a receiver. Determine the shortest multipath delay that can be tolerated to ensure that the cross-correlation of direct and multipath signal is less than 10% of the direct signal power. Assume $\omega_c T \gg 1$, and the multipath signal has the same amplitude as the direct signal when received. The cross-correlation of two signals is defined as

$$\gamma_{12} = \frac{1}{T} \int_0^T s_1(t) s_2(t)\, dt$$

CHAPTER 3

CARRIER RECEPTION

In Chapter 2 we examined carrier transmission and discussed the propagation effects that occur when a modulated carrier is sent over a communication channel as an electromagnetic field. To the communication engineer the parameter of primary interest is the actual amount of transmitted power that can be collected at the receiving station. This depends not only on the transmission effects but also on the structure of the receiving subsystem and its ability to intercept the electromagnetic field. Following field reception, the system designer is faced with the task of properly processing the resultant waveforms to achieve the most efficient communication link for useful information transfer. In this chapter we investigate the reception of the electromagnetic field and the filtering and other types of carrier processing that may be applied to the recovered waveform.

The receiver subsystem for collecting and processing the electromagnetic field is shown in Figure 3.1. If the transmitted field is sent as an unguided wave, a receiving antenna must be used to convert the impinging field to an electronic waveform. The resulting signal is then passed to the receiver *front end* electronics. In a guided system no receiving antenna is needed

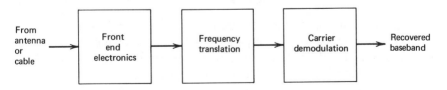

Figure 3.1. Receiver carrier processing subsystem.

and the field in the transmission line or cable need only be coupled directly into the front end. In the front end the carrier waveform is generally filtered to remove as much interference as possible without distorting the desired carrier waveform; it is then amplified to obtain a desired power level. During this front end processing other types of electronic interference may be imposed, and the effect of any front end filtering on the modulated carrier must be accounted for. In addition, the front end system may have to operate in the presence of other waveform channel degradations, similar to those discussed in Section 2.7.

Since modulation was introduced at the transmitter to allow the baseband information signal to be carried to the receiver, a corresponding demodulation operation must be performed at the receiver to extract the baseband. This carrier demodulation is perhaps the key operation of the receiver, and the overall success of the link depends to a large extent on the ability to demodulate the baseband waveform properly. Usually other forms of waveform processing, such as frequency translation, may be applied following front end filtering in order to prepare for the carrier demodulation. These receiver operations, from front end reception to demodulation input, are considered in this chapter. Primary emphasis is on subsystem modeling and design and on performance evaluation.

3.1. Receiving Antennas

Receiving antennas convert an impinging electromagnetic field to an electronic voltage waveform. Receiving antennas are best described in terms of an effective cross-sectional area $\mathscr{A}(\phi_l, \phi_z)$ associated with a ray direction (ϕ_l, ϕ_z) defined with respect to the receiving antenna coordinates. This area is defined by considering an electromagnetic plane wave, having a prescribed field power density, to impinge on the antenna from the direction (ϕ_l, ϕ_z), as in Figure 3.2, relative to the antenna orientation. We then determine the field power that will be observed at the receiving antenna output terminals under matched load conditions. The *effective receiving antenna area* in the direction (ϕ_l, ϕ_z) is then defined by

$$\mathscr{A}(\phi_l, \phi_z) = \frac{\text{power observed due to field from } (\phi_l, \phi_z)}{\text{power density of the field}} \quad (3.1.1)$$

Note that the antenna area is a function of the direction angles (ϕ_l, ϕ_z) and includes any antenna losses and polarization mismatch with the field. The effective area therefore describes a receiving pattern over all directions from the antenna coordinates. The maximum area

$$\mathscr{A}_m = \max_{\phi_l, \phi_z} \mathscr{A}(\phi_l, \phi_z) \quad (3.1.2)$$

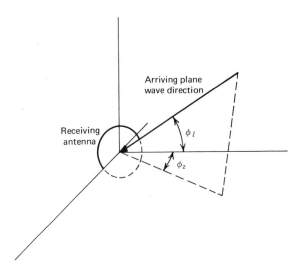

Figure 3.2. Receiving antenna geometry.

is the largest area presented to arriving fields. This maximum area is less than the physical antenna area $\mathscr{A}_p$ by the receiving antenna *aperture loss factor* ρ_{ap} in (2.5.14),

$$\mathscr{A}_m = \mathscr{A}_p \rho_{ap} \tag{3.1.3}$$

However, from the reciprocity condition [1, Chapter 4] of transmitting and receiving plane waves, it is known that

$$\frac{\mathscr{A}(\phi_l, \phi_z)}{\mathscr{A}_m} = \frac{g(\phi_l, \phi_z)}{g} \tag{3.1.4}$$

where $g(\phi_l, \phi_z)$ is the gain function of Section 2.5 if the same antenna had been used for transmission. This means that, except for scaling factors, an antenna's receiving and transmitting patterns are the same, and the ability of an antenna to collect field power from a given direction is identical to its ability to transmit. Thus the main transmission direction of a gain pattern while transmitting becomes the primary direction of field reception while receiving, and the sidelobe values represent reduced receiving capability in off-axis directions. If we relate (3.1.4) to (2.5.15) we see that for receiving antennas,

$$\frac{\mathscr{A}_m}{g} = \frac{\lambda^2}{4\pi} \tag{3.1.5}$$

Combining (3.1.4) and (3.1.5) shows

$$\mathscr{A}(\phi_l, \phi_z) = \frac{\lambda^2}{4\pi} g(\phi_l, \phi_z) \qquad (3.1.6)$$

in any direction (ϕ_l, ϕ_z). Equation (3.1.6) therefore relates antenna receiving area patterns directly to the gain pattern and the electromagnetic field frequency. Note that the receiving field of view is now identical to (2.5.14),

$$\Omega_{fv} = \frac{\lambda^2}{\rho_{ap} \mathscr{A}_p} \qquad (3.1.7)$$

associated with the gain pattern. This now becomes the solid angle associated with maximal power reception of the receiving antenna. This means high gain antennas have large effective receiving areas, as indicated by (3.1.5), and small fields of view for collecting power in only certain directions. This latter property is important for eliminating unwanted electromagnetic interference arriving from off-axis directions. This condition of high receiver selectivity therefore dictates high carrier frequencies in order to achieve reasonable antenna sizes. Thus the modulation of the baseband up to carrier frequencies aids us not only in carrier transmission but in carrier reception as well.

Note that (3.1.6) and (3.1.7) show that the design of receiving antennas can be equivalently stated in terms of designing transmitting gain patterns, and the use of transmitting design charts, as in Figure 2.11, can be applied here. Therefore, if we desire a receiving antenna with a given receiving field of view, we can design the antenna as if it were to be a transmitting antenna with the same field of view. For example, an antenna with the pattern shown in Figure 2.10 would therefore collect electromagnetic fields primarily over a 12° beamwidth.

3.2. Receiver Carrier Power

In analyzing or designing a communication system it is always necessary to know the actual amount of carrier power that can be transmitted to the receiver. As we shall see, this parameter plays a key role in assessing ultimate system performance. Based on our analysis to this point, we can determine the amount of carrier power theoretically to be recovered at the receiver antenna output terminals because of the power sent from the transmitter.

Consider first the carrier transmission channel in Figure 3.3 operating with unguided field propagation at a fixed carrier frequency f_c. We assume

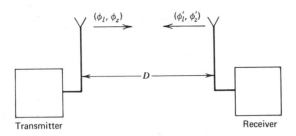

Figure 3.3. Transmitting and receiving channel.

the transmitter power amplifier produces the modulated carrier signal $c(t)$ with carrier power P_{amp} W. That is, P_{amp} is the value of P_c at the amplifier output, where P_c has one of the forms in Sections 2.1–2.3, depending on the type of modulation. The carrier power coupled into the antenna terminals is then given by

$$P_t = P_{\text{amp}} L_t \tag{3.2.1}$$

where L_t accounts for any guide or coupling losses at the transmitter [bracketed terms in (2.4.2) and (2.4.3)]. The EIRP of the transmitter in the direction of the receiver is obtained from (2.5.13) as

$$\text{EIRP} = P_t \rho_r g_t(\phi_l, \phi_z) \tag{3.2.2}$$

where ρ_r is radiation loss factor of the transmitting antenna, and $g_t(\phi_l, \phi_z)$ is the transmitting gain function in the direction of the receiver, the latter considered a point located a distance D from the transmitter. During propagation, this EIRP will be reduced by any added channel losses L_a (absorption, rainfall, etc.) due to the atmosphere. The field power density of the propagated plane wave arriving at the receiver antenna after propagating a distance D is obtained from (2.5.12) as

$$P_{\text{den}} = \frac{(\text{EIRP}) L_a}{4 \pi D^2} \tag{3.2.3}$$

Note that L_a accounts for all the additional propagation losses due to the atmospheric propagation channel ($L_a = 1$ for free space) and may also be a function of distance D as well as carrier frequency, as discussed in Section 2.6. The field plane wave power collected at the output of a receiving antenna presenting an effective area $\mathscr{A}(\phi_l', \phi_z')$ to the transmitter plane wave arriving from the transmitter direction ϕ_l', ϕ_z') is then

$$P_{\text{ant}} = P_{\text{den}} \mathscr{A}(\phi_l', \phi_z') \tag{3.2.4}$$

This effective receiving area can be related to its gain function $g_r(\phi'_l, \phi'_z)$ using (3.1.6), so that we can write instead

$$P_{\text{ant}} = P_{\text{den}}\left(\frac{\lambda^2}{4\pi}\right)g_r(\phi'_l, \phi'_z) \tag{3.2.5}$$

where λ is the carrier wavelength. The collected antenna power actually fed into the receiver front end must be reduced by any receiving antenna coupling losses L_r. Thus the total collected carrier power at the receiver front end terminals is

$$P_r = P_{\text{amp}}\left(\frac{L_t L_a \rho_r}{4\pi D^2}\right)g_t(\phi_l, \phi_z)\left(\frac{\lambda^2}{4\pi}\right)g_r(\phi'_l, \phi'_z)L_r \tag{3.2.6}$$

It is convenient to interpret

$$L_p = \left(\frac{\lambda}{4\pi D}\right)^2 \tag{3.2.7}$$

as an effective *free space propagation loss,* because of the manner in which it appears in (3.2.6). If the antennas are properly aligned, the transmitter and receiver directions are located at the peak of the gain functions, and (3.2.6) reduces to

$$P_r = P_{\text{amp}}\rho_r L_t L_a L_p L_r g_t g_r \tag{3.2.8}$$

where g_t and g_r are now the gains of the transmitting and receiving antennas at the carrier frequency. Equation (3.2.8) summarizes the effect of the complete carrier transmission subsystem as the power flows from transmitter amplifier to receiving antenna terminals and involves simply a product of the gains and losses along the way. It is often convenient to express this in decibels as

$$(P_r)_{\text{dB}} = (P_{\text{amp}})_{\text{dB}} + (\rho_r)_{\text{dB}} + (L_t)_{\text{dB}} + (L_a)_{\text{dB}} + (L_p)_{\text{dB}} + (L_r)_{\text{dB}} + (g_t)_{\text{dB}} + (g_r)_{\text{dB}} \tag{3.2.9}$$

where loss factors will appear as negative decibel values. The propagation loss factor in (3.2.7), when converted to decibels, reduces to the specific form

$$(L_p)_{\text{dB}} = -36.6 - 20 \log\left[D(\text{mi})f_c(\text{MHz})\right] \tag{3.2.10}$$

A nomogram of $(L_p)_{\text{dB}}$ is given in Figure 3.4 and aids in evaluating $(L_p)_{\text{dB}}$ from the preceding equation for typical carrier frequencies and transmission distances D. We see that computation of receiver power in decibels requires simply the addition of appropriate gain and loss decibels. This computation is often called a *power budget* and is a basic part of link analysis.

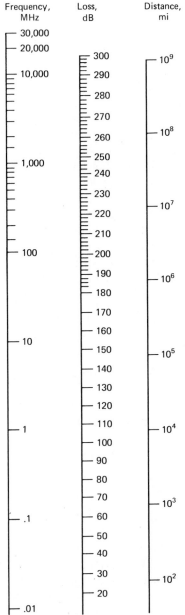

Frequency, MHz	Loss, dB	Distance, mi
30,000		
20,000	300	10^9
10,000	290	
	280	
	270	
	260	10^8
1,000	250	
	240	
	230	
	220	10^7
	210	
100	200	
	190	10^6
	180	
	170	
	160	
10	150	10^5
	140	
	130	
	120	
1	110	10^4
	100	
	90	
	80	
	70	10^3
.1	60	
	50	
	40	
	30	10^2
	20	
.01		

Figure 3.4. Nomogram of propagation loss.

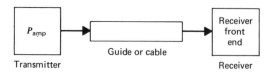

Figure 3.5. Cable transmission system.

When the propagation is by guided channels (Figure 3.5) the analysis is similar. The amplifier power coupled into the transmission cable is again given by (3.2.1), where L_t accounts for any transmitter coupling losses. Cable attenuation losses are typically stated in decibels per unit length of cable and include both the electromagnetic propagation loss and added losses due to cable walls, radiation, and internal reflections. If the cable has a length D and an attenuation loss factor of α dB/length, then the input decibel power is reduced by αD dB to give the cable output decibel power. If an amplifier of gain G_a is inserted a distance D_1 down the cable, the power after propagating a distance D_1 is now reduced by $\alpha D_1 - (G_a)$ dB. If a sequence of amplifiers is spaced along the cable, the propagating power is reduced by the loss of each section of cable and increased by the total inserted power gain. The final power coupled out of the cable and into the receiver front end is then further reduced by the coupling losses L_r. Thus for guided channels, the power flow equation becomes

$$(P_r)_{\mathrm{dB}} = (P_{\mathrm{amp}})_{\mathrm{dB}} + (L_t)_{\mathrm{dB}} - \alpha D + \sum_i (G_i)_{\mathrm{dB}} + (L_r)_{\mathrm{dB}} \qquad (3.2.11)$$

where D is the total guide length and the sum is over the number of cable amplifiers inserted. Again a power budget can be used for conveniently tabulating and summing the required terms.

3.3. Background Noise

In addition to collecting the desired carrier field from the transmitting antenna, a receiving antenna also collects noise energy from background sources in its field of view. This background energy is due primarily to random noise emissions from galactic, solar, and terrestrial sources, constituting the sky background. The amount of noise collected by the antenna is the ultimate limitation to the sensitivity of the receiving system, since it determines the weakest carrier signal that can be distinguished. The amount of pure background noise received depends on both the amount radiated from the sky background and the portion of that collected by the receiving antenna.

The background noise source is generally described by its *radiance* function, defined as

$$\mathcal{H}(\phi_l, \phi_z) = \begin{array}{l}\text{power received per Hz from the background} \\ \text{direction } (\phi_l, \phi_z) \text{ per unit area of receiving} \\ \text{antenna per unit solid angle of the source.}\end{array} \qquad (3.3.1)$$

Basically $\mathcal{H}(\phi_l, \phi_z)$ is the noise power spectral density level of the background one would receive from a unit solid angle source if a unit area receiving antenna is pointed at the sky at angle (ϕ_l, ϕ_z) from the antenna coordinates. The total background noise power in watts received from the sky background over a bandwidth B Hz is then

$$P_b = B \int_{\text{sphere}} \mathcal{H}(\phi_l, \phi_z) \mathcal{A}(\phi_l, \phi_z) \, d\Omega \qquad (3.3.2)$$

where $\mathcal{A}(\phi_l, \phi_z)$ is again the effective area the antenna presents to the direction (ϕ_l, ϕ_z). By substitution from (3.1.6), this can be rewritten in terms of the receiving antenna gain function $g_r(\phi_l, \phi_z)$ as

$$P_b = \frac{B\lambda^2}{4\pi} \int_{\text{sphere}} \mathcal{H}(\phi_l, \phi_z) g_r(\phi_l, \phi_z) \, d\Omega \qquad (3.3.3)$$

Thus by knowing the background radiance function and the antenna gain pattern, the receiver background noise power can be computed directly from (3.3.3). Radiance functions for the sky are obtained by employing a blackbody radiation model to describe the noise emissions. From Planck's law, we know that a body (in this case, the sky) at temperature $T°(\phi_l, \phi_z)$ Kelvin in the direction (ϕ_l, ϕ_z) produces a radiance function at frequency f of

$$\mathcal{H}(\phi_l, \phi_z) = \frac{\hbar f^3}{c^2} \left[\frac{1}{\exp\left[hf/\kappa T°(\phi_l, \phi_z)\right] - 1} \right] \qquad (3.3.4)$$

where c is the speed of light, and

$$h = \text{Planck's constant} = 6.624 \times 10^{-34} \text{ W-sec/Hz}$$

$$\kappa = \text{Boltzmann constant} = 1.379 \times 10^{-23} \text{ W/degree Kelvin-Hz}$$

If $hf/\kappa T° \ll 1$, then (3.3.4) is well approximated by

$$\mathcal{H}(\phi_l, \phi_z) \cong \kappa \left(\frac{f}{c}\right)^2 T°(\phi_l, \phi_z) = \left(\frac{\kappa}{\lambda^2}\right) T°(\phi_l, \phi_z) \qquad (3.3.5)$$

Note that when $T° = 300$ K, $hf/\kappa T° = 1$ at 6000 GHz, so that the preceding approximation is quite accurate for RF and microwave wavelengths, and for reasonable temperatures. Thus, for blackbody radiation, the radiance function $\mathcal{H}(\phi_l, \phi_z)$ is directly related to the temperature functions $T^\circ(\phi_l, \phi_z)$ describing the sky background. Such temperature functions are available as *radio temperature maps* [2], which indicate temperature contours of the sky when viewed from Earth. When (3.3.5) is used in (3.3.3) we have

$$P_b = \frac{B\kappa}{4\pi} \int_{\text{sphere}} T^\circ(\phi_l, \phi_z) g_r(\phi_l, \phi_z) \, d\Omega \qquad (3.3.6)$$

and background noise power can be determined by integrating the known sky temperature function directly. Note that the collected power P_b depends on the antenna orientation through the gain pattern $g_r(\phi_l, \phi_z)$. This equation suggests that an overall effective background temperature of the sky for a particular receiving antenna can be defined as

$$T_b^\circ = \frac{1}{4\pi} \int_{\text{sphere}} T^\circ(\phi_l, \phi_z) g_r(\phi_l, \phi_z) \, d\Omega \qquad (3.3.7)$$

This allows (3.3.6) to be written as simply

$$P_b = \kappa T_b^\circ B \qquad (3.3.8)$$

When expressed in this way the received background noise appears to have a flat, one-sided power spectral density level of κT_b° W/Hz over the bandwidth B. The effective temperature T_b° is determined by weighting the temperature function of the sky by the actual antenna gain pattern and integrating in (3.3.7). The integration is often simplified by using temperature function approximations. In general, the sky is composed of localized galactic sources (sun, moon, planets, stars, etc.), each having a temperature T_i° approximately constant over a spatial solid angle Ω_i when viewed from the antenna site, and a relatively constant sky background of temperature T_s°. The effective temperature of the overall background in (3.3.7) with this model simplifies to

$$T_b^\circ = T_s g_s + \sum_i T_i^\circ g_i \qquad (3.3.9)$$

where

$$g_i = \frac{1}{4\pi} \int_{\Omega_i} g_r(\phi_l, \phi_z) \, d\Omega$$

$$g_s = \frac{1}{4\pi} \int_{\text{rest of sky}} g_r(\phi_l, \phi_z) \, d\Omega \qquad (3.3.10)$$

Since the antenna function has a constant area of 4π, $(\sum_i g_i) + g_s = 1$. Thus the background temperature is collected as a weighted sum of individual galactic source temperatures and a constant sky background.

Note that each gain coefficient g_i in (3.3.9) depends on the location of the particular source within the gain pattern. Thus a particular galactic source does not contribute much noise to the antenna if it lies on the fringes of the gain pattern. Some discrete sources (called *radio stars*) appear as discrete points in the sky ($\Omega_i \approx 0$), and their effect is accounted for by treating them as spatial "delta functions" in (3.3.7). Hence a point source of temperature T_1° in direction (ϕ_{l_1}, ϕ_{z_1}) contributes a term $T_1^\circ g(\phi_{l_1}, \phi_{z_1})$ to the sum in (3.3.9).

Table 3.1 summarizes some common sources of galactic noise in the sky, listing their average temperature [2–7]. The sun is obviously the basic noise generator, having a beamwidth of about 0.5° as viewed from the Earth. Because of its relatively high temperature, the sun can contribute quite significantly to receiver noise if it appears in the antenna field of view. Most stars are relatively weak noise radiators and can be neglected. Some of the larger stars are more active noise sources with the temperatures listed, but since they occupy such small beamwidths relative to typical antenna patterns, their contribution is generally insignificant. The planets and the moon are passive reflectors of the sun's energy, with the moon being most dominant, since it has the large beamwidth ($\approx 0.5°$). The effect of the outer planets is lessened because they occupy extremely small fields of view. For passive radiators, the amount of noise energy contributed to the Earth is

Table 3.1. Typical Background Source Temperatures and Beamwidths

Source	Temp., °K	Beamwidth, degrees
Sun	6000	0.5
Moon	200	0.5
Stars (at 300 MHz)		
Cassiopeia	3700	$<10^{-3}$
Cygnus	2650	$<10^{-3}$
Taurius	710	$<10^{-3}$
Contaurus	460	$<10^{-3}$
Planets (10 MHz–10 GHz)		
Mercury	613	2×10^{-3}
Venus	235	6×10^{-3}
Mars	217	4.3×10^{-3}
Jupiter	138	1.3×10^{-3}
Saturn	123	5.7×10^{-3}
Earth (from Moon)	300	2

accounted for by modifying its incident sun energy by a constant called the *albedo* (ratio of reflected to incident energy). This albedo is a function of frequency, and received energy from these sources generally decreases with increasing frequency (i.e., at higher frequencies more sun energy is absorbed by them than is reflected). When the albedo is taken into account the planets appear as localized sources with the approximate temperatures listed.

The temperature T_s° in (3.3.9) is due to the clear sky background and appears as an accumulation of a continuum of cosmic and galactic noise sources, influenced by the frequency effects of the radiation albedo. As a result, the clear sky appears as an almost uniform noise source in all directions, with an equivalent temperature that decreases with frequency. Typical background clear sky temperature T_s° is shown in Figure 3.6 as a function of frequency.

When we view from the Earth, sky temperature is also affected by the atmosphere. The gases and vapors tend to absorb sun energy and reradiate a portion of this energy isotropically as noise. This atmospheric reradiation predominates at frequencies above 10 GHz and tends to increase beyond this point as the wavelengths approach particulate size. To a receiving

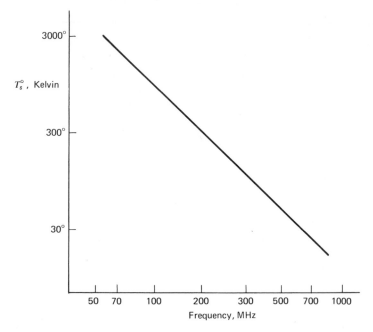

Figure 3.6. Sky background noise temperature versus frequency.

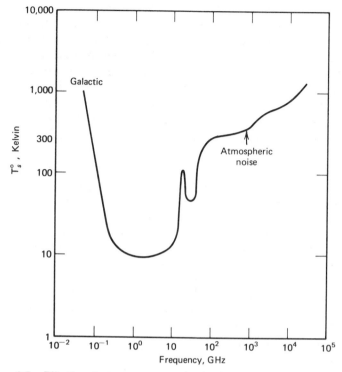

Figure 3.7. Effective sky temperature, galactic plus atmospheric reradiation.

antenna this reradiation appears as an extraneous source of background noise that tends to increase the sky temperature T_s° at higher frequencies. Thus the sky background appears as a combination of galactic effects that fall off with frequencies, as in Figure 3.6, and atmospheric effects that begin for short wavelengths and increases with frequency. When these two predominate effects are combined, one obtains an effective sky temperature behavior, as a function of frequency, shown in Figure 3.7. An important result of these dependencies is a region between 1 and 10 GHz, where an antenna aimed toward the sky receives less noise power than at a higher or lower frequency. This noise "window" is particularly advantageous for low power deep space and satellite communication systems and is the principal reason such systems utilize primarily carrier frequencies in the SHF band.

Although the sky background is the primary source of receiver noise, there are other types of interference that may appear during reception. One important type is *radio frequency interference* (RFI) due to other transmitting sources. This noise becomes important when the desired transmitter is located in an area dense with other operating transmitters, as, for example, with a satellite receiver aimed back toward the Earth. The amount of RFI can only be assessed with knowledge of these secondary source locations and their EIRP capability. An evaluation of the degree of interference then requires the computation of a link budget for each such source relative to the receiving antenna. RFI is particularly bothersome since it often is concentrated in frequency bands rather than being spectrally spread, as with pure background noise interference.

Another type of noise that appears with certain Earth based receivers is urban noise. Such noise is due to various electronic emissions associated with a typical urban environment, such as vehicles, communications, equipment, and so on. Such noise is usually impulsive, or burstlike, in nature and is generally modeled as a *shot noise* [8] process at the receiver front end. Shot noise interference takes the form of independently occurring random impulses obeying some prescribed distribution in time. When the rate of occurrence is high and fairly uniform in time, the shot noise pulses overlap into a continuous noise process with a well-defined power spectral density. When the rate of occurrence is low, however, the impulsive nature of the process is emphasized, preventing it from being treated mathematically as a continuous noise process, and care must be used in analytical treatments. Lastly, diffuse multipath noise, as in (2.7.2), may also contribute to the total receiver interference.

3.4. Receiver Front End Analysis

At the receiver the total antenna signal is coupled into the receiver front end circuitry, as shown in Figure 3.8a. Since the antenna may be located away from the actual front end electronics, a waveguide or cable may be used to provide this coupling. As stated in Section 2.4, such coupling devices inherently introduce attenuation into the antenna signal. The antenna produces at the front end input the carrier signal having the power level P_r, as computed in (3.2.8). In addition, any background noise and interference received by the antenna will produce a noise voltage that is superimposed on the receiver carrier signal. The combined carrier plus noise waveform is coupled through the cable to the receiver front end electronics.

The receiver front end circuitry has the prime objective of amplifying and filtering the relatively weak antenna signal. We model this front end filter-amplifier as having a filter transfer $\sqrt{G_c}H_c(\omega)$, Figure 3.8b, where G_c is the front end power gain and $H_c(\omega)$ represents a unit height carrier filter

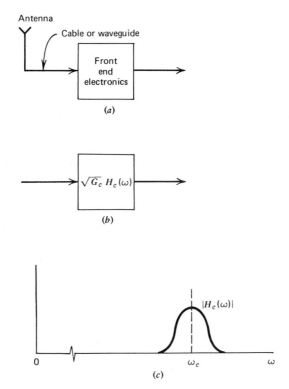

Figure 3.8. Receiver front end. (*a*) Block diagram, (*b*) filter model, (*c*) filter frequency function.

function. In reality the system may involve a cascade of several filters and amplifiers, so that $\sqrt{G_c}H_c(\omega)$ represents the combined effect of the cascade. The filter characteristic $H_c(\omega)$ represents the filtering that the front end provides for the received carrier signal. As such the filter is bandpass in nature, typically having a flat, unit level gain function over the bandwidth of the received carrier signal. Outside this band the filter should fall off rather quickly in frequency to remove as much noise and interference as possible. Thus the carrier filter function is a bandpass filter tuned to the receiver carrier signal, and therefore has the basic form sketched in Figure 3.8*c*. In many cases the carrier frequency (i.e., the actual location of the carrier spectrum) is not known exactly (e.g., because of Doppler), and the carrier filter may have to be several times larger in bandwidth to accommodate this uncertainty. The bandwidth actually needed in the front end filter also depends on the amount of distortion we are willing to accept in the modulated carrier. This topic is discussed in more detail in Section 3.5.

We might, however, point out that the construction of a suitable carrier filter for the receiver front end is not a trivial task. Difficulties are due to the following factors: (1) Filter bandwidths less than 0.1% of the center frequency often lead to unsatisfactory parameter values for practical realization. This means it is difficult to build arbitrarily narrow bandpass filters at the high RF carrier frequencies usually used. (2) The requirement of flat inband response and fast roll-off tends to be conflicting. In order to get sharp roll-off, one must often settle for nonflat inband response, as with Chebychev bandpass filters. On the other hand, flat inband behavior is generally accompanied by rounded-off band edges, as with Butterworth filters. (3) High order bandpass filters must generally be constructed as cascaded filter stages. Each stage, however, tends to introduce signal attenuation (*insertion loss*) that may severely reduce the overall front end gain. (4) Sharp cornered filters are invariably accompanied by severe phase distortion (nonlinear phase response) at the band edges. Such phase effects can seriously distort carrier waveforms and generally must be corrected by additional front end equalizing circuitry. Front end carrier amplification is generally produced by low-level tunnel diodes or low-noise cavity amplifiers.

Even after the carrier filter is properly designed for the front end, the carrier waveform must be amplified without distortion to provide the front end signal for the receiver processing that follows. For the model in Figure 3.8 the carrier power produced at the output of the front end is then

$$P_{\text{FE}} = P_r L_g G_c \qquad (3.4.1)$$

where G_c accounts for the front end amplifier power gain and L_g for any guide losses from antenna to front end input. Of course all antenna noise in the front end bandwidth is amplified as well.

When providing the amplification for the antenna signal, the receiver front end electronics invariably adds additional circuit noise to that already present as received background noise. his circuit is due both to thermal noise generated in passive resistors and semiconductors and to electronic noise resulting from flicker, shot, and tube noises in active amplifiers. Noise properties of any electronic device, whether passive or active, are accounted for by specification of its *noise figure*. Let us digress temporarily to discuss this parameter.

Consider an electronic device (filter, amplifier, etc.), as shown in Figure 3.9. We assume it is being fed by a source with output resistance Z_s that is matched to the input resistance Z_{in} of the device. We assume the source is at temperature T_0° Kelvin and the power gain of the device from input to output is G. It is well known [9–11] that a resistor Z_s at temperature T_0° generates a thermal noise voltage whose one-sided voltage spectral density

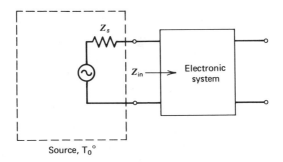

Figure 3.9. Electronic system with noisy source.

is $4\kappa T_0^\circ Z_s$ V^2/Hz, where κ is Boltzmann's constant. The noise power from the source in a bandwidth B that will be coupled into the electronic device under matched impedance conditions is then

$$\text{input noise power} = (4\kappa T_0^\circ Z_s)B\left(\frac{Z_{in}}{Z_s + Z_{in}}\right)^2 \frac{1}{Z_{in}}$$

$$= \kappa T_0^\circ B \qquad (3.4.2)$$

since $Z_s = Z_{in}$. Thus the power coupled in under matched conditions depends only on the source temperature and the prescribed bandwidth, but not on the value of the source resistance. Equation (3.4.2) implies that the input thermal noise power has a flat spectral density of level κT_0° W/Hz, one-sided, over the bandwidth B. The output noise power of the device due to this input thermal noise is given by the product of the device input power and the device power gain. Hence

$$\begin{array}{l}\text{noise power output of device in } B \text{ Hz} \\ \text{due to source thermal noise at } T_0^\circ \end{array} = \kappa T_0^\circ BG \qquad (3.4.3)$$

The noise figure F of the device is defined as

$$F \triangleq \frac{\text{total output noise in } B \text{ when input source temp} T_0^\circ = 290^\circ}{\text{output noise due to only the source at } T_0^\circ = 290^\circ}$$

$$(3.4.4)$$

Noise figure is therefore the ratio of total output noise to that due to the input noise alone, when the input source is at a specified temperature of 290°K. The total output noise is that due to the input source noise plus that due to internally generated noise within the device itself. Denoting the output noise power from internal sources as P_{int}, (3.4.4) becomes

$$F = \frac{\kappa(290°)BG + P_{int}}{\kappa(290°)BG}$$

$$= 1 + \frac{P_{int}}{\kappa(290°)BG} \tag{3.4.5}$$

We see that $F \geq 1$, and $F = 1$ only for a device with no internal noise (i.e., a noiseless device). Thus the noise figure of an electronic device indicates the amount by which the output noise will be increased over that of a noiseless device. Standard methods are known for measuring device noise figure (Problem 3.12). Typical noise figure values for receiver amplifiers are generally in the range 2–12 dB. From (3.4.5) we can solve for P_{int} as

$$P_{int} = \kappa[(F-1)290°]BG \tag{3.4.6}$$

which relates the output noise power due to internal sources to the device noise figure. When we compared it to (3.4.2) we see that internal noise is produced at the device output as if it were caused by an input source at temperature $(F-1)290°$. Thus we can account for internal device noise by simply adding to the input a matched source of temperature $(F-1)290°$ and treating the device as if it were noiseless. This equivalent source noise is added directly to any source noise already present, as shown in Figure 3.10. The temperature $(F-1)290°$ is called the *equivalent device temperature*. Note that it is not simply the physical temperature of the device, although the latter certainly affects the value of F. Low-noise receiver front ends using maser amplifiers or cryogenically cooled amplifiers can have noise figures as low as 1.1 (0.4 dB), and equivalent temperatures of about 30°K.

In many cases the electronic device may actually be composed of a cascade of devices, each with its own noise figure and power gain. One must therefore evaluate the noise figure of the cascade in order to assess the total output noise power that will occur. Consider the cascade of two devices as shown in Figure 3.11, each with noise figure F_1 and F_2 and

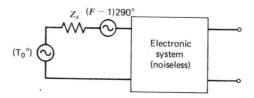

Figure 3.10. Equivalent noise input.

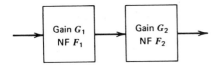

Figure 3.11. Cascade of two noisy systems.

power gain G_1 and G_2, respectively. We assume the devices are impedance matched at their input and output. The noise power at the output of the first device when a $290°$ matched source is placed at its input is $\kappa(290°)BG_1 + \kappa(F_1-1)290°BG_1$. This noise has the equivalent noise of the second stage, $\kappa(F_2-1)290°B$, added to it. The noise figure of the overall cascade is then

$$F = \frac{\kappa(290)°BG_1G_2 + \kappa(F_1-1)290°BG_1G_2 + \kappa(F_2-1)290°BG_2}{\kappa(290°)BG_1G_2}$$

$$= F_1 + \frac{F_2-1}{G_1} \tag{3.4.7}$$

We see that the noise figure of the cascade of the two devices is related to the individual noise figures by (3.4.7). Note that if the first stage gain G_1 is large, the noise figure of the cascade is approximately equal to that of the first stage alone. The result can be easily extended to more than two devices (Problem 3.8).

Any electronic device must necessarily have a noise figure. Consider a lossy transmission line, cable, or waveguide at temperature $T_g°$. Let L_g be the power loss factor of the line (ratio of output to input power). The output noise power due to a matched input source at $T_g°$ is then $\kappa T_g°BL_g$. However, the total noise at the output terminals under matched output conditions must be $\kappa T_g°B$, since the output (i.e., looking back into the guide) appears as a pure resistance at temperature $T_g°$. This means that $\kappa T_g°BL_g + P_{int} = \kappa T_g°B$, or

$$P_{int} = \kappa(1-L_g)T_g°B$$

$$= \kappa\left(\frac{1-L_g}{L_g}\right)T_g°BL_g \tag{3.4.8}$$

This immediately identifies the lossy waveguide as having an equivalent input noise temperature of $(1-L_g)T_g°/L_g$. Equating this to (3.4.6) shows that the lossy guide noise figure is

$$F_g = 1 + \left(\frac{1-L_g}{L_g}\right)\frac{T_g°}{290°} \tag{3.4.9}$$

when operated at temperature $T_g°$. When $T_g° = 290°$, $F_g = 1/L_g$. Note that the cascade connection of a lossy guide at 290° that is impedance matched to an amplifier with noise figure F_a is, from (3.4.7),

$$F = \frac{1}{L_g} + \frac{F_a - 1}{L_g}$$

$$= \frac{F_a}{L_g} \tag{3.4.10}$$

Thus attenuation circuitry immediately preceding amplifiers effectively increases the noise figure by the factor $1/L_g$.

Let us examine the noise properties of a typical receiver front end, using our earlier model of Figure 3.8, and inserting the noise figure definitions just derived. The antenna signal produces the carrier waveform along with the background noise having the equivalent temperature T_b°, as determined from (3.3.7). We assume the coupling waveguide or line has loss L_g and temperature T_g° and that the front end amplifier has transfer function $\sqrt{G_c} H_c(\omega)$, with noise figure F_a (Figure 3.12a). By using equivalent noise sources, the front end has the equivalent input noise sources shown in Figure 3.12b. If desired, all noise sources can be referred to the guide input, Figure 3.12c. The equivalent one-sided noise spectral level at the antenna output terminals can therefore be taken as κT_{eq}°, where

$$T_{eq}^\circ = T_b^\circ + \left(\frac{1 - L_g}{L_g}\right) T_g^\circ + \frac{(F_a - 1)290°}{L_g} \tag{3.4.11}$$

After waveguide attenuation and amplification, the equivalent antenna noise is filtered by the front end filter function, producing the resulting one-sided noise spectral density of the front end output as

$$\hat{S}_n(\omega) = \kappa T_{eq}^\circ L_g G_c |H_c(\omega)|^2 \tag{3.4.12}$$

The resulting output noise power, caused by background, thermal, and internal sources, is then

$$P_n = \kappa T_{eq}^\circ L_g G_c B_c \tag{3.4.13}$$

where

$$B_c = \frac{1}{2\pi} \int_0^\infty |H_c(\omega)|^2 \, d\omega \tag{3.4.14}$$

Here B_c represents the one-sided noise bandwidth of the carrier front end filter in Figure 3.8. If we again assume the front end subsystem has a bandwidth wide enough to pass the carrier waveform without distortion, the carrier power at the filter output is that given by (3.4.1). The resulting ratio of carrier power to total noise power at the front end filter output is

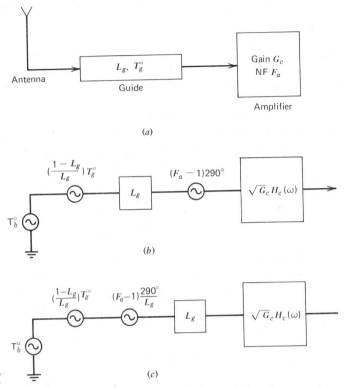

Figure 3.12. Front end noise model. (*a*) Block diagram, (*b*) equivalent noise model, (*c*) noise sources referred to input.

then

$$CNR = \frac{P_r L_g G_c}{\kappa T_{eq}^\circ L_g G_c B_c}$$

$$= \frac{P_r}{\kappa T_{eq}^\circ B_c} \qquad (3.4.15)$$

The preceding CNR parameter therefore serves as an indication of how well our front end subsystem has received the transmitted carrier. We see that this CNR depends only on the received carrier power P_r in (3.2.8), the carrier filter noise bandwidth in (3.4.15), and the equivalent temperature defined in (3.4.11). This latter parameter can be interpreted as the overall effective noise temperature of our receiver, taking into account all noise sources during reception and filtering. It is easy to see that T_{eq}° increases with T_b°, T_g°, F_a, and $1/L_g$, implying that receiver front ends should have as little noise and attenuation as possible, although which term actually does dominate in (3.4.11) depends on the specific elements involved. Background temperature T_b° can only be reduced by controlling the antenna pattern. Reduction of T_g° and F_a usually requires cooling or insulating the receiver front end. Reduction of front end attenuation requires close hardware tolerances.

It is interesting to note that CNR does not depend at all on the power gain G_c of the front end subsystem and depends on L_g only through T_{eq}°. This is due to the fact that both signal and noise are being power amplified during front end processing. Of course the signal and noise powers individually do depend on the system gain. We also see that CNR is related directly to B_c in (3.4.14), and obviously this should be as small as possible for best performance. Since $H_c(\omega)$ is normalized to one at band center, B_c can be reduced only by careful control of the filter characteristic, and using a filter that is no wider than necessary. Practical filter construction, however, often makes this a difficult task, as was pointed out earlier. In order to evaluate B_c we must perform the integration in (3.4.14), which requires specific knowledge of the function $H_c(\omega)$. For many such functions B_c can be directly related to the 3 dB bandwidth of the filter. Table 3.2 lists some common front end filter types and their corresponding B_c/B_{3dB} ratios.

It is common to determine CNR in decibels. When it is converted, (3.4.15) becomes

$$(\text{CNR})_{dB} = (P_r)_{dB} - [(\kappa)_{dB} + (T_{eq}^\circ)_{dB} + (B_c)_{dB}]$$

$$= (P_r)_{dB} + 228.6 - (T_{eq}^\circ)_{dB} - (B_c)_{dB} \qquad (3.4.16)$$

The carrier $(P_r)_{dB}$ is determined from the link budget in (3.2.8). Note that a CNR of 20 dB with a 300° receiver and a 1 MHZ noise bandwidth requires

Table 3.2.   Filter Noise Bandwidth B_c (B_{3dB} = 3dB Bandwidth)

Filter	Order	$\dfrac{B_c}{B_{3dB}}$	Filter	Order	$\dfrac{B_c}{B_{3dB}}$	Filter	Order	$\dfrac{B_c}{B_{3dB}}$
Butterworth	1	1.570	Chebychev	1	1.570	Chebychev	1	1.57
	2	1.220	$(\varepsilon = 0.1)$	2	1.15	$(\varepsilon = 0.158)$	2	1.33
	3	1.045		3	0.99		3	0.86
	4	1.025		4	1.07		4	1.27
	5	1.015		5	0.96		5	0.81
	6	1.010		6	1.06		6	1.26

a P_r of only $-123.6\,\text{dB}$, or about $4.3 \times 10^{-13}\,\text{W}$. It therefore does not require much received carrier power to generate a substantial CNR in (3.4.16).

When the transmission parameters constituting P_r in (3.2.8) are substituted into (3.4.15), we can rewrite

$$\text{CNR} = \left(\frac{P_{\text{amp}} L_t g_t \rho_r}{\kappa B_c}\right)(L_p L_a)\left(\frac{g_r}{T_{\text{eq}}^\circ}\right) \qquad (3.4.17)$$

When written in this form, the first bracket depends only on transmitter parameters (assuming the transmitted carrier bandwidth dictates the front end noise bandwidth), the middle bracket depends on the transmission channel, and the last bracket is determined purely by the receiver. Note that as far as *CNR* is concerned, the receiving system is characterized only by the ratio of gain to noise temperature. This ratio is often used as a figure of merit for assessing carrier receiving stations.

The $(g_r / T_{\text{eg}}^\circ)$ ratio implies that in achieving a desired level of performance, receiver antenna gain can be reduced if the front and noise temperature can be reduced. Hence, in the construction of receiving systems, antenna size can be traded off against the cost of low-noise front ends. This is particularly important in designing ground receiving stations for satellite transmitters, where T_{eq}° is sensitive to ground antenna pointing angle (recall Figure 2.16b). For example, suppose a receiving station uses low-noise front end circuitry so that its noise temperature is primarily due to background. Assume the ground antenna pointing is restricted to an elevation angle of about $5°$. At this angle the sky noise temperature is approximately $25°\text{K}$. If the minimum elevation angle could be increased to about $10°\text{C}$ (i.e., insuring that the satellite would never be seen from the ground station at a lower angle), the sky temperature would be reduced to about $12°\text{K}$. This reduces the receiver temperature by 50%, thereby allowing a 3 dB decrease in antenna gain, or a reduction in area by a half. Hence required ground station antenna size is directly related to satellite location relative to the station. By properly locating the satellite, all ground stations being served will have a favorable location (i.e., large elevation angles) and can therefore be designed at minimal size and cost.

When the field is propagated from transmitter to receiver by guided transmission, as for example, with long distance cables, there is no background noise involved. Instead, noise generated from the transmitter power amplifier and from any amplifiers inserted in the cable appears at the front end input and replaces the background noise power. This cable noise can be determined by computing the cable noise figure and inserting an equivalent noise temperature at the cable input, which in turn can be

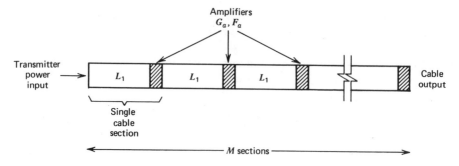

Figure 3.13. Cable model with amplifiers.

converted to an equivalent cable temperature at the front end input, replacing the background temperature T_b°. If the cable is simply a lossy line, its noise figure follows from (3.4.9). If the line contains inserted amplifiers, the noise figure must be calculated. This is most conveniently done by considering the line as a cascade of individual sections, each composed of lossy section of line feeding an amplifier having a specified noise figure. For example, consider the cable system in Figure 3.13 containing M identical amplifiers at equal distances. We can consider the cable to have M identical sections, each composed of a lossy line of gain L_1 and an amplifier of gain G_a and noise figure F_a. We assume the cable is at 290°K, and the amplifier gain G_a exactly compensates for the line loss L_1. That is, $L_a G_a = 1$ for each section. Using the result of Problem 3.8, the overall noise figure of the cable is

$$F_c = \frac{F_a}{L_1} + \frac{(F_a/L_a)-1}{L_1 G_a} + \frac{(F_a/L_1)-1}{(L_1 G_a)^2} \cdots$$

$$= M\left(\frac{F_a}{L_1}\right) - M + 1 \tag{3.4.18}$$

The effective noise temperature of the cable and source, referred to cable output terminals, is then $T_c^\circ = F_c$ (290°). This now represents the effective temperature of the noise that the cable transmissions channel inserts into the receiver front end. The carrier power at the same point was given in (3.2.11). Note that the overall noise figure F_c is directly proportional to the amplifier noise figure and to the number of amplifiers inserted. At first glance this might appear to indicate a disadvantage in using amplifiers at all. However, it must be remembered that the gain of the amplifier is canceling out the line loss. If no amplifiers were used, F_c would be equal to the reciprocal of the total cable loss; that is, $F_c = 1/L_1^M$. This means the effective noise would increase exponentially with M instead of only proportionally. Even if the amplifiers were quite noisy, the former effect is much more serious than the latter (see Problem 3.17).

3.5. Linear Filtering of Carrier Signals

The analysis of the previous sections has given us an indication of the power levels of the carrier at the front end output. It may still be necessary, however, to determine the actual effect of the front end filter on the carrier waveform. Simply constructing a filter with a bandwidth that passes the major portion of the carrier spectrum may not be adequate in terms of eventual waveform reconstruction. Of particular concern is the effect of the filter on the modulating baseband waveform embedded within the carrier, which will be subsequently extracted by the demodulation process. In this section we examine the basic effect of a bandpass carrier filter, such as the front end filter, on the various carrier waveforms $c(t)$.

Consider the system in Figure 3.14, where we let $c(t)$ be a modulated carrier at the input of a bandpass filter, we let $c_0(t)$ be the output carrier signal, and we represent the filter by its transfer function $H_c(\omega)$ and impulse response $h_c(t)$. We then have

$$c_0(t) = \int_{-\infty}^{\infty} h_c(x)c(t-x)\,dx \qquad (3.5.1)$$

For the class of modulated carrier signals of interest, we recall $c(t)$ can be written as

$$c(t) = a(t)\cos\left[\omega_c t + \theta(t)\right]$$
$$= \text{Real}\left\{a(t)e^{j\theta(t)}e^{j\omega_c t}\right\} \qquad (3.5.2)$$

where the form of $a(t)$ and $\theta(t)$ depends on whether it is amplitude or angle modulated. This means (3.5.1) takes the general form

$$c_0(t) = \int_{-\infty}^{\infty} h_c(x)\,\text{Real}\left\{a(t-x)\,e^{j\theta(t-x)}\,e^{j\omega_c t - j\omega_c x}\right\}dx$$

$$= \text{Real}\left\{e^{j\omega_c t}\int_{-\infty}^{\infty} h_c(x)a(t-x)\,e^{j\theta(t-x)}\,e^{-j\omega_c x}\,dx\right\} \qquad (3.5.3)$$

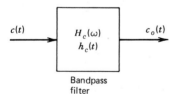

c(t) → | $H_c(\omega)$ $h_c(t)$ | → $c_0(t)$

Bandpass filter

Figure 3.14. Bandpass filtering of a carrier waveform.

Thus a study of filtering effects on carrier signals reduces to an examination of this particular equation. We examine the preceding for specific types of carrier waveforms.

Filtering AM Signals. Let $\theta(t) = \psi$, a constant, so that (3.5.2) represents an AM carrier. Then (3.5.3) becomes

$$c_0(t) = \text{Real}\,\{a_0(t)\,e^{\,j(\omega_c t + \psi)}\} \tag{3.5.4}$$

where

$$a_0(t) = \int_{-\infty}^{\infty} a(t-x)h_c(x)\,e^{-j\omega_c x}\,dx \tag{3.5.5}$$

Hence the output envelope function $a_0(t)$ appears as a complex time function obtained by filtering $a(t)$ with a complex impulse response $h_c(x)\,e^{-j\omega_c x}$. This impulse response corresponds to the equivalent transfer function

$$\tilde{H}_c(\omega) = \int_{-\infty}^{\infty} h_c(x)\,e^{-j\omega_c x}\,e^{-j\omega x}\,dx$$

$$= \int_{-\infty}^{\infty} h_c(x)\,e^{-j(\omega_c + \omega)x}\,dx$$

$$= H(\omega + \omega_c) \tag{3.5.6}$$

Thus the filter $\tilde{H}_c(\omega)$ is equivalent to the positive frequency filter function $H_c(\omega)$ shifted to $\omega = 0$, as shown in Figure 3.15. The filtered carrier $c_0(t)$ therefore has a complex envelope obtained by filtering $a(t)$, the amplitude modulation of the input carrier, with the equivalent filter $\tilde{H}_c(\omega)$. If $A(\omega)$ is the Fourier transform of $a(t)$, then the transform of $a_0(t)$ is

$$A_0(\omega) = A(\omega)\tilde{H}_c(\omega) \tag{3.5.7}$$

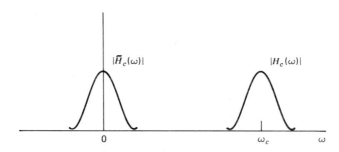

Figure 3.15. Bandpass filter and its low pass equivalent.

Note that $a_0(t)$ in (3.5.5) [i.e., the transform of (3.5.7)] is, in general, a complex time function, having both an amplitude and phase variation at each t, and the output carrier $c_0(t)$ is both amplitude and phase modulated. However, $a_0(t)$ is purely real (no phase variation) if its transform $A_0(\omega)$ has a magnitude even in ω and a phase odd in ω. (Recall Problem 1.3.) Since $a(t)$ is real, this even and odd symmetry occurs in $A_0(\omega)$ if $\hat{H}_c(\omega)$ is also even in magnitude and odd in phase about $\omega = 0$. This requires the filter $H_c(\omega)$ to have the same symmetry about ω_c. Thus if the carrier filter is symmetric about the carrier frequency, no phase modulation is induced on the output carrier, and the filtered carrier simply has an amplitude modulation obtained by filtering the input modulation with the equivalent filter $\tilde{H}_c(\omega)$. This means the effect on the modulation of bandpass filtering of the AM carrier can be determined by considering instead the effect of passing the modulation alone through the equivalent low pass filter. Thus to achieve a given distortion the bandpass filter must be as wide about $\omega = \omega_c$ to pass $c_0(t)$ as a low pass filter must be about $\omega = 0$ to pass $a(t)$ with the same distortion level. The problem of designing bandpass AM filters reduces therefore to a straightforward problem in linear low pass filter design. In fact, AM bandpass filters are often designed in exactly this manner—first designing the low pass equivalent, then using frequency translation (see Problem 1.15) to obtain the desired bandpass network.

Filtering Angle Modulated Carriers. In this case we reconsider (3.5.1) with $a(t) = A$, $\theta(t) = \Delta m(t)$, so that (3.5.3) becomes

$$c_0(t) = A \operatorname{Real}\left\{ e^{j\omega_c t} \int_{-\infty}^{\infty} h(x) e^{-j\omega_c x} e^{j\theta(t-x)} \, dx \right\} \qquad (3.5.8)$$

We recognize again the equivalent low pass filter impulse response in the integrand, but the desired modulation waveform $m(t)$ now appears in the exponential function. We therefore do not obtain the simple modulation-filtering interpretation as we did for AM. Analysis methods for (3.5.8) generally involve some type of expansion of the exponential modulation term. There are two basic procedures commonly used—the Carson–Fry [12] quasi-linear method and the Bedrosian–Rice expansion method [13].

In the quasi-linear approach the modulation is expanded in a Taylor series and the exponential expansion is terminated after several terms. Thus we write

$$\theta(t-x) = \theta(t) - \frac{\theta'(t)x}{1!} + \frac{\theta''(t)x^2}{2!} \cdots \qquad (3.5.9)$$

where the primes denote differentiations with respect to t, corresponding to a Taylor series expansion of $\theta(t-x)$ about $x = 0$. We then have for the integral in (3.5.8).

$$\int_{-\infty}^{\infty} h(x) e^{-j\omega_c x} \exp j\left[\theta(t) - \theta'(t)x + \frac{\theta''(t)x^2}{2} \cdots \right] dx \qquad (3.5.10)$$

If we retain the first three terms only, (3.5.8) can be written as

$$c_0(t) = \text{Real} \left\{ A \, e^{j\omega_c t + j\theta(t)} \int_{-\infty}^{\infty} h_c(x) e^{-j\omega(t)x} e^{j\theta''(t)x^2/2} \, dx \right\} \qquad (3.5.11)$$

where we have let $\omega(t) \triangleq \omega_c + \theta'(t)$, which is the instantaneous frequency of $c(t)$. If we further expand

$$\exp\left[j\frac{\theta''(t)x^2}{2} \right] \approx 1 + j\frac{\theta''(t)x^2}{2} \qquad (3.5.12)$$

and note from Appendix A, A.1,

$$\int_{-\infty}^{\infty} x^2 h_c(x) e^{-j\omega x} \, dx = -\frac{d^2}{d\omega^2} H_c(\omega) \qquad (3.5.13)$$

then

$$c_0(t) \approx \text{Real} \left\{ A \, e^{j\omega_c t + j\theta(t)} \left[H_c[\omega(t)] - j\frac{\theta''(t)}{2} H_c''[\omega(t)] \right] \right\}$$

$$\approx \text{Real} \left\{ A \, e^{j\omega_c t + j\theta(t)} H_c[\omega(t)] \left[1 - j\frac{\omega'(t)}{2}\left(\frac{H_c''[\omega(t)]}{H_c[\omega(t)]} \right) \right] \right\}$$

$$(3.5.14)$$

The preceding represents a somewhat complicated approximation to the exact filter output, but nevertheless does give some indication of the filtering effect. The first term is called the *quasi-linear* response term and has the following interpretation. Let us write the filter transfer function as $H_c(\omega) = |H_c(\omega)| \exp[j\angle H_c(\omega)]$ where $\angle H_c(\omega)$ is the phase function and $|H_c(\omega)|$ the magnitude. Then the first term in (3.5.14) yields an output component

$$A|H_c[\omega(t)]| \cos[\omega_c t + \theta(t) + \angle H_c[\omega(t)]] \qquad (3.5.15)$$

That is, at time t_0, the input amplitude is changed by the magnitude of the filter function at the instantaneous frequency $\omega(t_0)$, whereas the phase at that frequency, $\angle H_c[\omega(t_0)]$, is added to the input phase. Thus the quasi-linear term instantaneously evaluates the output amplitude and phase, as the frequency of the input carrier slides over the filter function. From the quasi-linear term we see that the bandpass filter should have essentially flat magnitude and linear phase over the maximum frequency excursions of

$c(t)$, if negligible amplitude and phase distortion are to occur. However, the quasi-linear term adequately represents $c(t)$ only if the remaining terms in (3.5.14) can be neglected. This means it is necessary that

$$\frac{1}{2}\left|\left(\frac{d\omega(t)}{dt}\right)\left[\frac{dH_c^2[\omega(t)]/d\omega^2}{H_c[\omega(t)]}\right]\right| \ll 1 \qquad (3.5.16)$$

or equivalently,

$$\left|\frac{dH_c^2[\omega(t)]/d\omega^2}{H_c[\omega(t)]}\right| \ll \frac{2}{|d\omega(t)/dt|} \qquad (3.5.17)$$

for all t. The term on the left depends on the specific function $H_c(\omega)$, whereas the term on the right depends only on the rate of change of the input carrier frequency. In essence, (3.5.17) requires that the frequency cannot change too rapidly relative to the filter function slope characteristics. Unfortunately, it is difficult to evaluate the left side of (3.5.17), although tabulations of its maximum values for several common types of filters have been presented [14].

A difficulty with this quasi-linear approach is the inability to assess the degree of accuracy of our approximation. We have terminated an infinite series without proper regard to the higher terms neglected. This means that although the quasilinear term has aided in filter design, the use of (3.5.14) alone to compute filter distortion accurately is somewhat suspect. Another disadvantage of the quasi-linear approach is that it accounts primarily for filter effects on the carrier waveform itself. In the typical analysis situation the system engineer is generally interested in the effect of the filter on the baseband modulation itself, which is imbedded in the carrier angle modulation. In this regard it is more advantageous to use the Bedrosian-Rise method to obtain an expansion of the phase of the output carrier $c_0(t)$ directly. If we let $\gamma(t)$ be the integral in (3.5.8),

$$\gamma(t) = \int_{-\infty}^{\infty} h_c(x)\, e^{-j\omega_c x}\, e^{j\theta(t-x)}\, dx \qquad (3.5.18)$$

representing the complex envelope of $c_0(t)$, we see immediately that

$$\text{phase of } c_0(t) = \text{phase of } \gamma(t)$$

$$= \text{Imag}\{\log \gamma(t)\} \qquad (3.5.19)$$

where $\text{Imag}\{\cdot\}$ means the imaginary part of $\{\cdot\}$. We now seek an expansion of $\log \gamma(t)$. If we denote

$$z(y, t) \triangleq \log \int_{-\infty}^{\infty} \tilde{h}_c(x)\, e^{jy\theta(t-x)}\, dx \qquad (3.5.20)$$

then we see that $\log \gamma(t) = z(1, t)$, with $\tilde{h}_c(x) = h_c(x) e^{-j\omega_c x}$. Thus an expansion of $\log \gamma(t)$ can be obtained from an expansion of $z(y, t)$ with the substitution of $y = 1$. The Taylor series expansion of $z(y, t)$ is

$$z(y, t) = \sum_{i=0}^{\infty} \left[\frac{d^i z(y, t)}{dy^i} \right]_{y=0} \frac{(jy)^i}{i!} \tag{3.5.21}$$

Therefore

$$\log \gamma(t) = \sum_{i=0}^{\infty} \left[\frac{d^i z(y, t)}{dy^i} \right]_{y=0} \frac{(j)^i}{i!} \tag{3.5.22}$$

and in (3.5.19)

$$\text{phase of } c_0(t) = \text{Imag} \left\{ \sum_{i=0}^{\infty} \frac{d^i z(y, t)}{dy^i} \bigg|_{y=0} \frac{(j)^i}{i} \right\} \tag{3.5.23}$$

The first imaginary term corresponds to $i = 1$ and has the form

$$\frac{dz(y, t)}{dy} \bigg|_{y=0} = \frac{1}{\tilde{H}_c(0)} \int_{-\infty}^{\infty} \tilde{h}_c(x)\theta(t - x) \, dx \tag{3.5.24}$$

Note that this term corresponds to simply the linear filtering of the phase modulation $\theta(t)$ by the low pass equivalent of the bandpass filter, just as in the AM case. This term is often called the *linear* term of the expansion and is considered the desired output signal component. The remaining terms of the series in (3.5.23) represent the distortion terms superimposed on the linear term. If we define $\Phi(t)$ as the linear term in (3.5.24) and let

$$d_n(t) \triangleq \frac{1}{n!} \int_{-\infty}^{\infty} \tilde{h}_c(x)[\theta(t - x) - \Phi(t)]^n \, dx \tag{3.5.25}$$

then the first such distortion term corresponds to $i = 3$ and is given by $d_3(t)$, whereas the term for $i = 5$ is given by $d_5(t) - (10/5!)d_3(t)d_2(t)$. The remaining terms of the series get exceedingly more complicated. Nevertheless, the series in (3.5.23) represents an exact converging expansion of the phase of the bandpass filtered carrier signal. In addition, the distortion terms are of a form relatively convenient for computer simulation. In general, the Rice-Bedrosian expansion is primarily applicable when the carrier modulation index is small and the total distortion is quite low. Reported studies [15] have shown that in these cases only the first several terms are necessary to predict the amount of distortion accurately by this method.

3.6. Receiver Noise Processes

In Section 3.4 we concentrated solely on the power values of the receiver noise. However, in subsequent discussions we must have a more definitive analytical model for the noise processes generated in our receivers. Recall that our primary sources of receiver noise were the blackbody radiation noise, the thermal circuit noise, and the flicker and tube noise associated with active elements. In the past each of these noises has been rigorously analyzed, both theoretically and experimentally, and each has been related to the accumulated effect of a large number of independent random movements of charged particles. When such models are developed, these noise sources invariably reduce to zero mean noise processes obeying Gaussian statistics, the latter obtained by application of some form of the central limit theorem (Section B.3). Thus noise processes in communication receivers are almost universally accepted as Gaussian random processes.

Earlier in this chapter we developed a receiver noise model in which the power spectral density of the receiver noise, referred to the receiving antenna output terminals, is given by κT_{eq}° W/Hz, where T_{eq}° was given in (3.4.11). Thus the receiver antenna noise has a spectrum that is flat over all microwave frequencies and has a two-sided spectral level (Figure 3.16a)

$$S_{ant}(\omega) = \kappa T_{eq}^{\circ}/2 \tag{3.6.1}$$

Thus receiver noise is inherently a Gaussian white noise process when referred to the antenna terminals. After front end filtering, this spectrum is shaped by the carrier filter function, producing the two-sided front end output noise spectrum in (3.4.12), which we write here as

$$S_n(\omega) = \tfrac{1}{2}\kappa T_{eq}^{\circ} G_{FE} |H_c(\omega)|^2 \tag{3.6.2}$$

where G_{FE} is the total front end gain (guide attenuation and amplifier gain). Note that this noise spectrum has been shaped by the filter function (Figure 3.16b) and limited to the spectral extent of the filter $H_c(\omega)$. Since this filter typically occupies a narrow symmetrical band about the filter center frequency, the filtered front end noise is a narrowband Gaussian process with this spectrum. Bandpass Gaussian noise processes of this type can be analytically written as (see Section B.6)

$$n(t) = n_c(t) \cos(\omega_c t + \psi) - n_s(t) \sin(\omega_c t + \psi) \tag{3.6.3}$$

where ψ is an arbitrary phase angle, ω_c is the filter frequency, and $n_c(t)$ and $n_s(t)$ are independent Gaussian noise processes, each having the two-sided spectrum

$$\tilde{S}_n(\omega) = (\kappa T_{eq}^{\circ}) G_{FE} |\tilde{H}_c(\omega)|^2 \tag{3.6.4}$$

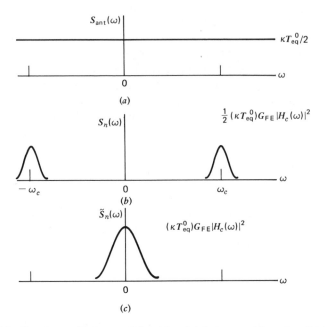

Figure 3.16. Receiver noise spectral densities. (*a*) Antenna white noise, (*b*) bandpass front end noise, (*c*) noise quadrature component spectrum.

where $\tilde{H}_c(\omega)$ is again the low frequency, shifted version of $H_c(\omega)$, as given in (3.5.6). This spectrum is shown in Figure 3.16*c*. We see that the two-sided low frequency spectrum $\tilde{S}_n(\omega)$ is simply a low frequency version of the bandpass noise spectrum about ω_c. That is, $\tilde{S}_n(\omega)$ is obtained by simply shifting the one-sided version of the spectrum $S_n(\omega)$ in (3.6.2) to the origin. The phase angle ψ is purely arbitrary and can be selected to be any desired (nonrandom) value without altering the statistical property of $n(t)$.

The representation in (3.6.3) is called a *quadrature expansion* of the bandpass noise $n(t)$, and $n_c(t)$ and $n_c(t)$ are its *quadrature components*. Note that the components themselves each have the same power as the noise $n(t)$ itself. That is,

$$\text{power of } n_c(t) = \text{power of } n_s(t)$$

$$= \frac{1}{2\pi} \int_{-\infty}^{\infty} \tilde{S}_n(\omega) \, d\omega$$

$$= \frac{1}{2\pi} \int_{-\infty}^{\infty} S_n(\omega) \, d\omega$$

$$= \text{power of } n(t) \tag{3.6.5}$$

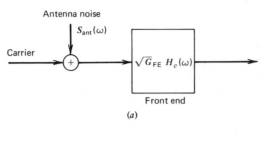

(a)

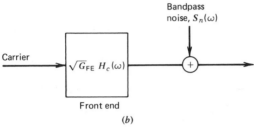

(b)

Figure 3.17. Receiver noise models (a) with antenna noise, (b) with bandpass front end noise.

We emphasize that each quadrature component does not have half the total power, as one might expect. This becomes a basic point in later analyses.

In summary, then, we see that receiver noise can be modeled in either of two ways. We can consider the noise as a Gaussian white noise process of spectral level κT_{eq}° appearing at the input to the receiver front end, as in Figure 3.17a, or we can consider the noise as a bandpass noise process, having the spectrum $S_n(\omega)$ in (3.6.2), appearing at the front end output, as in Figure 3.17b. Note the latter model is exactly equivalent to adding the white noise at the front end input and taking into account the filtering within. Of course both models produce exactly the CNR in (3.4.15) at the front end output.

3.7. Carrier Plus Noise Waveforms

We wish to examine in detail the front end output waveform. Recall that this output is composed of the desired carrier waveform $c(t)$ superimposed with the bandpass noise process $n(t)$ of the receiver front end. We denote this combined waveform as

$$x_c(t) = c(t) + n(t) \tag{3.7.1}$$

The carrier waveforms we have been discussing are of the general form

$$c(t) = a(t) \cos (\omega_c t + \theta(t) + \psi) \qquad (3.7.2)$$

where we remember that $a(t)$ or $\theta(t)$ may be constant, depending on the type of modulation. The noise $n(t)$ is the zero mean, narrowband, Gaussian process discussed in Section 3.6 and having the quadrature expansion in (3.6.3). Consider first the case of an AM carrier in which $\theta(t) = 0$ in (3.7.2). The combined waveform in (3.7.1) is then

$$x_c(t) = a(t) \cos (\omega_c t + \psi) + n_c(t) \cos (\omega_c t + \psi) - n_s(t) \sin (\omega_c t + \psi) \qquad (3.7.3)$$

where we have taken the arbitrary phase angle of the noise process to be identical to that of the carrier. Combining terms allows us to write this waveform as

$$x_c(t) = [a(t) + n_c(t)] \cos (\omega_c t + \psi) - n_s(t) \sin (\omega_c t + \psi)$$
$$= \text{Real} \{ \alpha(t) e^{jv(t)} e^{j(\omega_c t + \psi)} \}$$
$$= \alpha(t) \cos [\omega_c t + \psi + v(t)] \qquad (3.7.4)$$

where now

$$\alpha(t) = \{ [a(t) + n_c(t)]^2 + n_s^2(t) \}^{1/2} \qquad (3.7.5a)$$

$$v(t) = \tan^{-1} \left[\frac{n_s(t)}{a(t) + n_c(t)} \right] \qquad (3.7.5b)$$

Here $\alpha(t)$ is called the *envelope* of the combined signal $x_c(t)$ and $v(t)$ is referred to as the *phase noise*. Thus the effect of adding bandpass receiver noise to the AM carrier in (3.7.2) is to change the amplitude $a(t)$ to the envelope process $\alpha(t)$ and to introduce the additive phase noise $v(t)$. Note that the phase noise depends not only on the input noise, but also on the carrier amplitude function as well. Roughly speaking, the receiver interference has added noisy variations to the amplitude and phase of the original transmitted carrier.

For angle modulated carriers, $a(t) = A$ in (3.7.2) and

$$x_c(t) = A \cos [\omega_c t + \theta(t) + \psi] + n_c(t) \cos (\omega_c t + \psi) - n_s(t) \sin (\omega_c t + \psi) \qquad (3.7.6)$$

We cannot combine terms as we did in (3.7.3) because of $\theta(t)$. However, if we substitute the identities:

$$\cos (\omega_c t + \psi) = \cos (\theta(t)) \cos (\omega_c t + \theta(t) + \psi) + \sin (\theta(t)) \sin (\omega_c t + \theta(t) + \psi)$$
$$\sin (\omega_c t + \psi) = \cos (\theta(t)) \sin (\omega_c t + \theta(t) + \psi) - \sin (\theta(t)) \cos (\omega_c t + \theta(t) + \psi) \qquad (3.7.7)$$

then the total signal can be combined instead as

$$x_c(t) = [A + \hat{n}_c(t, \theta)] \cos (\omega_c t + \theta(t) + \psi) - \hat{n}_s(t, \theta) \sin (\omega_c t + \theta(t) + \psi) \quad (3.7.8)$$

where

$$\hat{n}_c(t, \theta) \triangleq n_c(t) \cos (\theta(t)) + n_s(t) \sin (\theta(t)) \quad (3.7.9a)$$

$$\hat{n}_s(t, \theta) \triangleq n_s(t) \cos (\theta(t)) - n_c(t) \sin (\theta(t)) \quad (3.7.9b)$$

The noise components $\hat{n}_c(t, \theta)$ and $\hat{n}_s(t, \theta)$ are modified forms of the quadrature components of the bandpass noise and involve the carrier phase modulation $\theta(t)$. Equation (3.7.8) can now be written as in (3.7.4) as

$$x_c(t) = \alpha(t) \cos [\omega_c t + \theta(t) + \psi + v(t)] \quad (3.7.10)$$

where again $\alpha(t)$ and $v(t)$ are given in (3.7.5) except with $\hat{n}_c(t, \theta)$ and $\hat{n}_s(t, \theta)$ replacing $n_c(t)$ and $n_s(t)$. Equation (3.7.10) again has the form of a random amplitude carrier with a phase noise $v(t)$ added to the original carrier phase. Thus again the receiver noise has distorted the amplitude and phase of the original transmitted carrier.

The statistical properties of the modified noise components in (3.7.9) can be derived from those of the quadrature components in Section 3.6. Their mean value is zero [since $n_c(t)$ and $n_s(t)$ have zero mean] and their autocorrelation functions are

$$R_{\hat{n}_c}(\tau) = \mathcal{E}[\hat{n}_c(t, \theta)\hat{n}_c(t + \tau, \theta)]$$

$$= R_{n_c}(\tau)[\cos (\theta(t)) \cos (\theta(t + \tau))]$$

$$+ R_{n_s}(\tau)[\sin (\theta(t)) \sin (\theta(t + \tau))] \quad (3.7.11)$$

where $R_{n_c}(\tau)$ and $R_{n_s}(\tau)$ are the autocorrelation functions of $n_c(t)$ and $n_s(t)$. However, these are identical, and (3.7.11) reduces to

$$R_{\hat{n}_c}(\tau) = R_{n_c}(\tau)[\cos (\theta(t) - \theta(t + \tau))] \quad (3.7.12a)$$

If $\theta(t)$ is a random process then the bracket must be further averaged over the joint statistics of $\theta(t)$ and $\theta(t + \tau)$. Similarly, the autocorrelation of $\hat{n}_s(t, \theta)$ and the cross-correlation of $\hat{n}_c(t, \theta)$ and $\hat{n}_s(t, \theta)$ are (Problem 3.24)

$$R_{\hat{n}_s}(\tau) = R_{n_s}(\tau)[\cos (\theta(t) - \theta(t + \tau))] \quad (3.7.12b)$$

$$R_{\hat{n}_c\hat{n}_s} = R_{n_c}(\tau)[\sin (\theta(t) - \theta(t + \tau))] \quad (3.7.13)$$

Thus the modified components $\hat{n}_c(t, \theta)$ and $\hat{n}_s(t, \theta)$ have their autocorrelation as that of $n_c(t)$ and $n_s(t)$, modified by the effect of the carrier phase function $\theta(t)$. We note this autocorrelation is, in general, not stationary. However, in most applications we can simplify (3.7.12) and (3.7.13). If the carrier phase modulation does not change significantly over the correlation time of the $n_c(t)$ process, we may state

$$\theta(t) \approx \theta(t + \tau) \qquad \text{over values of } \tau \text{ for which } R_{n_c}(\tau)$$
$$\text{is essentially nonzero} \qquad\qquad (3.7.14)$$

Then we may argue that

$$R_{\hat{n}_c}(\tau) = R_{\hat{n}_s}(\tau) \approx R_{n_c}(\tau)$$
$$R_{\hat{n}_c\hat{n}_s}(\tau) \approx 0 \qquad\qquad\qquad\qquad\qquad (3.7.15)$$

That is, the noise processes in (3.7.9) become approximately stationary and uncorrelated with autocorrelation given by $R_{n_c}(\tau)$. The condition in (3.7.14) is basically equivalent to the statement that the bandwidth of the spectrum of the bandpass noise is much larger than that of the carrier phase modulation $\theta(t)$. This means the front end bandwidth must exceed the phase modulation bandwidth. Since angle modulated systems are usually operated as wideband systems (i.e., given large phase or frequency deviations), this bandwidth condition is usually satisfied. When (3.7.15) is valid, the power spectral density of $\hat{n}_c(t, \theta)$ and $\hat{n}_s(t, \theta)$ are identical to those of $n_c(t)$ and $n_s(t)$. Thus, from (3.6.4),

$$S_{\hat{n}_c}(\omega) = S_{\hat{n}_s}(\omega) = \tilde{S}_n(\omega) \qquad\qquad (3.7.16)$$

where again $\tilde{S}_n(\omega)$ is the low frequency equivalent of the bandpass noise spectrum.

Although the autocorrelation and spectrum of the processes $\hat{n}_c(t, \theta)$ and $n_s(t, \theta)$ can be determined in this way, we often need to determine their actual probability densities for later analysis. If the carrier phase process $\theta(t)$ is a deterministic time function, then (3.7.9) corresponds to a sum of Gaussian variables at each t, and these modified components are therefore also Gaussian. If, however, $\theta(t)$ is itself a random process, then their individual probability density at any t is no longer obvious. To determine this we must investigate formally the transformation of the random variables $[n_c(t), n_s(t), \theta(t)]$ at any t over to the random variables $\hat{n}_c(t, \theta)$, $\hat{n}_s(t, \theta)$ at the same t, using (3.7.9) At a given t denote the random variable $n_c(t)$ by n_c, $\theta(t)$ by θ, and so on, and consider the conditional probability density $p(\hat{n}_c, \hat{n}_s | \theta)$. We write this as the formal transformation of densities, conditioned on θ (see Section B.2)

$$p(\hat{n}_c, \hat{n}_s | \theta) = \frac{1}{|J|} p(n_c, n_s | \theta) \Big|_{[n_c, n_s]} \qquad\qquad (3.7.17)$$

where J is the Jacobian of the transformation in (3.7.9)

$$J = \det \begin{bmatrix} \cos \theta & \sin \theta \\ -\sin \theta & \cos \theta \end{bmatrix} = 1 \qquad\qquad (3.7.18)$$

and $[n_c, n_s]$ represents the substitution for n_c and n_s in terms of $\hat{n}_c$ and $\hat{n}_s$ obtained by inverse solving (3.7.9). Therefore

$$p(\hat{n}_c, \hat{n}_s|\theta) = p(n_c, n_s|\theta)|_{[n_c, n_s]} \tag{3.7.19}$$

Now since $n_c(t)$ and $n_s(t)$ are independent of the carrier phase process $\theta(t)$, then the density on the right is jointly Gaussian in the variables n_c and n_s. After substituting $[n_c, n_s]$, (3.7.19) becomes

$$p(\hat{n}_c, \hat{n}_s|\theta) = \frac{1}{2\pi R_{n_c}(0)} \exp\left[-\frac{(n_c^2 + n_s^2)}{2R_{n_c}(0)}\right]\Bigg|_{\substack{n_c = \hat{n}_c \cos\theta + \hat{n}_s \sin\theta \\ n_s = \hat{n}_s \cos\theta - \hat{n}_c \sin\theta}}$$

$$= \frac{1}{2\pi R_{n_c}(0)} \exp\left[-\frac{(\hat{n}_c^2 + \hat{n}_s^2)}{2R_{n_c}(0)}\right] \tag{3.7.20}$$

The resulting density is that of a pair of independent Gaussian random variables. This establishes that $\hat{n}_c(t, \theta)$ and $n_s(t, \theta)$ are themselves jointly Gaussian at any t and independent of the probability density on $\theta(t)$. Therefore each process in (3.7.9) is itself Gaussian, no matter what the density on $\theta(t)$, and their correlation functions are given in (3.7.12) and (3.7.13). Under the condition of (3.7.14), they therefore become independent Gaussian processes each with spectral density given in (3.7.16). As such, they are then statistically identical to the quadrature components $n_c(t)$ and $n_c(t)$ and cannot be distinguished from them in any way. This means that without any loss of rigor we can replace the modified components $\hat{n}_c(t, \theta)$ and $\hat{n}_s(t, \theta)$ by their quadrature counterparts in subsequent analysis, so long as (3.7.14) is a fairly accurate approximation.

3.8. Frequency Translation and IF Processing

It is common in communication receivers to translate the front end waveform to a lower frequency range for further processing prior to demodulation. This lower frequency range is called the *intermediate frequency* (IF) of the receiver. The basic advantage of translating to a lower frequency (down conversion) is that electronic devices, such as filters, amplifiers, demodulators, and so on, are easier to construct at the lower frequencies. For example, a carrier bandwidth of 100 KHz at a carrier frequency of 1 GHz would require a front end filter whose width is 0.01% of center frequency. Such a filter would place severe constraints on parameter values. However, if this waveform can be shifted to a carrier frequency of 10 MHz, only a 1% filter bandwidth is required.

Frequency translation must, of course, be accomplished without introducing additional carrier distortion. Frequency translation is generally

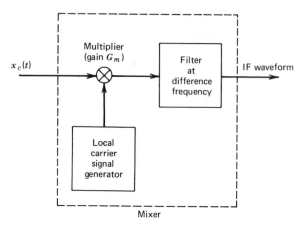

Figure 3.18. Frequency translator or mixer.

accomplished by frequency *mixing*. The mixing is produced by the multiplication of two carrier frequencies so as to produce a frequency shift. A typical mixer is shown in Figure 3.18. We assume the front end waveform $x_c(t)$ is again composed of the sum of the carrier and additive bandpass noise. A separate carrier signal (called the *local* carrier) is generated at the receiver to provide the mixing signal for the translation. The two signals are multiplied to produce sum and difference frequency terms. A filter following the multiplier filters off the terms of the difference frequency, the latter referred to as the *beat frequency*. The beat frequency terms now appear as the mixer output.

Let us write the signal $x_c(t)$ in the same form as in (3.7.1) and write the local carrier signal as $A_l \cos(\omega_l + \theta_l(t))$, where $\theta_l(t)$ represents any phase variation that may exist. We assume the mixer multiplier has a gain (or attenuation) G_m and contains a filter tuned to the difference frequency $(\omega_c - \omega_l)$ [or $(\omega_l - \omega_c)$, whichever is positive]. The multiplier performs the multiplication of $x_c(t)$ with the local carrier signal, while multiplying by the gain G_m, and the resulting signal is then filtered to provide the mixer output. The multiplier output is therefore

$$G_m x_c(t)\big[A_l \cos(\omega_l t + \theta_l(t))\big] = G_m[c(t) + n(t)][A_l \cos(\omega_l t + \theta_l(t))] \qquad (3.8.1)$$

We now substitute from (3.7.2), expanding the product of cosine terms into sum and difference frequencies. Noting that the filter eliminates sum frequency terms [assuming $(\omega_c + \omega_l) \gg |\omega_c - \omega_l|$] while passing only the beat frequency terms, the mixer output signal is then

$$x_{IF}(t) = K_m a(t) \cos[\omega_{cl}(t) + \theta(t) - \theta_l(t)] + n_m(t) \qquad (3.8.2)$$

where $K_m \triangleq G_m A_l/2$, $\omega_{cl} = \omega_c - \omega_l$, and $n_m(t)$ is the mixer output noise. The first term corresponds to carrier mixing, the second is due to the noise. Note that the mixer output appears as a linear summation, mixing the signal and noise individually. We see that the mixer carrier signal contains the same amplitude function as the front end carrier, but has the difference frequency ω_{cl} and difference phase of the front end and local carriers. Thus the front end carrier has been frequency translated to the frequency ω_{cl} while preserving the carrier amplitude (except for multiplication by K_m). Note that if the local carrier phase θ_l is constant, then phase modulation on the carrier is not affected by the mixer. By properly selecting ω_{cl} the carrier can be shifted to any desired IF for further processing.

Mixer output noise can be obtained by inserting the quadrature expansion of $n(t)$ in (3.8.1) and retaining the difference frequency terms, yielding

$$n_m(t) = K_m[n_c(t) \cos(\omega_{cl}t - \theta_l(t) + \psi) - n_s(t) \sin(\omega_{cl}t - \theta_l(t) + \psi)] \qquad (3.8.3)$$

If θ_l is a constant, then (3.8.3) is merely a shifted version of the front end noise. That is, $n_m(t)$ has a spectrum identical in shape to $S_n(\omega)$, except it is shifted to ω_{cl}. If $\theta_l(t)$ is a function of t, then (3.8.3) must be expanded using (3.7.7), resulting in

$$n_m(t) = K_m[\hat{n}_c(t, \theta_l) \cos(\omega_{cl}t + \psi) - \hat{n}_s(t, \theta_l) \sin(\omega_{cl}t + \psi)] \qquad (3.8.4)$$

where $\hat{n}_c(t, \theta_l)$ and $\hat{n}_s(t, \theta_l)$ are again the modified components in (3.7.9), only with the local carrier phase $\theta_l(t)$ involved instead of the phase modulation $\theta(t)$. If, however, $\theta_l(t)$ varies slowly with respect to the front end noise variations, the argument following (3.7.14) can be used to insert $n_c(t)$ and $n_s(t)$ in (3.8.4). In this case, $n_m(t)$ is again merely a shifted version of the front end noise. Thus we can conclude that as long as the local carrier phase variation is negligible relative to that of the noise variations, mixer output noise is simply a frequency shifted version of the bandpass noise appearing at its input.

The results in (3.8.3) are also directly applicable even if $\omega_{cl} = 0$ (i.e., $\omega_l = \omega_c$). In this case, we have

$$n_m(t) = K_m[n_c(t) \cos(\theta_l(t)) + n_s(t) \sin(\theta_l(t))], \quad \omega_{cl} = 0 \qquad (3.8.5)$$

and its corresponding autocorrelation function is

$$R_{n_m}(\tau) = K_m^2 R_{n_c}(\tau)[\cos[\theta_l(t) - \theta_l(t+\tau)]], \qquad \omega_{cl} = 0 \qquad (3.8.6)$$

When the conditions of (3.7.14) are true for θ_l, we approximate this by

$$R_{n_m}(\tau) \cong K_m^2 R_{n_c}(\tau), \qquad \omega_{cl} = 0 \qquad (3.8.7)$$

The mixer noise then has the power spectrum

$$S_{n_m}(\omega) = K_m^2 \tilde{S}_n(\omega), \qquad \omega_{cl} = 0 \qquad (3.8.8)$$

where $\tilde{S}_n(\omega)$ is the spectrum of the quadrature components. It can also be shown that the mixer noise in (3.8.5) is also a Gaussian noise process for any random phase process $\theta_l(t)$, by the same development leading to (3.7.20) (see Problem 3.26).

The requirement that θ_l be fairly constant implies that the local carrier should be produced from a highly phase stable oscillator, so that its output waveform is a pure carrier waveform. Phase variations in oscillators are caused by internal noise, which produces an inherent phase noise on the output carrier signal. Phase stability in oscillators is an important aspect of communication system design and often determines ultimate limitations to overall performance. A discussion of definitions and measurement procedures associated with oscillator phase noise is presented in Appendix C.

In actual system design, some consideration must be given to the selection of the IF frequency ω_{cl}. If further filtering is to be accomplished in the IF stage, then we generally desire ω_{cl} to be about 1000 times larger than the bandwidth of the modulated carrier. This makes it easier to construct IF bandpass filters that sufficiently reject out-of-band noise. On the other hand, the IF stage must interface with subsequent signal processors, such as demodulators, which may be preselected to operate at specific frequencies, thereby removing some freedom in IF selection. In addition, consideration must be given to the actual effect of the local carrier frequency on the other elements of the receiver. Since multiplier attenuation G_m may be significant, the amplitude of the local carrier A_l is generally quite large to produce an acceptable mixer gain K_m. This means a relatively high powered local carrier is being produced in the IF stage of the receiver, and its coupling into nearby elements may produce undesirable frequency interference. Of special concern is the possibility of the local carrier "leaking" back into the receiver front end. For this reason, IF is often selected such that the local frequency ω_l is not within the bandwidth of the front end filter. This requires $\omega_l < \omega_c - (2\pi B_c/2)$, or

$$\omega_{cl} > \pi B_c \qquad (3.8.9)$$

Mixers can also be used to translate a received carrier at one RF frequency to another RF frequency, instead of an IF frequency. This in fact represents the basic operation of the communication satellite relay. Here an RF carrier is received from the ground, frequency translated at the satellite to another RF carrier frequency, power amplified, and retransmitted back to Earth. Such a system is referred to as an RF *transponder*. Conditions similar to those in (3.8.9) are necessary to prevent the retransmitted carrier frequencies from coupling back into the transponder front end. In commercial satellite systems, carriers are transmitted up to the satellite in the 5.9–6.4 GHz frequency band, and then downconverted

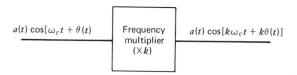

$a(t) \cos[\omega_c t + \theta(t)]$ → | Frequency multiplier ($\times k$) | → $a(t) \cos[k\omega_c t + k\theta(t)]$

Figure 3.19. Frequency multiplier.

in the satellite transponder to the 3.7–4.2 GHz downlink band (see Table 1.3).

We have considered only frequency translation to a lower frequency. Translation of a carrier waveform to a higher frequency (*up conversion*) can be accomplished by mixing and making use of the sum frequencies instead of the difference. Frequency up conversion can also be accomplished by direct frequency multiplication. A *frequency multiplier* is a device that multiplies the total phase and frequency of a carrier by a fixed factor. Figure 3.19 shows a block diagram model of typical multiplier and its effect on the input carrier. Frequency multipliers are constructed from either nonlinear devices or by digital frequency synthesizers. Nonlinear multipliers produce carrier harmonics of which one is selected (filtered off) for the output. The difficult task with nonlinear multipliers is to properly filter off the desired harmonic and its modulation without interference from adjacent harmonics. In addition it is difficult to get large frequency multiplication factors, since the higher harmonics are progressively weaker, and typical nonlinear multipliers are simply frequency doublers or triplers, which must be cascaded to get larger factors. Digital multipliers operate by using the original carrier to clock a fast digital readout that produces sequences of voltage samples of the multiplied frequency carrier. These are then filtered to synthesize a continuous version of the desired multiplied frequency carrier waveform.

3.9. Nonlinear Processing and Carrier Limiting

Often a communication receiver involves some type of nonlinearity during the processing of either the front end or IF carrier signal. This nonlinearity may be undesired, such as an amplifier having a nonlinear gain or saturation effects, or may be intentionally placed in the system, such as a limiter or a rectifier. It is therefore important in system design to understand the effects of such nonlinearities on the waveforms in the receiver and to account for them properly in system analysis. To consider this, we present some general results concerning nonlinear analysis of carrier waveforms. We concentrate only on the "memoryless" type of nonlinear device, in

$x(t)$ → | $g(x)$ | → $y(t)$

Figure 3.20.   Nonlinear device.

which the present value of the device output depends only on the present value of its input and not on any of its past.

Assume the nonlinearity of the system is described by the function $g(x)$. Thus if $x(t)$ is the time process at the nonlinearity input, the output is

$$y(t) = g[x(t)] \tag{3.9.1}$$

as shown in Figure 3.20. A common analytical procedure [16, 17] is to expand $y(t)$ in terms of the transform of the function $g(x)$. That is, if we denote $G(\omega)$ as Fourier transform of $g(x)$, then $y(t)$ can be rewritten as

$$y(t) = \frac{1}{2\pi} \int_{-\infty}^{\infty} G(\omega) e^{j\omega x(t)} d\omega \tag{3.9.2}$$

We point out that since typical nonlinearities $g(x)$ may have different characteristics for $x > 0$ and $x < 0$, $G(\omega)$ is actually a two-sided transform that often must be computed separately for $x > 0$ and $x < 0$ (Problem 3.29). This means that the inverse transform indicated in (3.9.2) may require separate contour integrations for the cases where $x(t) > 0$ and $x(t) < 0$. This point is discussed in more detail in References 16 and 17.

The use of (3.9.2) for analyzing nonlinearities in terms of their transforms is particularly convenient for the type of carrier waveforms we have been considering. Let us represent the general noisy carrier in the receiver (at either the front end or IF) as

$$x_c(t) = \alpha(t) \cos[\omega_c t + \theta(t)] \tag{3.9.3}$$

where $\theta(t)$ represents the combined phase variation due to the carrier modulation and any phase noise. We can then use the Jacobi–Anger identity (Section A.2)

$$e^{j\omega\alpha\cos\beta} = \sum_{i=0}^{\infty} \varepsilon_i I_i(j\omega\alpha) \cos(i\beta) \tag{3.9.4}$$

where $I_i(z)$ is the modified Bessel function of the first kind and order i[18], $\varepsilon_0 = 1$, and $\varepsilon_i = 2$, $i \neq 0$. Equation (3.9.2) becomes

$$y(t) = \sum_{i=0}^{\infty} c_i(\alpha(t)) \cos[i\omega_c t + i\theta(t)] \tag{3.9.5}$$

where

$$c_i(\alpha) = \frac{\varepsilon_i}{2\pi} \int_{-\infty}^{\infty} G(\omega) I_i[j\omega\alpha] \, d\omega \qquad (3.9.6)$$

Equation (3.9.5) is a general expression for the output of any nonlinearity in (3.9.1) when the input is given by (3.9.3). Note that the output always appears as the sum of harmonically related, modulated carriers with amplitude variations depending on the type of nonlinearity transform $G(\omega)$. We therefore see that harmonic generation is inherent in nonlinear processing. The effect of these harmonics must be carefully examined for their interference on later system components. Some common types of receiver nonlinearities and the resulting forms for $c_i(\alpha)$, evaluated from (3.9.6), are summarized in Table 3.3. Note that in certain cases only a finite number of harmonics appear, whereas in other cases only even (or odd) harmonics are generated. Clearly, the frequency multipliers described in the preceding section can be easily constructed from such nonlinear elements.

Since nonlinearities tend to produce harmonics, the nonlinear device is generally followed by a bandpass filter tuned to the carrier frequency. The combination of the nonlinear element followed by the bandpass filter is often called a *bandpass nonlinearity*. For such devices the only harmonic component appearing at the output is that corresponding to the index $i = 1$ in (3.9.5). For example, the output of the bandpass full wave odd νth power law devices in Table 3.3 would be

$$y(t) = C(\nu, 1)\alpha^{\nu}(t) \cos \left[\omega_c t + \theta(t) \right] \qquad (3.9.7)$$

The output is therefore a carrier with the same phase variation as the input, but with a modified amplitude variation.

One type of nonlinear operation that often occurs in receiver models is limiting. Limiting can be produced unintentionally by saturation effects in amplifiers, but in many systems limiters are purposely inserted in the IF channel to improve performance. The limiting prevents high peak voltages caused by noise spikes from exceeding allowable dynamical ranges in the receiver processing. In addition, limiters are useful for maintaining power and amplitude control during the subsequent demodulation operation.

An ideal *hard limiter* is a nonlinear device whose input-output function $g(x)$ is shown in Figure 3.21. Mathematically, the hard limiter is simply a $\nu = 0$, full wave, odd device listed in Table 3.3. A *bandpass hard limiter* (BPL) is a hard limiter followed by a bandpass filter tuned to the input carrier frequency, the latter allowing only the frequency components within its bandwidth to pass. If the input to a bandpass limiter is a noisy carrier waveform $c(t) = \alpha(t) \cos \left[\omega_c t + \theta(t) \right]$, the bandpass limiter output is

Table 3.3. Common Nonlinearities and Parameters

Device	Nonlinear Function $g(x)$	$c_i(\alpha)$
Half-wave νth power law		$\alpha^\nu C(\nu, i)$
Full wave (even) νth power law		$2\alpha^\nu C(\nu, i)$, $\quad i$ even 0, $\qquad\quad i$ odd
Full wave (odd) νth power law		$2\alpha^\nu C(\nu, i)$, $\quad i$ odd 0, $\qquad\quad i$ even
Half-wave linear bandpass		$\dfrac{\alpha}{\pi}, i = 1$
Full wave square law bandpass		$\dfrac{\alpha^2}{2}, i = 2$

134

Table 3.3. **(Continued)**

Device	Nonlinear Function $g(x)$	$c_i(\alpha)$
Fullwave odd, $\nu = 0$ bandpass	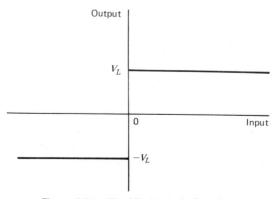	$\left(\dfrac{4}{\pi}\right), i = 1$

$$C(\nu, m) = \frac{\varepsilon_m(G_{am}[\nu+1])}{2^{\nu+1}\left(G_{am}\left[1-\dfrac{m-\nu}{2}\right]\right)\left(G_{am}\left[1+\dfrac{m+\nu}{2}\right]\right)}, \qquad G_{am}[z] = \int_0^\infty e^{-t} t^{z-1}\, dt$$

given by (3.9.7), with $\nu = 0$. Since $C(0, 1) = 4V_L/\pi$, the BPL output is

$$y(t) = \left(\frac{4V_L}{\pi}\right) \cos\left[\omega_c t + \theta(t)\right] \tag{3.9.8}$$

Note the output produces a RF carrier with the same phase and frequency modulation as the input but with a constant amplitude level. In effect, the BPL eliminates amplitude modulation, while preserving angle modulation. Obviously, limiters should not be used when AM carriers are involved. We also see that the power at the output of a bandpass limiter [power in $y(t)$ in (3.9.8)] is always given by

$$P_L = \frac{1}{2}\left(\frac{4V_L}{\pi}\right)^2 = \frac{8V_L^2}{\pi^2} \tag{3.9.9}$$

Figure 3.21. Hard limiter gain function.

for any input process. Thus by proper adjustment of the limiting level V_L, the amplitude levels and power output, of the BPL can be accurately adjusted. This allows control of the peak amplitude variations of the resulting carrier waveform during the subsequent receiver processing.

When the input to the BPL is composed of the sum of an angle modulated carrier plus additive receiver noise, it is often desirable to know the extent by which the carrier waveform has been preserved in passing through the BPL. This is difficult to determine from (3.9.8), since the limiter output noise is incorporated entirely into the phase noise of the carrier. However, it can be shown [16, 17] that the carrier power in the first harmonic term of the limiter output is given by

$$P_{c0} = \frac{2V_L^2}{\pi} \rho_i e^{-\rho_i} \left[I_0\left(\frac{\rho_i}{2}\right) + I_1\left(\frac{\rho_i}{2}\right) \right]^2 \tag{3.9.10}$$

where ρ_i is the limiter input CNR,

$$\rho_i \triangleq \frac{A^2/2}{R_n(0)} \tag{3.9.11}$$

Since the total power at the output of the BPL is P_L in (3.9.9), it follows that the noise and cross-product power terms must constitute the difference. Hence the BPL output noise power is

$$P_{n0} = P_L - (P_{c0}) \tag{3.9.12}$$

The resulting bandpass limiter output CNR is then

$$\text{CNR}_{BL} = \frac{P_{c0}}{P_{n0}}$$

$$= \frac{P_{c0}/P_L}{1 - (P_{c0}/P_L)} \tag{3.9.13}$$

where

$$\frac{P_{c0}}{P_L} = \left(\frac{\pi}{4}\right) \rho_i e^{-\rho_i} \left[I_0\left(\frac{\rho_i}{2}\right) + I_1\left(\frac{\rho_i}{2}\right) \right]^2 \tag{3.9.14}$$

A plot of the normalized ratio

$$\Gamma \triangleq \frac{\text{CNR}_{BL}}{\rho_i} \tag{3.9.15}$$

is shown in Figure 3.22 as a function of ρ_i. The result shows the way in which the CNR is altered in passing through the BPL. Note that CNR_{BL} has the asymptotic behavior

$$\text{CNR}_{BL} \cong \left(\frac{\pi}{4}\right) \rho_i, \qquad \text{for} \quad \rho_i \ll 1$$

$$\cong 2\rho_i, \qquad \text{for} \quad \rho_i \gg 1 \tag{3.9.16}$$

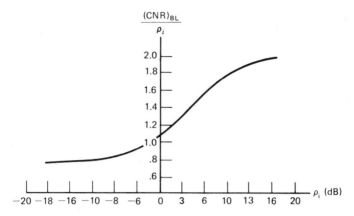

Figure 3.22. Bandpass limiter CNR suppression *ratio* (ρ_i = limiter input CNR).

Thus the effect of the BPL is to cause an increase in the CNR if the ratio is large but to cause a slight degradation (by about 2 dB) if the input CNR is low. We point out that CNR_{BL} is the output ratio of the carrier power to the total output interference power in the limiter bandpass filter bandwidth.

This anslysis of the BPL allows us to derive a relatively, simple model to account for the limiter effect on the input waveform. We consider the limiter input to be

$$x(t) = A \cos [\omega_c t + \theta(t)] + n(t) \qquad (3.9.17)$$

representing the combination of an angle modulated carrier with Gaussian noise of power $R_n(0)$. Since the limiter alters the effective CNR, we can account for this effect by writing the BPL output as

$$x_L(t) = (\alpha_s A) \cos [\omega_c t + \theta(t)] + \alpha_n n(t) \qquad (3.9.18)$$

where α_s and α_n represent the effective limiter *suppression factors* induced on the input carrier and noise by the limiter. The CNR of the BPL output in (3.9.13) is then

$$CNR_{BL} = \frac{\alpha_s^2 A^2 / 2}{\alpha_n^2 R_n(0)} = \left(\frac{\alpha_s^2}{\alpha_n^2}\right) \rho_i \qquad (3.9.19)$$

We see that (3.9.19) corresponds to the true CNR_{BL} in (3.9.15) if

$$\frac{\alpha_s^2}{\alpha_n^2} = \Gamma \qquad (3.9.20)$$

Thus, as far aș CNR is concerned, the BPL alters the carrier and noise amplitudes by the suppression factors in (3.9.18), where the ratio of these factors is given by the parameter Γ in Figure 3.22. Note that the suppression parameters themselves depend on ρ_i, and the manner in which the carrier and noise amplitudes are effectively modified by the BPL depends on their relative strength. We carefully emphasize that (3.9.18) is a model for the BPL output only as far as CNR is concerned. Note that it implies that the output noise is merely an amplitude adjusted version of the input noise, when in fact the limiter output interference is composed of both noise and carrier cross products. In "linearizing" the BPL in this way, we adjust the carrier and interference ratio so that the output CNR is correct, but we must use care in subsequent processing analysis when we model the output in this way.

We can solve for the individual suppression factors α_s and α_n in (3.9.18) by using the fact that $x_L(t)$ must have the power P_L. This means

$$\frac{\alpha_s A^2}{2} + \alpha_n^2 R_n(0) = P_L \tag{3.9.21}$$

Substituting from (3.9.20) and solving for $\alpha_s^2 A^2/2$ yields

$$\frac{\alpha_s^2 A^2}{2} = \frac{P_L}{1 + (R_n(0)/\Gamma A^2/2)}$$

$$= P_L \left[\frac{\text{CNR}_{BL}}{\text{CNR}_{BL} + 1} \right] \tag{3.9.22}$$

where CNR_{BL} is the limiter output CNR in (3.9.15). Similarly, we can solve for the output noise power as

$$\alpha_n^2 R_n(0) = P_L - \frac{\alpha_s^2}{A^2}$$

$$= P_L \left[\frac{1}{1 + \text{CNR}_{BL}} \right] \tag{3.9.23}$$

Thus the α_s and α_n suppression factors are such that they divide the available bandpass limiter output power P_L in accordance with the ratios (3.9.22) and (3.9.23). As CNR_{BL} increases, a greater portion of the limiter output power is associated with the carrier.

References

1. Angelakos, D. and Everhart, T. *Microwave Communications*, McGraw-Hill, New York, 1968.
2. Stephenson, R. G. "External Noise" *Space Communications*, edited by A. Balikrishnan, McGraw-Hill, New York, 1970.

3. Ko, H. "The Distribution of Cosmic Radio Background Radiation," *Proc. IRE*, vol. 46, no. 1, January 1958, pp. 218–225.

4. Cottony, H. and Johler, J. "Cosmic Radio Noise Intensity in the UHF Band," *Proc. IRE*, vol. 40, 1946, pp. 1487–1489.

5. Smith, A. "Extraterrestrial Noise in Space Communications," *Proc. IRE*, vol. 48, no. 4, April 1960, no. 4, pp. 594–603.

6. Mayer, C. "Thermal Radio Radiation from the Moon and Planets," *IEEE Trans.: On Antennas and Propagation*, vol. 12, no. 7, December 1964, pp. 902–913.

7. Pratt, W. *Laser Communication Systems*, Wiley, New York, 1969, Chap. 6.

8. Davenport, W. and Root, W., *Theory of Random Signals and Noise*, McGraw-Hill, New York, 1958.

9. Pierce, J. "Physical Sources of Noise," *Proc. IRE*, vol. 44, May 1956, pp. 601–608.

10. Bennett, W. *Electrical Noise*, McGraw-Hill, New York, 1960.

11. Van der Ziel, A. *Noise*, Prentice-Hall, Englewood Cliffs, N.J., 1954.

12. Panter, P. *Modulation, Noise, and Spectral Analyses*, McGraw-Hill, New York, 1965.

13. Bedrosian, E. and Rice, S. "Distortion and Crosstalk of Filtered Angle Modulated Signals," *Proc. IEEE*, vol. 56, January 1968, pp. 2–13.

14. Baghdady, E. (ed.) *Lecture on Communication Systems*, McGraw-Hill, New York, 1960.

15. Williams, O. "A Comparison of the Carson-Fry and Rice-Bedrosian Methods of FM Analysis," *IEEE Trans. Comm.*, vol. 52, August 1973, pp. 972–980.

16. Davenport and Root, loc. cit., Chap. 12.

17. Thomas, J. *Statistical Communication Theory*, Wiley, New York, 1969, Chap. 6.

18. Abromowitz, A. and Stegun, I. *Handbook of Mathematical Functions*, National Bureau of Standards, Washington, D.C., 1965, Chap. 9.

Problems

1. (3.1) A 2 m circular antenna operates at 1 GHz with an aperture loss factor of 0.5. (a) What is the maximum area it will present to an arriving plane wave? (b) Repeat if the carrier frequency was 100 MHz.

2. (3.1) Suppose the transmitted electromagnetic field is not an infinite plane wave but instead has a finite field cross-sectional area as it propagates. After propagating a distance D its wavefront has area $\mathscr{A}_f$. Assume it is transmitted with a given EIRP to a receiver of area $\mathscr{A}$ at D. Show that the receiver power collected in free space is

$$\left(\frac{\text{EIRP}}{4\pi D^2}\right)\mathscr{A}, \quad \text{if} \quad \mathscr{A} \leq \mathscr{A}_f$$

$$\left(\frac{\text{EIRP}}{4\pi D^2}\right)\mathscr{A}_f, \quad \text{if} \quad \mathscr{A}_f \leq \mathscr{A}$$

3. (3.2) A 100 W transmitter amplifier feeds a 1 GHz carrier into matched antenna terminals. The antenna has a 20 dB gain in the direction of the receiver and radiation loss of 0.9. The transmission channel is a 400 km link, of which half of the path involves atmospheric propagation with a 0.01 dB/km loss. The receiving antenna is 1 m with 0.5 aperture loss factor, a 0.95 radiation loss, and no coupling loss. How much power in watts will be collected at the receiver terminals?

4. (3.2)   A source couples 100 W into a 100 km cable having 2.0 dB/km loss. Ten amplifiers are to be inserted into the cable for carrier amplification. How much gain should each have in order to achieve 2 W of carrier power at the output? Neglect output coupling losses.

5. (3.3)   An Earth based antenna is pointed at the moon with gain function

$$g(\phi_l, \phi_z) = \left(\frac{2}{\Phi_b^2}\right) e^{-(\phi_l^2 + \phi_z^2)/2\Phi_b^2}$$

where $2\Phi_b$ is its beamwidth. Assume the temperature of the moon is 200°K over its surface area and the surrounding sky is 50°K. Neglect other background sources. Use the approximation $d\Omega = d\phi_l \, d\phi_z$.

(a) Determine the effective background temperature of the antenna when Φ_b is set equal to the beamwidth subtended by the moon.

(b) Sketch the behavior of the effective temperature as a function of Φ_b.

6. (3.3)   A passive planet occupies a 1° field of view with temperature T_0° K. A "black hole" is a discrete source (delta function in space), and assume it has the same temperature. Explain qualitatively what will happen in each case if we point an antenna directly at the source and begin to increase the antenna area.

7. (3.3)   Often a source is described by its *immittance function* $\mathcal{N}$, which is the product of its radiance and subtended field of view, $\mathcal{N} = \mathcal{H}\Omega_s$. Assume $\mathcal{N}$ is a constant and the antenna receiving area is $\mathcal{A}$ over its field of view, Ω_{fv}. Show that

$$P_b = B\mathcal{N}\mathcal{A} \qquad \text{if} \quad \Omega_s \le \Omega_{fv}$$

$$= B\mathcal{N}\mathcal{A}\frac{\Omega_{fv}}{\Omega_s} \qquad \text{if} \quad \Omega_{fv} < \Omega_s$$

8. (3.4)   (a) Show that a cascade of three devices, having gain and noise figure G_i, F_i, respectively, will have an overall noise figure of

$$F = F_1 + \frac{F_2 - 1}{G_1} + \frac{F_3 - 1}{G_1 G_2}$$

(b) Generalize the result to N stages.

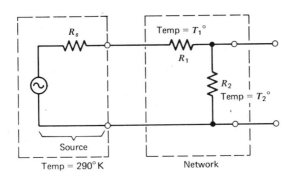

Figure P3.10.

9. (3.4) Given the parallel combination of two devices with noise figures F_1, F_2 and power gains G_1, G_2, determine the overall noise figure.

10. (3.4) Determine the noise figure of the resistor network in Figure P3.10. Assume $R_s = R_1 + R_2$. (*Hint:* Insert equivalent noise sources and compute total output power.)

11. (3.4) Given two lossy coupling lines in cascade, with power gain L_1 and L_2, and temperature T_1° and T_2°, respectively. Determine the overall noise figure and equivalent input temperature of the cascade. Assume matched impedances.

12. (3.4) Noise figure of a device can be measured in the following way. A resistor at 290°K is connected at the input and the output noise power is measured. The resistor is heated until the output noise power exactly doubles, and the resistor temperature T_2° is noted. Show that

$$F = \frac{T_2^\circ - 290^\circ}{290^\circ}$$

13. (3.4) The front end of an RF receiver is shown in Figure P3.13, where all elements are perfectly matched. How much power gain must the tunnel diode amplifier have to produce a front end noise figure of 10 dB?

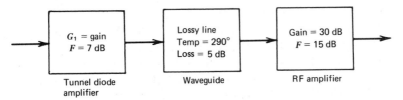

Figure P3.13.

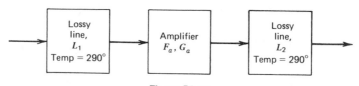

Figure P3.14.

14. (3.4) Find the necessary G_{amp} for the system in Figure P3.14 to obtain a NF of F_0.

15. (3.4) (a) A deep space link is to transmit an RF carrier with 10 kHz bandwidth at 5 GHz from Earth to a spacecraft near Venus over a free space channel (10^8 miles). A 200 ft transmitting antenna at the ground terminal is available. The spacecraft employs an isotropic antenna coupled to its front end with a line having 3 dB loss and operates at room temperature. The front end amplifier noise figure is 10 dB. The background sky presents an effective temperature of 10 dB below room temperature. How much transmitter power must be available to ensure transmission with a 10 dB SNR at the spacecraft in the carrier bandwidth?

(b) For downlink transmission (spacecraft to Earth) a 10 W transmitter is available. The ground receiver can be maintained at 7 dB below room temperature with a 2 dB noise figure and lossless coupling. The same sky background as in (a) is also present. The spacecraft antenna can be aligned to provide some directivity over isotropic transmission. Show a plot of the available RF bandwidth that can be theoretically provided at a 10 dB RF SNR in the downlink, as a function of the spacecraft antenna gain.

16. (3.4) A low level, noiseless power source of -103 dBW feed into a 100 km cable at 290°K. The cable has a loss of 0.2 dB/km. An amplifier of gain 70 dB at noise figure of 4 dB is to be inserted into the cable. An output SNR of 30 dB is desired in a 1 MHz bandwidth. How far down the cable should the amplifier be inserted (or does it make any difference)? Neglect coupling losses.

17. (3.4) A long cable transmission line is composed of M sections, each section having a line of loss L, followed by an internal amplifier. The gain of the amplifier is selected so that it just cancels the line loss. The amplifier has a noise figure of F_a. The line is fed by a power of P_1 W from a source of temperature 290°K. The cable is also at 290°K. (a) Determine the SNR of output if there was just one section. (b) Determine the output SNR for M sections. (c) Determine the output SNR if no amplifiers were used. (d) Let $F_a = 10$ dB, $M = 10$, and $L = 3$ dB, and comment on whether long lines should have noisy amplifiers inserted.

18. (3.4) Shot noise is often modeled mathematically as

$$x(t) = \sum_{i=0}^{k} h(t - t_i)$$

where $h(t)$ is a known response function, $\{t_i\}$ is a sequence of independent random location points, each uniformly distributed over $(0, T)$. The number k is also random, independent of the $\{t_i\}$, and having Poisson density, Prob $(k = j) = ((\alpha T)^j / j!) \exp(-\alpha T)$, where α is the rate of occurrence. Determine the autocorrelation and power spectra density of $x(t)$. Assume $h(t)$ is much narrower in width than T.

19. (3.5) The front end of a receiver has the filter function

$$H_c(\omega) = \frac{1}{1 + j((\omega - \omega_c)/2\pi B_3)^2}$$

(a) Determine the power in the output signal when receiving the AM signal

$$c(t) = a(t) \sin(\omega_c t + \psi)$$

with

$$a(t) = a_0 + b_0 \sin(\omega_m t)$$

(b) Repeat when the filter input signal is a random waveform having a bandlimited white noise power spectrum of one-sided level S_0 and bandlimited to $2 B_n$ Hz about the carrier frequency.

(c) Determine how wide B_3 should be in (b) in order to pass 96% of the input waveform power.

20. (3.5) Given the carrier $H_c(\omega) = \exp[-(\omega - \omega_c)^2 / 2\sigma^2]$. Find the maximum rate of change of carrier frequency that can be tolerated for quasi-linear analysis.

21. (3.5) Determine the first $(i = 3)$ distortion term in the Rice–Bedrosian expansion of (3.5.23).

22. (3.6) Given the quadrative noise expansion in (3.6.3). Assume $n_c(t)$ and $n_s(t)$ are uncorrelated, and each has the same spectral density $\tilde{S}_n(\omega)$. Determine the correlation function of $n(t)$ and compute its power spectrum. (*Hint:* See Section B.6.)

23. (3.6) Show that if white noise (level $N_0/2$) is to be represented as in (3.6.3), then $n_c(t)$ and $n_s(t)$ will not be uncorrelated and each will have spectrum $N_0, |\omega| \le \omega_c$ and $N_0/2, |\omega| > \omega_c$.

24. (3.7) Derive (3.7.12) and (3.7.13) from (3.7.9), for the modified noise components.

25. (3.7) Verify the Jacobian in (3.7.18) for the transformation defined in (3.7.9).

26. (3.8) Prove the mixer noise $n_m(t)$ in (3.8.5) is always Gaussian, even if θ_l is random. [*Hint:* Define an auxiliary process $n'_m(t) = n_s(t)$ $\cos \theta_l - n_c(t) \sin \theta_l$ and determine the joint density of n_m and n'_m.]

27. (3.8) A carrier occupies a 1 MHz RF bandwidth at 1 GHz. What local oscillator frequency is required to generate an IF with a 1% bandwidth?

28. (3.8) A carrier squaring device is used to obtain frequency doubling. Determine the carrier power loss in decibels for the doubled frequency carrier. Repeat for the tripled frequency carrier when a cubic device is used.

29. (3.9) Let $G_+(\omega)$ be the Fourier transform of $g(x)$ over $x > 0$. Show that for an even nonlinear device $(g(-x) = g(x))$ we have $G(\omega) = G_+(\omega) + G_+(-\omega)$, whereas for an odd device $(g(-x) = -g(x))$ we have $G(\omega) = G_+(\omega) - G_+(-\omega)$.

30. (3.9) A receiver uses an RF band at three times the IF frequency band. There is the possibility of IF signals leaking back into the RF band. Explain the events, in terms of RF interference, that will occur if the IF contains an amplifier that is (a) a full wave odd device, (b) a full wave even device, (c) a squaring device. (*Hint:* Refer to Table 3.3.)

CHAPTER 4

CARRIER DEMODULATION

Following RF and IF processing, the communication receiver must demodulate the carrier to recover the baseband information signal. In this chapter we consider this demodulation operation when the latter is performed in the presence of receiver noise. We assume all front end processing of the carrier waveform has been accounted for in the description of the carrier, and the interfering IF noise is the narrowband Gaussian process discussed in Section 3.7. We concentrate only on the cases where the carrier may be of the AM, FM, or PM type. In each case we are primarily interested in the properties of the demodulated signal and its relation to the corresponding transmitter baseband signal. An outgrowth of our study is an attempt to model the entire carrier subsystem from transmitter baseband to demodulator output. A result of this type allows us to concentrate separately on baseband and carrier subsystem design.

4.1. The Demodulating Subsystem

The demodulation portion of the carrier subsystem at the receiver is shown in Figure 4.1. The antenna signal is processed in the RF and IF stages, as discussed. The output of the IF filter stage serves as the input to the receiver carrier demodulator. This input is the sum of the IF carrier and the IF noise waveform. We again write this combined waveform as

$$x_{\text{IF}}(t) = c(t) + n(t) \tag{4.1.1}$$

where the carrier and noise are referred to the demodulator input. The carrier $c(t)$ is at the IF frequency ω_{IF} and the narrowband IF noise $n(t)$

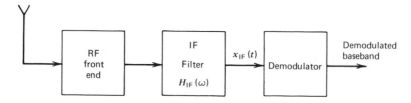

Figure 4.1.   The receiver demodulating subsystem.

again has the quadrature expansion

$$n(t) = n_c(t) \cos(\omega_{IF}t + \psi) - n_s(t) \sin(\omega_{IF}t + \psi) \qquad (4.1.2)$$

relative to this frequency. The quadrature components $n_c(t)$ and $n_s(t)$ are independent Gaussian processes, each having spectral density

$$S_{n_c}(\omega) = S_{n_s}(\omega) = \tilde{S}_n(\omega)$$
$$= N_0 |\tilde{H}_{IF}(\omega)|^2 \qquad (4.1.3)$$

where

$$N_0 = \kappa T^\circ_{eq} G_1 \qquad (4.1.4)$$

and $\tilde{H}_{IF}(\omega)$ is the low frequency equivalent of the IF bandpass filter function $H_{IF}(\omega)$. The gain G_1 accounts for the total receiver gain from antenna to demodulator input. Since the power of the received carrier signal at the antenna output terminals is P_r, the power in the carrier at the demodulator input is

$$P_c = G_1 P_r \qquad (4.1.5)$$

where P_r was computed in Section 3.2. We are now interested in determining the performance of various demodulating systems that operate on the IF waveform in (4.1.1). The structure and performance of these demodulators depend on the type of modulation involved.

4.2.   Amplitude Demodulation

We first consider demodulation of an AM carrier. We write the AM carrier at the demodulator in the standard form

$$c(t) = C[1 + m(t)] \cos(\omega_{IF}t + \psi) \qquad (4.2.1)$$

where $m(t)$ is the baseband modulation, ψ is the arbitrary carrier phase shift, and the amplitude coefficient C satisfies (4.1.5); that is,

$$P_c = \frac{C^2}{2}(1 + P_m) = G_1 P_r \qquad (4.2.2)$$

where P_m is the power of the modulation $m(t)$. We assume all filtering effects of the RF and IF stages have been incorporated into the description of $m(t)$. The input to the demodulator in Figure 4.1 is then the sum of the carrier in (4.2.1) and the noise process in (4.1.2). This combined waveform can be expanded as in (3.7.4),

$$x_{IF}(t) = \alpha(t) \cos[\omega_{IF}t + \psi + v(t)] \tag{4.2.3}$$

where

$$\alpha(t) = \{[C(1+m(t))+n_c(t)]^2+n_s^2(t)\}^{1/2} \tag{4.2.4}$$

and $v(t)$ is the phase noise. Let us first consider an AM demodulator that consists of an ideal envelope detector followed by a low pass filter $H_m(\omega)$, as shown in Figure 4.2. This filter is selected to have a bandwidth large enough to pass the modulation $m(t)$ without distortion, and therefore has a bandwidth of approximately B_m Hz, the latter the bandwidth of $m(t)$. The ideal envelope detector is considered to be a device that produces an output voltage directly proportional to the envelope of the input. In practice, envelope detectors are constructed from rectifier circuits with long time constants [1, 2], so as to follow the envelope function but be immune to the high frequency carrier variations. When the IF signal $x_{IF}(t)$ in (4.2.3) is the input, the envelope detector yields the output signal

$$x_1(t) = K_d[\text{envelope of } x_{IF}(t)]$$

$$= K_d\alpha(t)$$

$$= K_d\{[C(1+m(t))+n_c(t)]^2+n_s^2(t)\}^{1/2} \tag{4.2.5}$$

where K_d is the proportionality (gain) constant of the detector. Thus the envelope detector produces an output that contains the desired baseband $m(t)$ in combination with the noise components. In particular, we note it does not yield the baseband as a separate, distinct output unless the noise

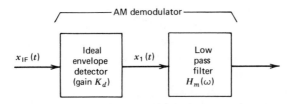

Figure 4.2. The AM demodulator (ideal envelope detector).

processes are zero. However, if we can assume that*

$$C[1+m(t)] \gg |n_c(t)|, |n_s(t)| \tag{4.2.6}$$

we can approximate

$$x_1(t) \cong K_d C(1+m(t)) + K_d n_c(t)$$
$$= K_d C + K_d C m(t) + K_d n_c(t) \tag{4.2.7}$$

The envelope detector output now appears as the sum of the baseband signal $K_d C m(t)$ plus an additive noise $K_d n_c(t)$ (the constant $K_d C$ can be theoretically removed by subtraction or ac coupling). The output filter $H_m(\omega)$ is designed to pass the baseband modulation without distortion while filtering the noise process $K_d n_c(t)$. This latter noise has the spectrum $K_d^2 \tilde{S}_n(\omega)$, where $\tilde{S}_n(\omega)$ is given in (4.1.3). The total noise power appearing at the demodulator filter output is therefore

$$P_n = \frac{K_d^2}{2\pi} \int_{-\infty}^{\infty} \tilde{S}_n(\omega)|H_m(\omega)|^2 \, d\omega$$
$$= \frac{K_d^2 N_0}{2\pi} \int_{-\infty}^{\infty} |\tilde{H}_{IF}(\omega)|^2 |H_m(\omega)|^2 \, d\omega \tag{4.2.8}$$

Assuming the filter does not distort the baseband signal $m(t)$, the SNR at the demodulator output is then

$$\mathrm{SNR}_d = \frac{K_d^2 C^2 P_m}{K_d^2 N_0 (2\tilde{B}_m)}$$
$$= \frac{C^2 P_m}{2 N_0 \tilde{B}_m} \tag{4.2.9}$$

where $\tilde{B}_m$ is the one-sided noise bandwidth

$$\tilde{B}_m = \frac{1}{2\pi} \int_0^{\infty} |H_{IF}(\omega)|^2 |H_m(\omega)|^2 \, d\omega \tag{4.2.10}$$

Thus the ideal envelope detector in Figure 4.2 demodulates with the output SNR in (4.2.9), provided that (4.2.6) is valid. The latter corresponds to a condition that the carrier envelope be strong enough relative to the noise. This occurs with a high probability if the power in the carrier is

---

* This condition requires that the signal envelope exceed the noise processes at every t. Since the noise is random, one can only interpret this statement in a statistical sense. Hence we mean "for most of the time" the envelope will exceed the noise.

sufficiently greater than that of the noise components in the IF stage. That is,

$$\text{CNR}_{\text{IF}} \triangleq \frac{G_1 P_r}{N_0 B_{\text{IF}}} = \frac{P_r}{\kappa T^\circ_{\text{eq}} B_{\text{IF}}} \gg Y \qquad (4.2.11)$$

where Y is a suitably selected threshold value (see Problem 4.1) and B_{IF} is the noise bandwidth of the IF filter stage. In practical systems Y is generally taken as a value in the range 10–20 dB. We therefore conclude that an ideal envelope detector will demodulate the baseband with the SNR in (4.2.9), provided that the demodulator input has a carrier to noise power ratio in (4.2.11) that satisfies an appropriate threshold. If the IF and low pass filter are taken to be ideal filters over the bandwidths $2B_m$ and B_m, respectively, corresponding to the modulated and baseband bandwidths, the noise bandwidth in (4.2.10) then becomes

$$\tilde{B}_m = \frac{1}{2\pi} \int_0^{2\pi B_m} d\omega = B_m \qquad (4.2.12)$$

and the output SNR is identical to (4.2.9) with $\tilde{B}_m$ replaced by B_m. In this case the output SNR can be written in terms of the demodulator input CNR by using $B_{\text{IF}} = 2B_m$ and substituting (4.2.11) into (4.2.9), yielding

$$\text{SNR}_d = \frac{C^2 P_m}{2N_0 B_m} = 2\left(\frac{P_m}{1+P_m}\right)\frac{C^2(1+P_m)/2}{2N_0 B_m}$$

$$= 2\left(\frac{P_m}{1+P_m}\right)\text{CNR}_{\text{IF}} \qquad (4.2.13)$$

Thus the demodulator output SNR is directly related to the input CNR and depends on the power that can be put into the modulation. In standard AM we required $|m(t)| \leq 1$ to avoid overmodulation, which forces $P_m \leq 1$. If $P_m = 1$, its largest possible value, then $(\text{SNR})_d = (\text{CNR})_{\text{IF}}$ and the demodulated SNR is equal to that of its carrier input. Therefore to obtain an increase in SNR_d with ideal envelope AM demodulation, we can only attempt to increase the IF CNR, which requires transmitting higher power P_r, or reducing the receiver noise temperature T°_{eq}.

If overmodulation occurs with envelope detection, the demodulator will produce a distorted baseband during these time periods, since it does not respond to the carrier phase that reflects these sign changes. Thus envelope detection of nonstandard AM carriers, such as suppressed carrier or SSB-AM waveforms, is not advisable in AM demodulation. There are, however, other methods for demodulation in these cases, as is considered in Section 4.3.

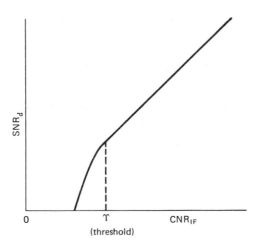

Figure 4.3. Typical AM SNR_d curve.

Note from (4.2.4) that if the threshold condition is satisfied, so that (4.2.6) is valid, the demodulator output appears as the sum of the desired baseband $m(t)$ and the filtered version of the quadrature noise. Thus, above threshold, AM demodulation with ideal envelope detection effectively recovers the carrier modulation without distortion while simple adding low pass noise to the waveform. However, below threshold, where (4.2.6) has reversed inequalities, (4.2.5) instead becomes

$$x_1(t) \approx [2C(1+m(t))n_c(t)+n_c^2(t)+n_s^2(t)]^{1/2} \qquad (4.2.14)$$

and the baseband $m(t)$ appears only with a multiplication by the noise. That is, the signal disappears and the noise totally dominates. Thus, as the IF CNR decreases below threshold, the noise effect increases and the usable signal is suppressed, causing a rapid deterioration in the demodulation process. The overall effect is to produce a typical AM SNR_d characteristic, as shown in Figure 4.3. Above threshold Y, (4.2.13) is valid, producing a linear increase with $(CNR)_{IF}$. Below threshold, SNR degrades quickly, as the IF noise dominates the demodulation operation. As stated, there is always some question regarding the actual value of the threshold Y, as measured from empirical data used to generate such curves, since the "knee" of the SNR_d curves tends to be somewhat broad.

4.3. AM Coherent Demodulation

A second way to demodulate an AM carrier is by the use of the system in Figure 4.4, where frequency mixing is used for demodulation. We replace the ideal envelope detector by a carrier mixer as shown, and use the low

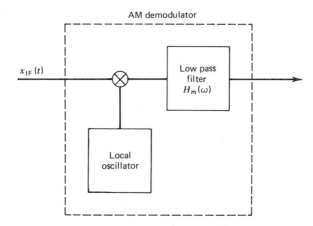

Figure 4.4. AM coherent demodulator.

pass filter $H_m(\omega)$ as the mixer filter. The mixer uses a local carrier generated at the receiver having the same frequency ω_{IF} as the IF carrier. We write this local oscillator signal as the pure carrier

$$A_l \cos (\omega_{IF}t + \psi_l) \tag{4.3.1}$$

where ψ_l represents its phase angle. The mixer filter output is that discussed in Section (3.8) and has the form

$$K_m C[1 + m(t)] \cos (\psi - \psi_l) + n_m(t) \tag{4.3.2}$$

where K_m is the mixer gain and $n_m(t)$ is the mixer noise in (3.8.4) with $\omega_{cl} = 0$. Thus the AM product demodulator produces the baseband signal term

$$K_m C \cos (\psi - \psi_l) m(t) \tag{4.3.3}$$

which appears as the desired baseband modulation multiplied by the cosine of the phase error between input carrier and local oscillator. If $\psi_l = \psi$, the local oscillator is said to be *phase synchronized* or *phase coherent* with the input carrier, and the demodulated output contains the undistorted baseband signal component. Since the mixer noise has the same spectral shape as $\tilde{S}_n(\omega)$ in (4.1.3), the mixer noise power at the output filter is again given by (4.2.8). The resulting demodulator output SNR is then the ratio of the signal power to the noise power. Hence

$$\begin{aligned} \text{SNR}_d &= \frac{K_m^2 C^2 P_m}{K_m^2 N_0(2\tilde{B}_m)} \\ &= \frac{C^2 P_m}{2 N_0 \tilde{B}_m} \end{aligned} \tag{4.3.4}$$

Note this is identical to that of the ideal envelope detector system operating above threshold. Hence an AM product demodulator operated in phase synchronism produces the same output SNR_d as an ideal envelope detector. When a phase error exists between the carrier signals, the output SNR_d instead becomes

$$(SNR)_{dem} = \frac{C^2 P_m}{2N_0 \tilde{B}_m} (\cos^2 \psi_e) \qquad (4.3.5)$$

where ψ_e is the phase error ($\psi - \psi_l$). In this case the output SNR_d is suppressed from the synchronized case by the squared cosine error. Since $\cos \psi_e \to 0$ as $\psi_e \to 90°$, (4.3.5) shows the importance of maintaining phase coherence during AM mixer demodulation. If ψ_e is considered a random variable (because either ψ or ψ_l may be random), $\cos^2 \psi_e$ must be replaced by $\mathscr{E}[\cos^2 (\psi_e)]$, where $\mathscr{E}$ is the expectation operator over the density of the phase error. It is important to note that (4.3.5) was obtained without requiring any explicit threshold condition, as in (4.2.11). In essence, the CNR condition is replaced by a condition of maintaining small phase error between IF and demodulator oscillator carriers. Later we see that this phase synchronism can be achieved with tracking loops, and the small phase error condition does in fact imply an inherent threshold condition on the IF carrier power.

If the phase error ψ_e is a function of time, then (4.3.3) corresponds to a multiplication of the desired baseband signal by the cosine error time function. This latter effect can cause complete baseband distortion. For example, if the local oscillator differs by frequency ω_d from the IF carrier, then $\psi_e(t) = \omega_d t$ and (4.3.3) becomes instead $K_m C m(t) \cos \omega_d t$. This will appear as a shift of the spectrum of $m(t)$ to the carrier offset frequency ω_d. This resulting signal corresponds to an offset baseband spectrum that may be distorted by the baseband output filter if ω_d is large. That is, the frequency error may shift the baseband spectrum outside the bandpass of the output filter.

The notion of maintaining phase coherency between a received and local carrier signal is fundamental throughout much of our subsequent study. Such coherence requires the local carrier to learn and "keep up with" the incoming carrier phase, an operation that can be ahcieved with a tracking system that forces the local carrier phase to follow that of the received carrier. This operation is discussed later in this chapter and in more detail in Chapter 9. Maintaining phase coherence, which for many years was considered a difficult electronic task, has been shown over the past decade to be both feasible and in fact relatively simple to achieve. For this reason phase coherent systems have become an integral part of modern communication systems.

Since (4.3.4) follows from (4.3.2) even if overmodulation occurred, we see that coherent AM demodulation can be used to obtain undistorted demodulation of suppressed carrier AM, provided that the local oscillator remains phase coherent during the overmodulation. That is, the local oscillator must change phase each time the overmodulated carrier changes phase. There are also other methods for demodulating such waveforms (Problem 4.4). In the suppressed carrier AM case, P_c in (4.2.2) is given instead by $C^2 P_m/2$ and (4.2.13) becomes

$$\text{SNR}_d = 2\text{CNR}_{\text{IF}} \qquad (4.3.6)$$

Thus a factor of 2 is gained in output SNR by using suppressed carrier AM with coherent demodulation. However, the task of achieving the necessary phase coherence is now made more difficult without the presence of the carrier component.

4.4. Frequency Demodulation

In this section we consider the demodulation of an FM carrier. The demodulator input waveform is assumed again to have the form in (4.1.1) when now $c(t)$ corresponds to a general FM carrier signal having the form

$$c(t) = A \cos [\omega_{\text{IF}} t + \theta(t)] \qquad (4.4.1)$$

where now

$$\theta(t) = \Delta_\omega \int m(t)\, dt$$

$$\frac{A^2}{2} = P_r G_1 \qquad (4.4.2)$$

and Δ_ω is the frequency deviation in radians. The baseband signal $m(t)$ occupies a bandwidth of B_m Hz, and we assume $m(t)$ is normalized so that $P_m = 1$. The carrier signal $c(t)$ occupies an IF bandwidth given approximately by the Carson rule bandwidth

$$B_{\text{IF}} = 2(\beta + 1)B_m \text{ Hz} \qquad (4.4.3)$$

where $\beta = \Delta_\omega/2\pi B_m$ is the carrier modulation index. We are primarily interested in the case where $\beta \gg 1$, so that the B_{IF} is several times the modulating bandwidth B_m, correspnding to wideband FM. The waveform to be demodulated is the sum of (4.4.1) and the narrowband Gaussian noise process of (4.1.2). As discussed in Section 2.6, the IF signal can then be written as in (3.7.4):

$$x_{\text{IF}}(t) = \alpha(t) \cos [\omega_{\text{IF}} t + \theta(t) + v(t)] \qquad (4.4.4)$$

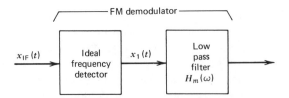

Figure 4.5. FM demodulator (ideal frequency detector).

where the phase noise is

$$v(t) = \tan^{-1}\left[\frac{n_s(t)}{A + n_c(t)}\right] \qquad (4.4.5)$$

Here we have replaced the modified noise components $\hat{n}_c(t, \theta)$ and $\hat{n}_s(t, \theta)$ with $n_c(t)$ and $n_s(t)$ by using the wideband condition of (3.7.14). Consider the processing of the preceding IF signal by the FM demodulator in Figure 4.5, composed of the cascade connection of an ideal frequency detector followed by the low pass filter $H_m(\omega)$. An ideal frequency detector is a device whose output voltage is proportional to the instantaneous frequency variation of the input. Such devices are often called *frequency discriminators* and can be constructed by combining tuned circuits and using the slope of the gain function, or can be implemented by zero crossing counters and integrators (Problem 4.7). In the absence of noise, the ideal FM demodulator responds to he instantaneous frequency of $c(t)$ about ω_{IF}, and therefore yields an output directly proportional to the baseband signal $m(t)$. When noise is present, however, the frequency detector responds to the frequency of the IF signal in (4.4.4). Its output is therefore

$$x_1(t) = K_d[\omega(t) - \omega_{IF}]$$

$$= K_d \frac{d}{dt}[\theta(t) + v(t)] \qquad (4.4.6)$$

where K_d is the detector gain. Expanding the derivative of the phase noise process yields

$$x_1(t) = K_d\left[\Delta_\omega m(t) + \frac{(A + n_c(t))n_s'(t) - n_s(t)n_c'(t)}{(A + n_c(t))^2 + n_s^2(t)}\right] \qquad (4.4.7)$$

where the primes denote differentiation with respect to t. The first term is the desired baseband, and the second term is the effect of the phase noise, the latter causing undesired deviations of the detector output. An understanding of this noise interference is hindered by the rather complex manner in which the noise components appear. If, however, we apply an assumption similar to (4.2.6)

$$A \gg |n_c(t)|, \ |n_s(t)| \qquad (4.4.8)$$

we can then expand (4.4.7) as

$$x_1(t) \cong K_d\left[\Delta_\omega m(t) + \frac{n_s'(t)}{A} + \frac{n_c(t)n_s(t)}{A^2} - \frac{n_s(t)n_c'(t)}{A^2} + O\left(\frac{1}{A^3}\right)\right] \qquad (4.4.9)$$

where $O(1/A)^3$ represent terms of order $(1/A)^3$ or higher. If we neglect all terms of second order or higher in $1/A$, we simplify to

$$x_1(t) = K_d\Delta_\omega m(t) + K_d\left(\frac{n_s'(t)}{A}\right) \qquad (4.4.10)$$

Thus, under the condition of (4.4.8), the FM detector output noise is

$$n_1(t) = \frac{K_d n_s'(t)}{A} = \left(\frac{K_d}{A}\right)\frac{dn_s(t)}{dt} \qquad (4.4.11)$$

The quadrature noise component $n_s(t)$ has the power spectrum $\tilde{S}_n(\omega)$. Hence the noise in (4.4.11) has the corresponding spectrum (recall Problem 1.29)

$$S_{n_1}(\omega) = \left(\frac{K_d}{A}\right)^2 \omega^2 \tilde{S}_n(\omega) = \left(\frac{K_d}{A}\right)^2 \omega^2 N_0 |\tilde{H}_{IF}(\omega)|^2 \qquad (4.4.12)$$

Thus the output of the ideal frequency detector is the sum of the desired modulating signal plus a noise process whose spectral density is given by (4.4.12). This spectrum is sketched in Figure 4.6, assuming an ideal IF filter over the bandwidth B_{IF}. Since the output filter has a bandwidth B_m much less than B_{IF} when wideband FM is used, the demodulator output noise is considerably reduced. If the output filter has transfer function $H_m(\omega)$ and

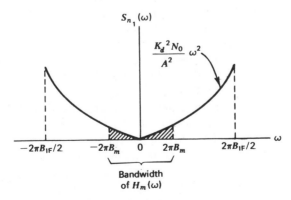

Figure 4.6. FM noise spectrum after demodulation.

passes the baseband signal $m(t)$ without distortion, the FM demodulator output has baseband signal power

$$P_s = K_d^2 \Delta_\omega^2 P_m = K_d^2 \Delta_\omega^2 \tag{4.4.13}$$

and noise power

$$P_n = \left(\frac{K_d}{A}\right)^2 \frac{N_0}{2\pi} \int_{-\infty}^{\infty} \omega^2 |\tilde{H}_{\text{IF}}(\omega)|^2 |H_m(\omega)|^2 \, d\omega \tag{4.4.14}$$

If the output filter has a rectangular filter function over B_m Hz, then

$$\begin{aligned}
P_n &= \left(\frac{K_d}{A}\right)^2 \frac{N_0}{2\pi} \int_{-2\pi B_m}^{2\pi B_m} \omega^2 \, d\omega \\
&= \frac{2K_d^2 N_0 (2\pi)^2 B_m^3}{3A^2}
\end{aligned} \tag{4.4.15}$$

The FM demodulator output SNR is then

$$\begin{aligned}
\text{SNR}_d &= \frac{P_s}{P_n} = \frac{3A^2 K_d^2 \Delta_\omega^2}{2K_d^2 N_0 (2\pi)^2 B_m^3} \\
&= 3\beta^2 \left[\frac{A^2/2}{N_0 B_m}\right]
\end{aligned} \tag{4.4.16}$$

where again $\beta = \Delta_\omega/2\pi B_m$. The preceding is called the *FM improvement* formula and reveals the basic advantage of FM communications. It relates the output SNR to the parameter β and $(A^2/2N_0 B_m)$. The latter parameter is the ratio of the IF carrier power $(A^2/2)$ to input IF noise power in the bandwidth of the baseband $m(t)$. Note that if the parameters A, B_m, and N_0 are considered fixed, the output SNR_d in (4.4.16) increases with the parameter β, which in turn depends on the transmitted deviation Δ_ω. This improvement in demodulated SNR appears as a "quieting" effect on observations of the output. For large output SNR, FM systems should therefore operate with a large frequency deviation relative to their base-bandwidth. However, recall that the IF carrier (Carson rule) bandwidth is given by $2(\beta + 1)B_m)$, and therefore also increases with β. Hence FM improvement in output SNR is achieved at the expense of required FM bandwidth. This latter effect illustrates the advantage of using wideband FM systems. By increasing the transmitted carrier bandwidth we obtain an increase in the demodulated output SNR. Recall no such trade-off was possible in AM systems. Of course, physical limitations on the amount by which a carrier can be deviated, or on the available carrier bandwidth, will ultimately constrain the achievable SNR_d predicted by (4.4.16).

The preceding results are of course predicated on (4.4.8). We can again relate this to a condition on the probability of the noise components having values less than the carrier amplitude, as in Problem 4.1. Thus the condition that this probability be sufficiently high is equivalent to a condition that

$$\text{CNR}_{\text{IF}} = \frac{A^2/2}{N_0 B_{\text{IF}}} \geq Y \tag{4.4.17}$$

for some Y. Equation (4.4.17) plays the role of an FM threshold condition on the demodulator input (IF) carrier power, just as in the AM case in (4.2.11). Note that as the index β increases, the carrier power needed to satisfy the preceding theshold also must increase, since B_{IF} increases also with β. In many communication systems, transmitter power is at a premium, whereas bandwidth is expendable (e.g., deep space links). For these systems FM is extremely suitable, since it allows this trade-off of power for bandwidth. That is, if we desire a given SNR_d, then (4.4.16) indicates that bandwidth (β) can be traded off for carrier power ($A^2/2$) by increasing β and decreasing A^2, while keeping SNR_d fixed. This trade-off can be made until A^2 reaches a lower value such that (4.4.17) is no longer satisfied. Equation (4.4.16) can be rewritten directly in terms of CNR_{IF} by substitution from (4.4.3),

$$\text{SNR}_d = 6\beta^2(\beta + 1)\text{CNR}_{\text{IF}} \tag{4.4.18}$$

This allows a direct relation between demodulator input and output SNR in FM receivers.

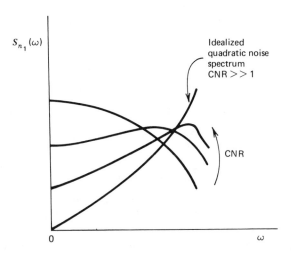

Figure 4.7. Variation in output noise spectrum as CNR_{IF} decreases.

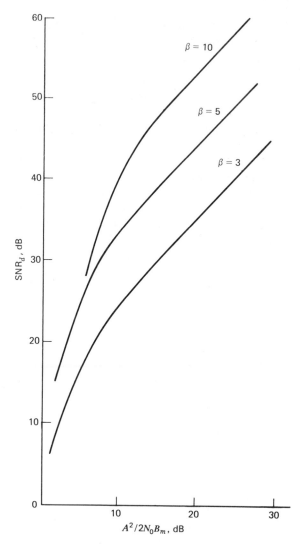

Figure 4.8. FM improvement curve. Output SNR_d versus CNR in the modulation bandwidth B_m. β = modulation index.

Operation below threshold renders the approximation in (4.4.8) no longer valid, and the effect of the higher order noise terms in (4.4.9) must be considered. These remaining terms cause an increase in the output noise and a rapid deterioration of the output SNR occurs, just as in AM. Note that the first second order term [third term in (4.4.9)] has a power spectrum given by the convolution of the spectra of the quadrature components, and therefore has a spectrum concentrated about zero frequency. The effect of including this term is to add more low frequency noise, converting the output noise spectrum from the quadratic form in Figure 4.6 to a more flattened spectrum, as shown in Figure 4.7. Thus as FM carrier power is decreased, the addition of the higher order terms in (4.4.9) causes an increase in the output inband noise and a reduction of the output SNR from that predicted by the FM improvement formula in (4.4.16). Figure 4.8 shows typical FM output SNR_d curves, exhibiting the manner in which the FM improvement formula in (4.4.16) is not longer valid below threshold. The "knee" of the curves involved is typically on the order of $CNR_{IF} \approx 12\text{-}16$ dB. Below threshold operation of FM detectors has been studied rigorously by Blackman [3], Middleton [4], and others [5, 6] from a spectral analysis point of view, and more recently by modeling the higher order effects in terms of phase "jumps" and applying an appropriate "click" analysis [7]. Each of these approaches adequately predicts the changing demodulator noise spectrum as the input carrier power is reduced, as depicted in Figure 4.7.

4.5. Phase Demodulation

When the carrier is phase modulated with the baseband signal $m(t)$, demodulation is accomplished by recovering the phase of the received IF signal. In PM systems the IF carrier is given by

$$c(t) = A \sin [\omega_{IF}t + \theta(t)]$$

$$= A \sin [\omega_{IF}t + \Delta m(t)] \qquad (4.5.1)$$

where again $A^2/2$ is the IF carrier power and Δ is the phase modulation index. When the preceding carrier is received in the presence of additive Gaussian IF noise, the resulting signal can be combined into the IF signal, as in (4.1.1),

$$x_{IF}(t) = c(t) + n(t)$$

$$= \alpha(t) \sin [\omega_{IF}t + \Delta m(t) + v(t)] \qquad (4.5.2)$$

where $v(t)$ is the phase noise in (4.4.5). Let us consider the behavior of the preceding phase modulated IF signal when acted on by an ideal phase detector (i.e., a device whose output is proportional to the phase deviation of the input IF signal). The overall PM demodulator is shown in Figure 4.9. The output low pass filter has a bandwidth corresponding to the bandwidth B_m of the baseband $m(t)$. From (4.5.2) the phase detector output will be

$$x_1(t) = \Delta m(t) + v(t) \qquad (4.5.3a)$$

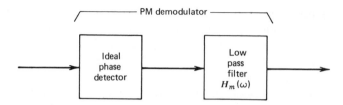

Figure 4.9. The PM demodulator (ideal phase detector).

where

$$v(t) = \tan^{-1}\left[\frac{n_s(t)}{A + n_c(t)}\right] = \frac{n_s(t)}{A} + O\left(\frac{1}{A^2}\right) \qquad (4.5.3b)$$

If we again require

$$\frac{A^2/2}{N_0 B_{IF}} \geq Y \qquad (4.5.4)$$

we can approximate $v(t)$ by dropping the second order terms in $1/A$ in the expansion of the inverse tangent function. We then have

$$x_{IF}(t) = \Delta m(t) + \frac{n_s(t)}{A} \qquad (4.5.5)$$

The output of the demodulator is then the modulation component with power $\Delta^2 P_m$ and a Gaussian noise component of power $2\tilde{B}_m N_0/A^2$, where $\tilde{B}_m$ is the noise bandwidth in (4.2.10). Assuming $P_m = 1$, we can write the idealized output SNR for PM demodulation as

$$SNR_d = \frac{\Delta^2}{2N_0\tilde{B}_m/A^2} = \Delta^2\left(\frac{A^2/2}{N_0\tilde{B}_m}\right) \qquad (4.5.6)$$

which is similar to the FM result with $3\beta^2$ replaced by Δ^2. The threshold condition in (4.5.4) again requires a suitable CNR in the IF bandwidth, the latter given by $B_{IF} = 2(\Delta + 1)B_m$. Thus the PM system also trades off bandwidth for carrier power, and the discussion for FM improvement is appropriate here for PM as well. However, the problem of obtaining wideband PM improvements in practice is hindered by the fact that Δ, the transmitter phase deviation, cannot be arbitrarily increased. This would require then that the ideal phase detector assumed in Figure 4.9 be able to detect absolute phase deviations rather than module 2π detection (i.e., distinguish between θ degrees and $(n2\pi + \theta)$ degrees). Since practical phase detectors [8] involve sine wave phase comparisons, this 2π multiple is often lost, and the absolute value of the phase deviation cannot be detected.

4.6. FM and PM Demodulation With Feedback Tracking

The demodulation of FM and PM considered in Sections 4.4 and 4.5 was accomplished by detection circuits that recovered the carrier modulation. Angle modulated carriers can also be demodulated by the use of feedback tracking. In the latter schemes, an output waveform is fed back to control an oscillator phase that is mixed with the modulated carrier to perform the demodulation. The entire feedback subsystem now plays the role of a carrier demodulator, with the output of the feedback system representing the demodulated waveform. Because of the feedback operation, such forms of demodulators are less susceptible to circuit parameter variations, and have the basic advantage that the demodulation can be achieved with less received carrier power than with the standard discriminator circuits of the previous sections.

An FM demodulator using feedback tracking is depicted in Figure 4.10. The input is the IF FM carrier to be demodulated. The mixer beats the IF carrier with the output of a local oscillator. The local oscillator is at the same frequency as the IF carrier, and the mixing operation produces a baseband waveform at the mixer output. This waveform is differentiated and low pass filtered to produce the demodulated output, and reduce the higher frequency output noise. The output is also fed back to control the frequency variation of the local oscillator. Thus the local oscillator is itself a form of voltage controlled oscillator (VCO), having its output frequency varied proportionally to the demodulated waveform.

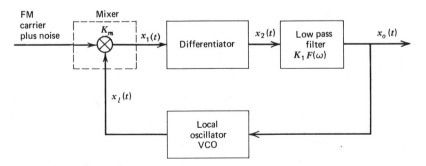

Figure 4.10. Feedback tracking FM demodulator.

To understand the way feedback causes FM demodulation to occur, let us write the FM carrier at the loop input in Figure 4.10 as in (4.4.1)

$$c(t) = A \cos [\omega_{IF} t + \theta(t)] + n_{IF}(t) \tag{4.6.1}$$

where ω_{IF} is the IF frequency, $n_{IF}(t)$ is the IF noise process, and again

$$\theta(t) = \Delta_\omega \int m(t) \, dt \tag{4.6.2}$$

We write the local oscillator waveform as

$$x_l(t) = \sin [\omega_{IF} t + \theta_l(t)] \tag{4.6.3}$$

where $\theta_l(t)$ accounts for the phase variation inserted by the feedback. Note that while (4.6.1) involves a cosinusoidal waveform, $x_l(t)$ uses a sinusoid at that same frequency, and thereby assumes a fixed 90° phase shift (aside from the modulation) between these carriers. The output of the mixer is then

$$x_1(t) = K_m A \sin \theta_e(t) + n_m(t) \tag{4.6.4}$$

where K_m is the mixer gain, $n_m(t)$ the mixer output noise, and

$$\theta_e(t) = \theta(t) - \theta_l(t) \tag{4.6.5}$$

The mixer baseband signal therefore involves the sine of the instantaneous phase error between the IF and the local carrier waveform. If we assume this phase error $\theta_e(t)$ is small enough so that

$$\sin \theta_e(t) \approx \theta_e(t) \tag{4.6.6}$$

then (4.6.4) simplifies to

$$x_1(t) = AK_m[\theta(t) - \theta_l(t)] + n_m(t) \tag{4.6.7}$$

This signal is then differentiated to produce

$$x_2(t) = AK_m \frac{d\theta_e(t)}{dt} + \frac{dn_m(t)}{dt} \tag{4.6.8}$$

The waveform $x_2(t)$ is low pass filtered by $K_1 F(\omega)$ to generate the demodulator output $x_o(t)$. The constant K_1 represents the gain level of the filter. The Fourier transform of the output waveform is

$$X_o(\omega) = K_1 F(\omega)[j\omega[AK_m \Phi_e(\omega) + N_m(\omega)]] \tag{4.6.9}$$

where

$$\Phi_e(\omega) = \Phi(\omega) - \Phi_l(\omega) \tag{4.6.10a}$$

$$\Phi(\omega) = \Delta_\omega \frac{M(\omega)}{j\omega} \tag{4.6.10b}$$

and capitals denote Fourier transforms of the corresponding lowercased time functions. We have symbolically denoted $N_m(\omega)$ as the transform of the noise process $n_m(t)$ in order to determine the effect of the loop filtering, without being concerned with its specific meaning. The signal $x_o(t)$ serves as the output and is also fed back to control the frequency of the local oscillator. Since the frequency of the VCO is controlled by its input waveform, its phase $\theta_l(t)$ is proportional to the integral of $x_o(t)$. Hence we can write

$$\Phi_l(\omega) = K_2 \frac{X_o(\omega)}{j\omega} \qquad (4.6.11)$$

where K_2 is a proportionality constant associated with the oscillator, and indicates the radian phase shift per volt of control signal. Combining (4.6.10) and (4.6.11) with (4.6.9) yields

$$X_o(\omega) = K_1 F(\omega) j\omega A K_m \left[\frac{M(\omega)}{j\omega} - K_2 \frac{X_o(\omega)}{j\omega} \right] + K_1 F(\omega) j\omega N_m(\omega) \qquad (4.6.12)$$

Solving for $X_o(\omega)$ shows

$$X_o(\omega) = \left[\frac{A K_m K_1 j\omega F(\omega)}{1 + A K F(\omega)} \right] \Delta_\omega \frac{M(\omega)}{j\omega} + \left[\frac{K_1 F(\omega)}{1 + A K F(\omega)} \right] j\omega N_m(\omega)$$

$$= \left[\frac{A K_m K_1 F(\omega)}{1 + A K F(\omega)} \right] \Delta_\omega M(\omega) + \left[\frac{K_1 F(\omega)}{1 + A K F(\omega)} \right] j\omega N_m(\omega) \qquad (4.6.13)$$

where

$$K = K_m K_1 K_2 \qquad (4.6.14)$$

is the total component gain around the loop. Thus the output of the feedback tracking loop appears as a filtered version of the carrier frequency modulation in (4.6.2), plus an added filtered noise term. Except for this filtering, the demodulator has recovered the frequency modulation of the IF carrier. We note that the output in (4.6.13) is identical to that which would be produced by the linear feedback system in Figure 4.11. (Recall Figure 1.16c.) The latter is called the *baseband equivalent* system to the FM feedback tracking demodulator, and is equivalent in the sense that it generates the same output waveform $x_o(t)$. Note that the input to the equivalent baseband loop is the phase modulation $\theta(t)$ of the FM carrier, while the mixer noise is effectively added at the point shown. In the loop, the mixer becomes a phase subtractor, the loop VCO represents a phase integrator, while the loop filtering (differentiator and low pass filter) is identical to the filtering in the actual system. We emphasize that the baseband equivalent loop is only to aid analysis, and does not indicate the

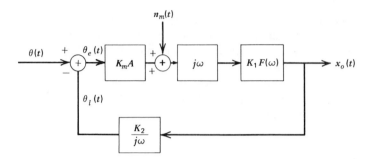

Figure 4.11. Equivalent baseband linear loop model of the phase tracking FM demodulator.

actual system implemented for demodulation, the latter shown instead in Figure 4.10. However any design of the filters using the equivalent loop is of course directly applicable to the actual feedback loop. From (4.6.13) we see that if $F(\omega) = 1$ for all $|\omega| \leq 2\pi B_m$, and 0 elsewhere, $X_o(\omega)$ becomes

$$X_o(\omega) = \left[\frac{AK_mK_1}{1+AK}\right]\left[\Delta_\omega M(\omega) + j\omega \frac{N_m(\omega)}{AK_m}\right]$$

$$= 0 \qquad \text{elsewhere} \qquad\qquad (4.6.15)$$

The output appears as an undistorted baseband signal, modified by the gain constant $AK_mK_1/(1+AK)$, plus additive noise. The output noise has its spectrum confined to frequencies within a bandwidth B_m Hz, and has a multiplication by the same gain constant. Since the gain constant is always less than one, the signal portion of the output is reduced in amplitude but the noise portion is similarly reduced, while noise components outside the signal band are eliminated. Recall the mixer noise has the power spectrum $K_m^2\tilde{S}_n(\omega)$ given in (3.8.8). The output noise in (4.6.15) will therefore have the spectrum

$$S_{n_o}(\omega) = \frac{\omega^2 N_0}{A^2}\left(\frac{AK_mK_1}{1+AK}\right)^2 \qquad |\omega| \leq 2\pi B_m \qquad (4.6.16)$$

similar to (4.4.12). The output signal power due to the first term in (4.6.15) will be given by $(AK_mK_1\Delta_\omega)^2/(1+AK)^2$. Hence the output SNR_d is identical to (4.4.16) when feedback tracking is used, and therefore performs identically to an ideal FM discriminator. This occurs without having to apply any threshold condition as we did in (4.4.17). Instead we had to apply the small phase error condition in (4.6.6). This is reminiscent of the coherent AM demodulation operation, and suggests again the notion of phase synchronization between the local and IF carrier. However here the

IF carrier phase is varying according to the frequency modulation, and the local carrier phase must continually maintain the synchronism throughout the variations. We say the two carriers are "locked" in phase, and the loop is "in lock," if the phase error continually remains small. When the phase error gets so large that (4.6.6) is no longer valid, we say the loop is "out of lock," and the preceding linear analysis is no longer applicable. (We must then resort to nonlinear loop theory for further analysis, as we shall do in Chapter 9.) Thus the feedback demodulating loop in Figure 4.10 will demodulate the FM carrier as long as the VCO phase "tracks" (remains close to) the input carrier phase variation. For this reason feedback demodulators of this type are often referred to as *phase tracking* demodulators.

Note of the role of the differentiator in the loop in Figure 4.10. The output of the mixer produces a waveform proportinal to phase error. The differentiator is necessary to produce a waveform proportional to the frequency error, which is then fed back to produce the modulation waveform at the output. Note also that in Figure 4.11 the IF carrier amplitude A appears as an effective loop gain parameter in the baseband equivalent system. This means that loop performance will necessarily depend on the value of this amplitude. Furthermore, it means that any loop design based on the equivalent system will require a particular value of A, and performance will vary if the IF carrier has a different amplitude. For this reason amplitude control devices, such as limiters or automatic gain control subsystems, are often inserted prior to phase tracking demodulators.

A similar development applies to the demodulation of PM carriers. A phase tracking PM demodulator is shown in Figure 4.12. The input PM carrier is first mixed to baseband by the VCO. Since the mixer output is proportional to phase error, a differentiator is not needed in the forward path, and the low pass filter produces the output phase signal. This phase

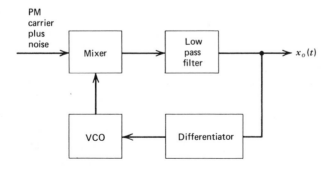

Figure 4.12. The phase tracking PM demodulator.

signal is fed back to control the VCO phase and, since a VCO has its output frequency proportional to its input voltage, we require a differentiator in the feedback path to force the VCO phase to follow the output phase voltage, as shown. To examine the PM demodulator, we let the loop input again be given by the sum of a PM carrier $c(t)$ plus IF noise, as in (4.5.2). If the VCO signal is written as in (4.6.3) the mixer output is again given by (4.6.4) and (4.6.5), except $\theta(t)$ is proportional to the phase modulation and has transform

$$\Phi(\omega) = \Delta M(\omega) \tag{4.6.17}$$

If we apply the phase error assumption in (4.6.6), the loop output has transform

$$X_o(\omega) = K_1 F(\omega)[AK_m(\Phi(\omega) - \Phi_l(\omega))] + K_1 F(\omega)N_m(\omega) \tag{4.6.18}$$

The output waveform is differentiated and used to control the VCO. Since the VCO has its frequency controlled by this differentiated output, the VCO phase $\theta_l(t)$ has transform

$$\Phi_l(\omega) = j\omega\left(\frac{K_2}{j\omega}\right)X_o(\omega) = K_2 X_o(\omega) \tag{4.6.19}$$

where K_2 is again the VCO proportionality constant. Substituting (4.6.19) into (4.6.18), and solving for $X_o(\omega)$ yields

$$X_o(\omega) = \left[\frac{AK_1 K_m F(\omega)}{1 + AKF(\omega)}\right]\Delta M(\omega) + \left[\frac{K_1 F(\omega)}{1 + AKF(\omega)}\right]N_m(\omega) \tag{4.6.20}$$

The output in (4.6.20) can be generated by a linear feedback system identical to Figure 4.11, except the $j\omega$ box is moved to the feedback path. The latter is the baseband equivalent of the PM phase tracking demodulator, and provides the same aid to analysis as Figure 4.11 did for the FM demodulator. The noise is again inserted as mixer noise at the point shown. If we again assume an ideal $F(\omega)$, with large enough bandwidth to pass the modulation, the output signal power in (4.6.20) is $(AK_1 K_m \Delta)^2/(1 + AK)^2$. The mixer noise has spectrum $K_m \tilde{S}_n(\omega)$, $|\omega| \leq 2\pi B_m$, and its output noise power is $(AK_1 K_m)^2 2N_0 B_m/(1 + AK)^2$. The resulting phase tracking demodulator output SNR_d, is identical to (4.5.6), corresponding to that of an ideal phase detector. Hence the PM phase tracking demodulator in Figure 4.12, under the phase locked condition of (4.6.6), performs as an ideal phase demodulator.

Mathematically PM phase tracking demodulators require the differentiator in the feedback path in order to achieve the proper loop output filtering function. However the use of feedback differentiation is generally undesirable since it tends to cause stability problems and produce large

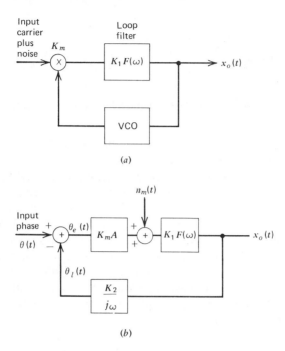

Figure 4.13. The phase lock loop. (*a*) The basic loop, (*b*) the equivalent linear model.

dynamical noise variations, making it difficult to achieve the small error condition. For this reason phase tracking PM demodulators are usually designed with no return path differentiators. The loop is reduced to a simple combination of a mixer, low pass loop filter, and VCO as shown in Figure 4.13*a*. The latter loop is referred to as a *phase lock loop* (PLL) and its equivalent baseband system is shown in Figure 4.13*b*. Its operation is identical to the previous phase tracking loops, with the mixer output generating an error signal that is filtered and fed back to control the VCO output. However, since the PLL no longer contains feedback differentiation, its overall phase transfer function from modulation input to demodulated output no longer corresponds to (4.6.20). For this reason PLL demodulators are generally followed by external filters (Figure 4.14) placed outside the loop to achieve the desired overall response. The PM demodulator in Figure 4.14 has the baseband transfer function

$$H_m(\omega) = \left[\frac{AK_mK_1F(\omega)}{1 + (AKF(\omega)/j\omega)} \right] H_{of}(\omega) \qquad (4.6.21)$$

The first bracket is the linear (small phase error) transfer function of the

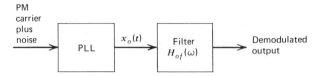

Figure 4.14. The PLL PM demodulator with output filter.

PLL alone, and depnds only on the loop filter and loop gain. The function $H_{of}(\omega)$ accounts for the output filtering. The advantage of using a PLL in cascade with an output filter as a phase tracking PM demodulator is that the loop can be designed for best phase tracking performance, while the output filter can be desiged to produce the desired demodulation transfer function. Since the output filter is not part of the feedback loop, it can be designed to have any desired shape without affecting loop tracking or stability. Because of its simplicity of design, the PLL has emerged as a basic component in modern communication system design [9–12]. As we shall see in later discussions, its basic operation (i.e., causing an oscillator to phase track a carrier waveform) makes it advantageous in several other important communication applications as well.

4.7. Linear PLL Modulation Tracking

The PLL in Figure 4.13 must operate as a linear system, which means its error in tracking the input phase variation must be relatively small for all t. Let us examine the requirements for constructing a PLL to achieve the small error condition and determine the parameters that will influence the magnitude of the phase error. It is first convenient to define

$$H(\omega) \triangleq \frac{AK[F(\omega)/j\omega]}{1+[AKF(\omega)/j\omega]} \tag{4.7.1}$$

as the loop gain function of the PLL, where $F(\omega)$ is the loop filter and $K = K_1K_2K_m$ is the total component gain of the loop. From Figure 4.13b we see that $H(\omega)$ is actually the transfer function from the baseband loop input to VCO output of the baseband equivalent system. The error in tracking, $\theta_e(t) = \theta(t) - \theta_l(t)$, is then the signal appearing at the subtractor output. Using linear feedback analysis, its transform is then

$$\Phi_e(\omega) = \Phi(\omega) - \Phi_l(\omega)$$

$$= \left[\frac{1}{1+AK[F(\omega)/j\omega]}\right]\Phi(\omega) - \left[\frac{AK[F(\omega)/j\omega]}{1+AK[F(\omega)/j\omega]}\right]\frac{N_m(\omega)}{K_mA} \tag{4.7.2}$$

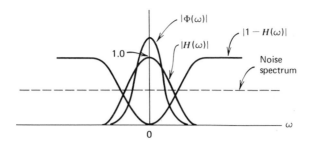

Figure 4.15. Signal and noise filtering in PLL design.

Using (4.7.1), this can be written as

$$\Phi_e(\omega) = [1 - H(\omega)]\Phi(\omega) - H(\omega)\left(\frac{N_m(\omega)}{K_m A}\right) \qquad (4.7.3)$$

The first term is the contribution to the phase error due to the carrier phase modulation $\theta(t)$. The second is due to the mixer noise entering the loop. The problem of designing modulation tracking loops therefore reduces to proper design of the loop gain function $H(\omega)$, which in turn depends on the loop filter function $F(\omega)$ and the loop gain K. The filtering operation indicated by (4.7.3) is sketched in Figure 4.15, representing a filtering of $\theta(t)$ with the transfer function $(1 - H(\omega))$ and a filtering of $n_m(t)/K_m A$ with $H(\omega)$. Since the modulation $\theta(t)$ is expected to have a low pass spectrum, increasing the bandwidth of $H(\omega)$ [letting $H(\omega) \approx 1$ over a wider frequency range] reduces the phase error due to modulation but increases the mean square phase error due to noise. Thus the design of modulation tracking loops basically involves a trade-off of modulation error versus noise error. Note that if no extraneous phase shifts or noise were involved, then adequate signal error reduction requires that $|H(\omega)| = 1$ over the bandwidth of the modulation process $m(t)$. That is, the loop transfer function $H(\omega)$ should have a bandwidth roughly equivalent to the modulation bandwidth. When extraneous phase modulation and noise is present, this bandwidth may have to be larger or smaller, depending on the dominant contributor to the total error. This trade-off is examined in detail in Section 4.8.

Specific forms of the loop gain function $H(\omega)$ for several common type of loop filter functions $F(\omega)$ are listed in Table 4.1. The *order* of the loop is an indication of the degree of filtering provided by the loop, and therefore specifies loop complexity. Most PLL are designed to be of the first or second order to avoid stability problems prevalent with more complex loops. Third order loops are used only in special situations where severe

Table 4.1. Tabulation of Loop Filters and Loop Bandwidths

Loop filter	Loop order	$H(\omega)$	B_L
$F(\omega) = 1$ (no loop filter)	1	$\dfrac{AK}{j\omega + AK}$	$\dfrac{AK}{4}$
$F(\omega) = \dfrac{j\omega\tau_2 + 1}{j\omega\tau_1 + 1}$ $\omega_n^2 = \dfrac{AK}{\tau_1}$ $2\zeta\omega_n = \dfrac{1 + AK\tau_2}{\tau_1}$	2	$\dfrac{1 + j\left(\dfrac{2\zeta}{\omega_n} - \dfrac{1}{AK}\right)\omega}{-\left(\dfrac{\omega}{\omega_n}\right)^2 + j\left(\dfrac{2\zeta}{\omega_n}\right)\omega + 1}$	$\dfrac{\omega_n}{8\zeta}\left[1 + \left(2\zeta - \dfrac{\omega_n}{AK}\right)^2\right]$
$F(\omega) = \dfrac{j\omega\tau_2 + 1}{j\omega\tau_1}$ $\omega_n^2 = \dfrac{AK}{\tau_1}$ $2\zeta\omega_n = \dfrac{AK\tau_2}{\tau_1}$	2	$\dfrac{1 + j\left(\dfrac{2\zeta}{\omega_n}\right)\omega}{-\left(\dfrac{\omega}{\omega_n}\right)^2 + j\left(\dfrac{2\zeta}{\omega_n}\right)\omega + 1}$	$\dfrac{\omega_n}{8\zeta}(1 + 4\zeta^2)$
$F(\omega) = \dfrac{(j\omega)^2 + aj\omega + b}{(j\omega)^2}$	3	$\dfrac{AK((j\omega)^2 + j\omega a + b)}{(j\omega)^3 + AK(j\omega)^2 + AKaj\omega + bAK}$	$\dfrac{AK(aAK + a^2 + b^2)}{4(aAK - b)}$

phase modulation dynamics are involved. First order loops involve no loop filtering, and are the easiest to design and analyze. Second order loops have loop filters that can be constructed as relatively simple RC networks (Problem 4.20). The second order, high gain PLL has the general loop gain function

$$H(\omega) = \frac{1 + 2\zeta(j\omega/\omega_n)}{-(\omega/\omega_n)^2 + 2\zeta(j\omega/\omega_n) + 1} \qquad (4.7.4)$$

The parameter ζ is called the loop *damping factor* and ω_n is called the loop *natural frequency*. The damping factor is typically in the range $0.5 \leq \zeta \leq 1.5$, and is an indication of the stability of the loop. As the damping factor approaches zero, the loop becomes less stable, and a much tighter tolerance must be placed on the mixer, oscillator, and filter in terms of their time delay contribution within the loop. The loop natural frequency ω_n is an indication of the loop bandwidth and its reciprocal specifies loop

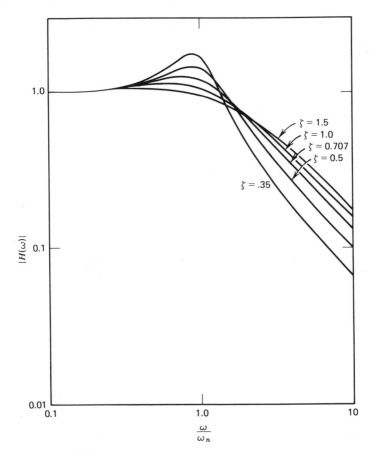

Figure 4.16. Loop gain function for second order PLL. $\zeta =$ damping factor, $\omega_n =$ loop natural frequency.

response time. The relation between these loop parameters and the parameters of the loop components and filter coefficients is also listed in Table 4.1. Figure 4.16 shows plots of $|H(\omega)|$ in (4.7.4) as a function of ω/ω_n for several values of ζ. Most loops are operated with $\zeta = 0.707$, which is referred to as a *critically damped* loop.

From (4.7.3) we see that for any PLL the corresponding phase error in the time domain is given by

$$\theta_e(t) = \theta_{em}(t) + \theta_{en}(t) \tag{4.7.5}$$

where $\theta_{em}(t)$ is the error time function associated with the inverse transform of the first term, and $\theta_{en}(t)$ is that due to the noise. Since the mixer noise is a Gaussian noise process, $\theta_{en}(t)$ itself evolves as a Gaussian process, and the first term depends on the type of modulation model used. For any $H(\omega)$ the mean squared value of the noise process $\theta_{en}(t)$ at any t can be obtained by integrating over the spectral density of the last term in (4.7.3). Since the mixer noise has spectral density $K_m^2 \tilde{S}_n(\omega)$, this is given by

$$\sigma_n^2 = \frac{1}{2\pi} \int_{-\infty}^{\infty} |H(\omega)|^2 \frac{\tilde{S}_n(\omega)}{A^2} \, d\omega$$

$$= \frac{N_0}{A^2} \left[\frac{1}{2\pi} \int_{-\infty}^{\infty} |\tilde{H}_{IF}(\omega)|^2 |H(\omega)|^2 \, d\omega \right] \quad (4.7.6)$$

If we denote

$$\tilde{B}_L \triangleq \frac{1}{2\pi} \int_0^{\infty} |\tilde{H}_{IF}(\omega)|^2 |H(j\omega)|^2 \, d\omega \quad (4.7.7)$$

as the one-sided *loop noise bandwidth*, we can write simply

$$\sigma_n^2 = \frac{N_0 2 \tilde{B}_L}{A^2} \quad (4.7.8)$$

Noting that $A^2/2$ is the IF carrier power and that $N_0 \tilde{B}_L$ is the noise power in the tracking loop noise bandwidth in (4.7.7), we can denote

$$\text{CNR}_L \triangleq \frac{A^2/2}{N_0 \tilde{B}_L} \quad (4.7.9)$$

as the loop CNR in the loop noise bandwidth. Hence

$$\sigma_n^2 = \frac{1}{\text{CNR}_L} \quad (4.7.10)$$

Thus the mean square tracking error contributed by noise is given by the reciprocal of this CNR. It is again important to note that $H(\omega)$, and therefore $\tilde{B}_L$, is itself a function of A, as is evident from (4.7.1), and the loop noise bandwidth actually depends on the received carrier power. Therefore in specifying a particular loop bandwidth, indication of the carrier power level at which it must be measured must also be given.

In most operating situations the IF noise bandwidth is much larger than the modulation bandwidth. In this case the bandwidth of $\tilde{H}_{IF}(\omega)$ greatly exceeds that of the loop function $H(\omega)$, and (4.7.7) reduces to

$$B_L \triangleq \frac{1}{2\pi} \int_0^{\infty} |H(\omega)|^2 \, d\omega \quad (4.7.11)$$

This means $\tilde{B}_L$ can be replaced by B_L above, with the latter depending only on the parameters of the loop itself. Table 4.1 also lists the relations for one-sided loop noise bandwidths B_L obtained by evaluating (4.7.11) using (4.7.1). Note that increasing the order of the loop always widens the loop noise bandwidth, while also increasing its tracking capability (shorter response time).

The modulation term $\theta_{em}(t)$ of the loop phase error has a transform given by the first term in (4.7.3). Its effect therefore depends on the form of the carrier modulation $m(t)$. If $m(t)$ is considered to be a deterministic time function, then $\theta_{em}(t)$ can be computed by inverse transforming and will contribute a deterministic time response to the loop phase error. For example, if we consider the carrier phase modulation to be a sinusoid of frequency ω_m and amplitude Δ, then $\theta_{em}(t)$ in (4.7.5) is also a sinusoid with amplitude $|1 - H(\omega_m)|\Delta$. Note that this modulation error term is approximately zero if $H(\omega_m) \approx 1$ or if ω_m is well within the bandwidth of the loop transfer function $H(\omega)$. For other types of deterministic modulations, $\theta_{em}(t)$ can be determined by straightforward linear system theory.

Alternatively, the carrier phase modulation $\theta(t)$ can be modeled itself as a Gaussian modulation process with spectrum $\Delta^2 S_m(\omega)$. The time process $\theta_{em}(t)$ is now a Gaussian response process having a zero mean value and a mean square value at any t given by

$$\sigma_m^2 = \frac{\Delta^2}{2\pi} \int_{-\infty}^{\infty} |1 - H(\omega)|^2 S_m(\omega) \, d\omega \qquad (4.7.12)$$

The error process $\theta_e(t)$ in (4.7.5) is then the sum of two uncorrelated Gaussian processes, which is itself a Gaussian process, having a total mean square value at any t of

$$\sigma_e^2 = \sigma_n^2 + \sigma_m^2 \qquad (4.7.13)$$

with σ_n^2 given in (4.7.10). We see that the phase error in the modulation tracking PLL always appears as a random Gaussian time process, the latter a direct consequence of our assumption that the noise processes are Gaussian and the loop is linear. If the carrier modulation is considered a deterministic time function, then $\theta_{em}(t)$ is a known time function that serves as the mean of the process $\theta_e(t)$, the latter having mean square value σ_n^2 about this mean at each value of t.

With these models in mind, let us investigate our principal assumption that the loop is linear, which means $\theta_e(t)$ must be small (<1) for almost all t. At any instant of time, $\theta_e(t)$ is a Gaussian random variable with mean m and variance σ_e^2. Therefore the probability that this error process is less

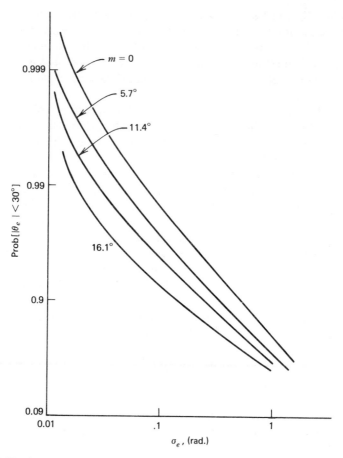

Figure 4.17. Loop phase error probabilities (m = mean phase error, σ_e = rms phase error.

than any desired phase error value θ_0 at any particular time t is

$$\text{Prob}\,[|\theta_e(t)| < \theta_0] = \int_{-\theta_0}^{\theta_0} \frac{1}{\sqrt{2\pi}\sigma_e} \exp\,[-(x-m)^2/2\sigma_e^2]\,dx$$

$$= \tfrac{1}{2}\,\text{Erf}\left(\frac{\theta_0 - m}{\sqrt{2}\sigma_e}\right) + \tfrac{1}{2}\,\text{Erf}\left(\frac{\theta_0 + m}{\sqrt{2}\sigma_e}\right) \qquad (4.7.14)$$

where $\text{Erf}\,(a)$ is tabulated in Appendix B. In general, we can consider $\sin\theta_e \approx \theta_e$ if $|\theta_e| \le \pi/6$ rad ($= 30°$). Figure 4.17 shows a plot of (4.7.14) for $\theta_0 = 30°$ and for several values of m, as a function of σ_e. One can consider this probability as equivalent to the percentage of time the loop remains in linear operation. Note that increasing m, the mean error caused by modulation tracking, greatly reduces the probability of linear lock-in. If the mean error is zero, the ability to remain linear depends only on the noise. That is, if $m = 0$, then

$$\text{Prob } [|\theta_e(t)| \le 30°] = \text{Erf}\left(\frac{\pi/6}{\sqrt{2}\sigma_e}\right) \qquad (4.7.15)$$

For the loop to remain linear 95% of the time we require

$$\frac{\pi/6}{\sigma_e} \ge 2$$

or

$$\sigma_e^2 \le \left(\frac{\pi}{12}\right)^2 \qquad (4.7.16)$$

Hence the mean square value of the phase error at any t must be sufficiently small to guarantee linear operation. For deterministic modulation and zero modulation tracking error, $\sigma_e^2 = \sigma_n^2$, and (4.7.10) and (4.7.16) imply

$$\text{CNR}_L \ge \left(\frac{12}{\pi}\right)^2 = 12.45 = 12.2 \text{ dB} \qquad (4.7.17)$$

For random modulation, σ_e^2 is given in (4.7.13), and we have instead

$$\text{CNR}_L \ge \frac{1}{(\pi/12)^2 - \sigma_m^2} \qquad (4.7.18)$$

which requires a higher value of CNR_L. We therefore conclude that a PLL phase tracking demodulator is linear only if the CNR in its loop noise bandwidth is sufficiently large. The right-hand side of (4.7.17) and (4.7.18) serve as a threshold for determining just how large this CNR_L must be, depending on the assumptions made concerning the modulation function $m(t)$. Thus the condition that the PLL be linear is equivalent to a threshold condition on the CNR in the B_L bandwidth. If the threshold condition is violated, then the probability of the loop operating in the linear region is decreased, as is evident from (4.7.15). If the loop does not track the modulation in a linear manner, then the equivalent baseband filtering model developed earlier is no longer valid, and the demodulated baseband waveform is further distorted.

It should be emphasized here that the threshold depends on the noise in the loop bandwidth, which must be only wide enough to track the carrier phase modulation. In general, this modulation bandwidth is on the order of B_m Hz, and the loop noise bandwidths are much less than typical IF noise bandwidths. This fact demonstrates the advantage of using feedback tracking demodulation instead of standard detector-type demodulation, discussed earlier. The latter system must transmit enough carrier power to combat the noise in the entire IF bandwidths to satisfy their threshold condition [recall (4.5.4) and (4.4.17)]. However the feedback systems must only overcome the noise in the loop bandwidth $B_L \approx B_m$ to satisfy its threshold in (4.7.17). Hence tracking demodulators can be operated with less carrier power than detector demodulators. It is precisely this fact that has fostered an increasing interest in phase tracking demodulation in modern systems.

4.8. Design of Modulation Tracking Loops

In the previous section we have shown how we can begin with a particular phase demodulating loop and, knowing the properties of the IF signal to be demodulated (i.e., knowing the modulation waveform and noise spectrum), we can determine the ability of the loop to remain in lock and act as a phase demodulator. In particular, we found that this condition was related to the loop tacking error. We can, instead, approach the problem from the opposite point of view. Given the waveform properties, we can determine the best loop design for maintaining a locked condition (i.e., minimizing the mean squared tracking error). In this section we consider this approach to the design of a modulation tracking PLL. We concentrate on two particular cases of interest.

Sinusoidal Modulation, Second Order PLL.   Consider first the case where

$$\theta(t) = \Delta \sin(\omega_m t + \psi_m) \tag{4.8.1}$$

and ψ_m is a uniformly distributed random phase angle over $(0, 2\pi)$. Since the modulation process has a power spectrum composed of delta functions at $\pm\omega_m$, the mean squared loop tracking error, from (4.7.12) and (4.7.13), is then

$$\sigma_e^2 = \tfrac{1}{2}|1 - H(\omega_m)|^2 \Delta^2 + \sigma_n^2 \tag{4.8.2}$$

We assume a second order PLL is to be used for the modulation tracking, represented by the general loop gain function in (4.7.4). Hence,

$$|1 - H(\omega_m)|^2 = \left| \frac{(\omega_m/\omega_n)^2}{(\omega_m/\omega_n)^2 + (2\zeta\omega_m/\omega_n) + 1} \right|^2 \tag{4.8.3}$$

and

$$\sigma_n^2 = \frac{N_0 B_L}{A^2/2} = \frac{N_0}{A^2}\left[\frac{\omega_n}{4\zeta}(1 + 4\zeta^2) \right] \tag{4.8.4}$$

Thus the loop mean squared error depends explicitly on the loop parameters ω_n and loop damping factor ζ. If we assume the modulation frequency ω_m lies well within the loop natural frequency so that $\omega_m \ll \omega_n$, we can approximate

$$\sigma_e^2 \cong \left(\frac{\omega_m}{\omega_n}\right)^4 \frac{\Delta^2}{2} + \left(\frac{N_0}{A^2}\right)\left[\frac{\omega_n}{4\zeta}(1+4\zeta^2)\right] \qquad (4.8.5)$$

The mean squared error σ_e^2 can now be minimized with respect to ω_n and ζ by solving the simultaneous equations:

$$\frac{d\sigma_e^2}{d\zeta} = 0 = \frac{N_0\omega_n}{4}\left(\frac{-1}{\zeta^2}+4\right)$$

$$\frac{d\sigma_e^2}{d\omega_n} = 0 = \frac{\omega_m^4\Delta^2}{2}\left(\frac{-4}{\omega_n^5}\right) + \frac{N_0}{A^2 4\zeta}(1+4\zeta^2) \qquad (4.8.6)$$

The solution follows immediately as

$$\zeta = \tfrac{1}{2}$$

$$\omega_n = \left[\frac{2\Delta^2\omega_m^4 A^2}{N_0}\right]^{1/5} \qquad (4.8.7)$$

Note that a loop damping factor of $\tfrac{1}{2}$ is specified, whereas the required loop natural frequency ω_n is directly related to the index and frquency of the modulation and to the parameters of the IF noise. This, unfortunately, is a property of optimal design procedures associated with a particular type of waveform—the resulting design is directly related to the specific waveform parameters. This means that these parameters values must be known precisely before the optimal loop can be constructed. The parameters in (4.8.7) can be used in conjunction with Table 4.1 to determine the loop gain and design the appropriate loop filter function.

Random Modulation. Now consider the case where $\theta(t)$ is a Gaussian random modulating process with zero mean and power spectrum $S_m(\omega)$. The mean square error is now given by

$$\sigma_e^2 = \frac{1}{2\pi}\int_{-\infty}^{\infty}\left[|1-H(\omega)|^2 S_m(\omega) + |H(\omega)|^2\left(\frac{N_0}{A^2}\right)\right] d\omega \qquad (4.8.8)$$

The first term does not integrate in closed form as it did in the previous case. Hence (4.8.8) must be minimized directly. By straightforward application of the calculus of variations (Section A.3) we define

$$f[H(\omega)] = |1-H(\omega)|^2 S_m(\omega) + |H(\omega)|^2\left(\frac{N_0}{A^2}\right) \qquad (4.8.9)$$

We then replace $H(\omega)$ by $H_0(\omega)+\varepsilon\eta(\omega)$, and compute the solution function $H_0(\omega)$ that satisfies

$$\frac{\partial f[H_0(\omega)+\varepsilon\eta(\omega)]}{\partial\varepsilon}\bigg|_{\varepsilon=0} = 0 \qquad (4.8.10)$$

for any arbitrary function $\eta(\omega)$. This yields

$$0 = -[[1-H_0(\omega)]\eta^*(\omega)+[1-H_0^*(\omega)]\eta(\omega)]S_m(\omega)$$

$$+[H_0(\omega)\eta^*(\omega)+H_0^*(\omega)\eta(\omega)]\frac{N_0}{A^2}$$

$$= 2\,\mathrm{Real}\left\{\left[-(1-H_0(\omega))S_m(\omega)+H_0(\omega)\left(\frac{N_0}{A^2}\right)\right]\eta^*(\omega)\right\} \quad (4.8.11)$$

where * denotes complex conjugates. The optimal solution $H(\omega)=H_0(\omega)$ follows as

$$H_0(\omega) = \frac{S_m(\omega)}{S_m(\omega)+(N_0/A^2)} \qquad (4.8.12)$$

The preceding equation yields the PLL loop gain function for minimizing (4.8.8). However, (4.8.12) may not always produce a realizable loop functin (such a constraint was not part of the initial problem formulation). A realizable form of $H(\omega)$ is obtained by utilizing only the upper ω plane roots of the function on the right, and (4.8.12) must be interpreted as a solution in this sense. We mention that (4.8.12) is, in reality, a form of *Wiener filter* [13], about which much has been written in terms of realizable and unrealizable filter solutions [14, 15]. Thus the design of modulation tracking loops can be readily obtained by making use of well known results from the theory of Wiener filtering.

As an example, let

$$S_m(\omega) = \frac{\Delta^2}{1+(\omega/\omega_m)^2} \qquad (4.8.13)$$

Then in (4.8.12),

$$H_0(\omega) = \frac{C_1}{1+C_2\omega^2} \qquad (4.8.14)$$

where $C_1 = \Delta^2/(\Delta^2+N_0)$ and $C_2 = N_0/\omega_m^2\Delta^2$. The denominator in (4.8.14) must now be factored to determine the upper half ω plane roots. Since $(1+C_2\omega^2)=(1+j\sqrt{C_2}\omega)(1-j\sqrt{C_2}\omega)$, we have

$$H_0(\omega) = \frac{C_1}{1+j\sqrt{C_2}\omega} \qquad (4.8.15)$$

is the realizable loop gain function. Note that the desired loop is in this case a first order loop with loop time constants dependent on the modulation and noise parameters. In particular, note that as the noise is weakened, $N_0 \to 0$, the required loop bandwidth increases, whereas the opposite is true when the noise is made stronger. Equation (4.7.1) identifies the required loop filter function as

$$AKF(\omega) = \frac{j\omega H_0(\omega)}{1 - H_0(\omega)} = C_1 \frac{j\omega}{(1 - C_1) + j\sqrt{C_2}\,\omega} \tag{4.8.16}$$

which can easily be constructed as an RC network.

We see therefore, from these examples, that there are well defined theoretical approaches available for aiding in the design of modulation tracking phase demodulators. The relationship of loop design and Wiener filter theory has been shown, which generates a plethora of loop design procedures and interpretations for optimal demodulator construction. The interested reader may wish to pursue this point further in the work of Stiffler [16], Lindsey [9], Lindsey and Simon [10], and Van Trees [17].

4.9. Carrier Subsystem Models

We have investigated carrier demodulation techniques associated with the various modulation formats presented in earlier chapters. This demodulation operation has the primary objective of extracting the baseband modulating waveform from the carrier and is hindered by the presence of receiver noise that contaminates the carrier signal at the receiver IF. We have seen, however, that if the strength of the noise is weak relative to that of the carrier signal then, to a first order approximation, the demodulation operation recovers the carrier baseband waveform but with an additive noise interference. The condition on the relative signal strengths was developed in the form of a threshold condition on the ratio of the IF carrier to noise power levels. The characteristics of the demodulated noise depend on the type of demodulation used. In all cases we were able to determine a demodulated output signal-to-noise ratio, which serves as an overall indicator of the performance of the entire carrier subsystem. Below threshold, when the receiver noise processes become significant relative to the carrier, we invariably noted a rapid degradation of this SNR, indicating a deterioration of the link performance.

These conclusions allow us to identify a basic carrier subsystem model representing the entire baseband transmission from transmitter to demodulator output. This model is shown in Figure 4.18. In Figure 4.18*a* the carrier subsystem is shown in block diagram form, with the baseband

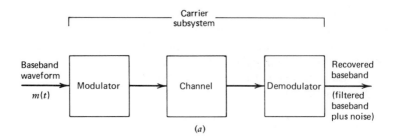

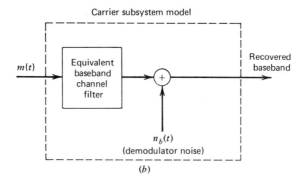

Figure 4.18. Carrier subsystem model. (*a*) Subsystem block diagram, (*b*) equivalent baseband additive noise channel model.

waveform to be modulated appearing at the transmitter modulator. This waveform is then modulated onto the carrier, transmitted over the carrier channel, received and processed in the receiver front end, and finally demodulated to the recovered baseband at the receiver. Each of these operations has been discussed at length in the past chapters. The demodulated baseband contains the equivalent filtered version of the transmitter modulation, where the filtering accounts for any baseband filtering, or effective baseband filtering imposed by the bandpass carrier filtering. To this is added the demodulation baseband noise $n_b(t)$. Thus, as far as the baseband waveform is concerned, it appears as if the entire carrier subsystem has merely filtered the modulating waveform and added in the demodulating noise. This leads to the block diagram representation shown in Figure 4.18*b*, which is often referred to as the *additive noise channel model* for the carrier subsystem. Implicit in this model is the basic assumption that the transmitter carrier power is sufficient to satisfy demodulation thresholds. In addition, other carrier interference effects, such as those discussed in Section 2.7, have been neglected.

Table 4.2. Summary of Modulation Formats

Modulation	Threshold noise bandwidth Hz	SNR_d	$S_{nb}(\omega)$ (pos. freq. only)
AM	$2B_m$	CNR_{IF}	
FM	$2(\beta+1)B_m$	$6\beta^2(\beta+1)CNR_{IF}$	
FM (modulation tracking)	B_L	$3\beta^2\,CNR_L$	
PM	$2(\Delta+1)B_m$	$2\Delta^2(\Delta+1)CNR_{IF}$	
PM (modulation tracking)	B_L	$\Delta^2 CNR_L$	

P_c = IF carrier power, $CNR_{IF} = P_c/N_0 B_{IF}$, β = modulation index, $CNR_L = P_c/N_0 B_L$.

Note that in this additive noise channel model, the form of the carrier modulation enters only through the characteristics of the additive Gaussian baseband noise $n_b(t)$. This noise will have a power spectrum, $S_{nb}(\omega)$, that reflects the type of demodulation that was used. Summarizing the results of this chapter, we see that $S_{nb}(\omega)$ will have one of the forms:

$$
S_{nb}(\omega) = \begin{cases}
N_0|\tilde{H}_{\text{IF}}(\omega)|^2|H_m(\omega)|^2, & \text{for AM} \\[2ex]
\left(\dfrac{N_0}{A^2}\right)\omega^2|\tilde{H}_{\text{IF}}(\omega)|^2|H_m(\omega)|^2, & \text{for FM} \\[2ex]
\left(\dfrac{N_0}{A^2}\right)|\tilde{H}_{\text{IF}}(\omega)|^2|H_m(\omega)|^2, & \text{for PM} \\[2ex]
\left(\dfrac{N_0}{A^2}\right)|H_m(\omega)|^2, & \text{for modulation tracking PM}
\end{cases}
\tag{4.9.1}
$$

where $N_0 = \kappa T_{\text{eq}}^{\circ} G_1$, $A^2/2$ is the IF carrier power, $\tilde{H}_{\text{IF}}(\omega)$ is the equivalent IF filtering applied to the modulation, and $H_m(\omega)$ is the demodulator low pass filtering. When the filters involved are ideal filters, (4.9.1) simplifies to give the demodulated SNR_d values computed earlier. These results are summarized in Table 4.2, showing threshold bandwidths, the output SNR, and the functional shapes of $S_{nb}(\omega)$. In the next chapters we make use of the additive noise model, and the results of Table 4.2, to investigate in detail the baseband portion of the transmitter and receiver subsystems.

References

1. Black, H. *Modulation Theory*, Van Nostrand, New York, 1953.

2. Panter, P. *Modulation, Noise, and Spectral Analysis*, McGraw-Hill, New York, 1965, Chap. 6.

3. Blackman, N. *Noise and Its Effect on Communications*, McGraw-Hill, New York, 1966.

4. Middleton, D. *Statistical Communication Theory*, McGraw-Hill, New York, 1960.

5. Goldman, S. *Frequency Analysis, Modulation, and Noise*, McGraw-Hill, New York, 1948.

6. Stumpers, F. "Theory of FM Noise," *Proc. IRE*, September 1948.

7. Taub, H. and Schilling, D. *Principles of Communication Systems*, McGraw-Hill, New York, 1971, Chap. 10.

8. Terman, F. *Electronic and Radio Engineering*, McGraw-Hill, New York, 1955.

9. Lindsey, W. *Synchronization Theory in Control and Communications*, Prentice-Hall, Englewood Cliffs, N.J., 1971.

10. Lindsey, W. and Simon M. *Telecommunication Systems Engineering*, Prentice-Hall, Englewood Cliffs, N.J., 1973.

11. Viterbi, A. *Principles of Coherent Communications*, McGraw-Hill, New York, 1966.

12. Gardner, F. *Phaselock Techniques*, Wiley, New York, 1966.

13. Wiener, N. *Extrapolation, Interpolation, and Smoothing of Stationary Time Series*, MIT Press, Cambridge, Mass., 1949.

14. Lee, Y. *Statistical Theory of Communication*, Wiley, New York, 1960, Chaps. 14–19.

15. Thomas, J. *Introduction to Statistical Communication Theory*, Wiley, New York, 1969, Chap. 5.

16. Stiffler, J. *Theory of Synchronous Communications*, Prentice-Hall, Englewood Cliffs, N.J., 1971, Chap. 5.

17. Van Trees, H. *Detection, Estimation, and Modulation Theory, Part II*, Wiley, New York, 1971, Chap. 2.

Problems

1. (4.2) Let $n_c(t)$ be a zero mean Gaussian random process with power σ^2. (a) Write the probability that $|n_c(t)| \leq A$ at any t. (b) Show that (a) will be satisfied with a probability greater than P if $A^2/2\sigma^2 \geq (\mathrm{Erf}^{-1}[P])^2$.

2. (4.2) An AM demodulator uses the envelope detector of Figure 4.2. (a) Derive an expression for the output SNR_d when the AM modulation $m(t)$ has power spectrum $S_m(\omega) = [1 + [\omega/2\pi B_m)^2]^{-1}$ and $H_m(\omega) = 1$, $|\omega| \leq 2\pi B_m$. Assume preceding threshold operation, white noise, and infinite IF bandwidths. (b) Repeat when $H_m(\omega)$ is replaced by the low pass filter $H_m(\omega) = [1 + (j\omega/2\pi B_m)]^{-1}$.

3. (4.3) Show that a phase coherent AM detector that is phase coherent with the sinusoidal term of a SSB-AM signal will demodulate the SSB-AM signal.

4. (4.3.) Show that a suppressed carrier AM waveform can be demodulated by adding to it a phase coherent pure carrier and squaring the resulting combined signal.

5. (4.3) (a) Derive the expression for the output SNR_d for an ideal suppressed carrier AM system, assuming phase coherent demodulation. Assume the IF carrier has power P_c, modulation has unit power, and bandwidth is B_m Hz. (b) Show that $\mathrm{SNR}_d = 2\,\mathrm{CNR}_{\mathrm{IF}}$, where CNR is the IF CNR.

6. (4.4) Assume the IF noise is white over the IF bandwidth and sketch the shape of the power spectrum of the next $(1/A^2)$ terms in (4.4.9).

7. (4.4) A device produces a unit impulse every time a positive going zero crossing occurs. Show that placing a T sec integrator after the device produces an approximation to a FM detector. What is the required relation between T and the bandwidth of the modulation?

8. (4.4)   An FM system has a receiver RF threshold of 20 dB. How much received carrier power and RF bandwidth is needed to transmit a 1 kHz baseband signal with a demodulated SNR of 40 dB ($N_0 = 10^{-10}$ W/Hz)?

9. (4.4)   (a) An FM system uses a modulating sine wave at 10 kHz and a deviation of 100 kHz. If the demodulator operates at a threshold of 15 dB, compute the demodulated SNR. (b) If the deviation was limited to 1% of the carrier frequency, what is the maximum SNR_d attainable with a 3 MHz RF carrier.

10. (4.4)   Two sine waves at frequency f_1 and f_2, each with deviation Δ_ω, are added and transmitted by FM modulation on an RF carrier. An ideal FM receiver is operated above threshold with received carrier power P_r and additive RF noise of level N_0. The demodulated output is filtered by the filters shown in Figure P4.10. (a) Determine the output SNR of each filter. (b) Repeat for an ideal PM system, with mod index Δ.

11. (4.4)   A pure carrier of amplitude A at 1 GHz and a white noise process of level N_0 are received at the input to a system composed at a bandpass filter at 1 GHz with bandwidth 20 MHz, followed by an ideal FM detector and low pass filter with 20 MHz bandwidth. (a) Determine the output spectral density. (b) Determine the output power. (c) Repeat (a) with the carrier shifted to 1.005 GHz (system remains the same).

12. (4.4)   (a) Derive an expression for the output SNR_d of an FM demodulator in which the low pass output filter has transfer function $|H(\omega)|^2 = [1 + \omega/2\pi B_m)^4]^{-1}$. Assume that $P_m = 1$, wideband FM ($\Delta_\omega \gg$

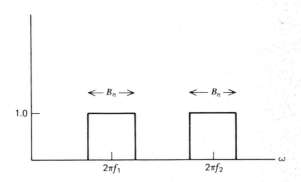

Figure P4.10.

$2\pi B_m$), white noise, above threshold operation, and that the filter passes $m(t)$ without distortion. (b) Compare the result to the ideal filter case.

13. (4.4) An FM system uses a standard demodulator with a 15 dB threshold. It is to achieve an output SNR of 50 dB when transmitting a 10 kHz modulating signal. The carrier system parameters are: range loss -165 dB, isotropic transmitting antenna, receiving antenna gain 40 dB, and receiver noise equivalent spectral level $N_0 = -191$ dBW/Hz. Neglecting pointing, coupling, and atmospheric losses, determine the required carrier bandwidth and transmitter power to maintain the system.

14. (4.6) Consider again the system in Problem 4.11, but with the FM detector replaced by an ideal phase detector. Determine the output spectral density of the low pass filter output when the pure carrier and noise of Problem 4.11 is inserted at the input.

15. (4.6) A phase tracking PM demodulator is to have the overall equivalent filter function in (4.6.21) that corresponds to an ideal filter of bandwidth of B_m Hz. Determine the required output filter function to achieve this when a first order tracking loop is used with loop bandwidth B_m.

16. (4.7) A phase lock loop having a loop noise bandwidth of 1 kHz has an unmodulated carrier plus white Gaussian noise of level $N_0 = 10^{-10}$ W/Hz fed into it. Using linear theory, determine the required carrier power to guarantee a mean squared tracking error of 0.01 rad^2.

17. (4.7) Verify the bandwidth B_L of the first order loop in Table 4.1.

18. (4.7) A carrier with a phase offset ψ is fed into a first order PLL having the loop gain function $H(\omega) = 1/1 + j2\omega$. Gaussian noise

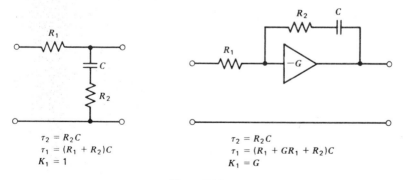

$$\tau_2 = R_2 C$$
$$\tau_1 = (R_1 + R_2)C$$
$$K_1 = 1$$

$$\tau_2 = R_2 C$$
$$\tau_1 = (R_1 + GR_1 + R_2)C$$
$$K_1 = G$$

Figure P4.20.

producing a CNR_L in the loop bandwidth of 10 dB is also inserted. What is the probability that the loop will remain linear (i.e., have a phase error of less than 30°)?

19. (4.7) A PLL tracks a carrier in additive noise (N_0) using a loop with filter $F(\omega)$ and loop gain K. Determine the loop noise bandwidth as the carrier amplitude $A \to 0$.

20. (4.7) Show that the loop filter function can be constructed as either of the RC networks in Figure P4.20. Show that the filter parameters and time constants are related as shown.

21. (4.8) Complete the design problem in (4.8.7). That is, design the loop filter and gain using Table 4.1 and problem 4.20.

22. (4.9) A PM carrier with a phase deviation $\Delta \ll 1$ is passed through a bandpass carrier filter function $H_c(\omega)$. It is then ideally phase demodulated to produce the modulation. Show that the equivalent filter function $H_{bc}(\omega)$ in Figure 4.18 corresponds to $\tilde{H}_c(\omega)$, the low frequency version of $H_c(\omega)$.

CHAPTER 5

BASEBAND WAVEFORMS, SUBCARRIERS, AND MULTIPLEXING

Our study to this point has concentrated on the carrier portion of the communication system. We have shown that under proper operating conditions, the entire carrier transmission can be represented by a basic additive noise model operating on the transmitter baseband waveform. The baseband waveform was considered to be an arbitrary waveform specified by a known bandwidth B_m and power level P_m. We now wish to examine more specifically the actual characteristics of typical baseband waveforms, and the manner in which the desired source information can be extracted at the receiver. Recall from Figure 1.2a that the baseband is formed from the source signal by means of the baseband conversion, and therefore contains, in some context, the desired information. The specific form of the baseband waveform depends on the type of source, the manner in which the source information is baseband converted, and whether a single source or a group of sources is involved. In this chapter the various baseband formats are investigated.

5.1. The Baseband Subsystem

The baseband subsystem represents the portion of the communication link involving the source and baseband waveforms. As such, the baseband subsystem contains components at both the transmitter and receiver ends,

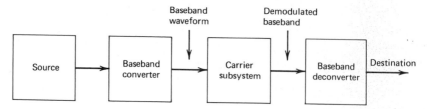

Figure 5.1.   The baseband subsystem block diagram.

which are interconnected by the carrier subsystem previously considered. A typical baseband subsystem model is shown in Figure 5.1. The source signal is baseband converted to the baseband waveform, which is then transmitted over the carrier subsystem via modulation and demodulation. The latter subsystem can be modeled by the additive noise and filtering channel of Figure 4.18. The baseband waveform recovered at the receiver must then be baseband deconverted to regenerate the source waveform. It is evident that a study of the baseband subsystem reduces primarily to a study of the converting and deconverting operations.

The form of the baseband waveform $m(t)$ depends on the conversion format and the type of information source. When a single source is involved, its waveform may be used directly as the baseband signal (i.e., the baseband conversion is eliminated), or it can be converted by an encoding, filtering, or modulation format. When many sources are involved, the source set must be multiplexed (combined together) to form the baseband. In all cases the resulting baseband waveform is determined by the inherent properties of the source signal itself and the processing imposed by the converter.

Sources are classified as analog or digital sources. An analog source produces a continuous waveform in time, and its time variation represents the information sent to the receiver destination. Examples of such sources are the electronic signals from telephones (*audio* sources) or television cameras (*video* sources). A digital source produces a sequence of binary symbols, which may correspond to digital commands or computer words, or may be produced by digital conversion of an analog source. With digital sources the baseband signal is formed by encoding the source symbols into a baseband waveform. In this chapter we concentrate primarily on analog sources, and defer discussion of digital sources and baseband encoding to Chapter 6.

If the source waveform is used directly as the baseband signal, then clearly $m(t)$ has the same characteristics as the source itself. However, in many cases it may be advantageous to filter the source waveform prior to carrier modulation, or perhaps to modulate the source waveform onto a secondary carrier (subcarrier) to form the baseband signal. The advantage of filtering is that we can shape the baseband spectrum prior to carrier modulation, which allows us to emphasize or deemphasize specific portions of the source spectrum so as best to combat channel distortion. Subcarrier modulation allows us to control the location of the baseband spectrum along the frequency axis, and thereby shift the baseband away from undesired interference. In the next sections we examine these forms of baseband conversion operations.

5.2. Baseband Filtering Systems

With baseband filtering, the source signal is simply filtered prior to carrier modulation. At the receiver the demodulated baseband may then be further filtered to attempt to reconstruct the original source frequency characteristics. Using the additive noise model to represent the carrier subsystem, the overall baseband link from source to destination will have the form shown in Figure 5.2. The filter function $H_t(\omega)$ represents the premodulation filtering inserted at the transmitter for source waveform shaping. The filter function $H_{bc}(\omega)$ represents the channel filtering in Figure 4.18 effectively applied to the baseband by the channel during carrier transmission. The filter $H_b(\omega)$ represents baseband filtering inserted at the receiver following carrier demodulation. The spectrum $S_{nb}(\omega)$ is the spectral density of the added demodulator baseband noise, and $S_d(\omega)$ is the source spectrum. Typically, $H_t(\omega)$ and $H_b(\omega)$ are in the hands of the system designer, whereas $H_{bc}(\omega)$, $S_d(\omega)$, and $S_{nb}(\omega)$ are fixed by system constraints. The baseband design problem is then to select the best receiver and/or transmitting filters for the baseband link model of Figure 5.2. We assume our prime objective is to achieve as large a destination SNR, or signal distortion ratio SDR, as possible.

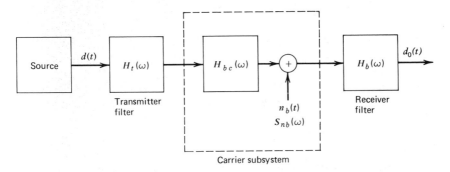

Figure 5.2. Baseband filtering model.

Consider first the case where no transmitter filtering is used (i.e., $H_t(\omega) = 1$), and the source waveform is used directly as the baseband. The receiver baseband filter $H_b(\omega)$ must then filter the baseband noise without significant distortion of the baseband signal. If we denote $d_0(t)$ as the filtered baseband signal at the destination, then a signal error term, $d(t) - d_0(t)$, can be defined, and a signal to distortion power can be determined as in (1.7.7). The total output distortion is then the sum of the error distortion power and the output noise power. Thus, as in Problem 1.32, we write

$$\text{SDR} = \frac{P_d}{\dfrac{1}{2\pi} \displaystyle\int_{-\infty}^{\infty} |1 - H_{bc}(\omega)H_b(\omega)|^2 S_d(\omega) \, d\omega + \dfrac{1}{2\pi} \displaystyle\int_{-\infty}^{\infty} |H_b(\omega)|^2 S_{nb}(\omega) \, d\omega}$$

(5.2.1)

where P_d is the source waveform power. With a given P_d, it is clear that SDR is maximized if the denominator in (5.2.1) is minimized. Hence we seek a baseband filter function $H_b(\omega)$ that minimizes this denominator for a specified channel filter $H_{bc}(\omega)$. Equivalently, we seek the function $G(\omega) \triangleq H_{bc}(\omega)H_b(\omega)$ that minimizes

$$\int_{-\infty}^{\infty} \left[|1 - G(\omega)|^2 S_d(\omega) + |G(\omega)|^2 \left(\frac{S_{nb}(\omega)}{|H_{bc}(\omega)|^2} \right) \right] d\omega \qquad (5.2.2)$$

When stated in this way, the problem is identical to the minimization carried out in Section 4.8, and by straightforward substitution we obtain the desired filter function for $H_b(\omega) = G(\omega)/H_{bc}(\omega)$, or

$$H_b(\omega) = \frac{1}{H_{bc}(\omega)} \left[\frac{S_d(\omega)}{S_d(\omega) + (S_{nb}(\omega)/|H_{bc}(\omega)|^2)} \right] \qquad (5.2.3)$$

Note that the optimal filter function inverts the channel filtering, and filters according to the relative strengths of the signal spectrum and the effective noise spectrum $S_{nb}(\omega)/|H_{bc}(\omega)|^2$. This latter filtering is again a form of Wiener filter, similar to that derived in (4.8.12). At frequencies where $S_d(\omega) \gg S_{nb}(\omega)$, $H_b(\omega) \approx 1/H_{bc}(\omega)$, and the filter concentrates on compensating for the channel filtering over these frequencies. At frequencies where $H_{bc}(\omega)S_d(\omega) \ll S_{nb}(\omega)$, we have $H_b(\omega) \approx S_d(\omega)/S_{nb}(\omega) \ll 1$, and the filter effectively rejects transmission at frequencies where the noise dominates. Thus the optimal receiver baseband filter in (5.2.3) attempts to achieve a satisfactory compromise between reducing signal distortion caused by the channel filter and filtering the baseband noise from the output.

When the optimal filter function of (5.2.3) is inserted into (5.2.1), we obtain the maximum SDR that can be attained in the baseband channel.

The denominator becomes

$$\frac{1}{2\pi}\int_{-\infty}^{\infty}\left\{\left|1-\frac{S_d(\omega)}{S_d(\omega)+S'_{nb}(\omega)}\right|^2 S_d(\omega)+S'_{nb}(\omega)\left[\frac{S_d(\omega)}{S_d(\omega)+S'_{nb}(\omega)}\right]^2\right\}d\omega$$

$$=\frac{1}{2\pi}\int_{-\infty}^{\infty}\frac{S'_{nb}(\omega)S_d(\omega)}{S_d(\omega)+S'_{nb}(\omega)}\,d\omega \qquad (5.2.4)$$

where $S'_{nb}(\omega)\triangleq S_{nb}(\omega)/|H_{bc}(\omega)|^2$. That is, $S'_{nb}(\omega)$ is the baseband noise referred back to the source and represents the equivalent noise that would have to be inserted at the source to produce $S_{nb}(\omega)$. The maximum achievable baseband SDR at the receiver is then

$$\text{SDR}_{\max}=\frac{P_d}{\dfrac{1}{2\pi}\displaystyle\int_{-\infty}^{\infty}\dfrac{S'_{nb}(\omega)S_d(\omega)}{S_d(\omega)+S'_{nb}(\omega)}\,d\omega} \qquad (5.2.5)$$

Given the source and equivalent source noise spectra, we can therefore immediately determine the attainable SDR of the baseband subsystem. This is extremely convenient in analysis, since it allows us to assess the subsystem without having first to compute the required optimal receiver filter. Note that the SDR depends only on the integrated spectral densities in (5.2.5). Since it involves the product of the noise and source spectra at each ω, we see that we can never completely remove the baseband noise (i.e., obtain SDR$\to\infty$) as long as there is frequency overlap of the two spectra. Although we would intuitively expect this conclusion, we have here proven it to be rigorously true, and, in addition, have determined the precise manner in which the degree of overlap will influence performance.

Consider now the case where a baseband filter $H_t(\omega)$ is inserted at the transmitting end. We now have the advantage of having two separate baseband filters to adjust for best performance. We intuitively expect that the receiver filter $H_b(\omega)$ should be used primarily for noise rejection, whereas $H_t(\omega)$ can be used to compensate for signal filtering. Hence we can set

$$H_t(\omega)H_{bc}(\omega)H_b(\omega)=1 \qquad (5.2.6)$$

while defining the transmitted baseband power

$$P_m=\frac{1}{2\pi}\int_{-\infty}^{\infty}|H_t(\omega)|^2 S_d(\omega)\,d\omega \qquad (5.2.7)$$

With (5.2.6) no source signal distortion occurs in the baseband link, and the output interference is due only to output noise. Hence the baseband output SNR is

$$\text{SNR} = \frac{P_d}{\dfrac{1}{2\pi} \displaystyle\int_{-\infty}^{\infty} |H_b(\omega)|^2 S_{nb}(\omega)\, d\omega} \qquad (5.2.8)$$

We again seek the minimization of the denominator of (5.2.8), subject to a fixed P_m in (5.2.7) and the condition of (5.2.6). From Section A.3, this requires minimization of

$$f[|H_b(\omega)|] = |H_b(\omega)|^2 S_{nb}(\omega) + \lambda \frac{S_d(\omega)/|H_{bc}(\omega)|^2}{|H_b(\omega)|^2} \qquad (5.2.9)$$

with respect to $|H_b(\omega)|$. The solution follows easily as

$$|H_b(\omega)|^4 = \frac{K^2 S_d(\omega)}{|H_{bc}(\omega)|^2 S_{nb}(\omega)} \qquad (5.2.10a)$$

where K is an arbitrary filter gain constant. The corresponding transmitter filter function, from (5.2.6), is then

$$|H_t(\omega)|^4 = \frac{S'_{nb}(\omega)}{K^2 S_d(\omega)} \qquad (5.2.10b)$$

with K adjusted to satisfy (5.2.7). Thus the filter pair in (5.2.10), for the transmitter and receiver baseband subsystems, is required for maximization of SNR. Note that now only the receiver baseband filter magnitude is specified, and is markedly different from the previous result of (5.2.3) when no transmitter filtering was included. Nevertheless, the filter still tends to have large gain at frequencies where the source spectrum dominates and low gain when the noise dominates. The corresponding transmitter filter essentially inverts the baseband operation, emphasizing the frequencies where the subsequent channel noise dominates and attenuating the source frequencies where the signal dominates. Such transmitter filters are often called *preemphasis filters*. Their basic role is purposely to predistort the signal spectrum so that the subsequent baseband filtering will regenerate the correct spectral shape, as indicated in (5.2.6), while producing the maximal output SNR.

5.3. Subcarrier Modulated Baseband Systems

A baseband system using subcarrier modulation for baseband conversion is shown in Figure 5.3. The source waveform is modulated directly onto the subcarrier to form the baseband (Figure 5.3a), which is therefore itself a

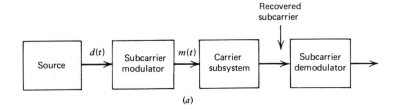

(a)

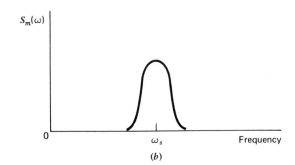

(b)

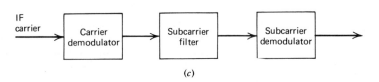

(c)

Figure 5.3. Baseband subcarrier system. (a) Block diagram, (b) baseband spectrum, (c) receiver system.

modulated carrier, having a baseband spectrum similar to that in Figure 5.3b. The advantage of this is that we control the location of the spectrum of the baseband signal, allowing us to shift the source spectrum away from the lower frequencies and avoid undesired baseband interference. The selection of the subcarrier frequency ω_s is arbitrary, although there may be physical constraints that limit the modulation parameters of an oscillator for a given subcarrier frequency. For example, in FM subcarriers there is a limit to the amount by which a subcarrier frequency can be deviated, as given in (1.5.19).

After the source signal is modulated onto the subcarrier the baseband is then modulated onto the main (RF) carrier for transmission, producing the RF carrier $c(t)$. Note there are two modulation stages—the source waveform onto the subcarrier, and the subcarrier onto the RF carrier. Each modulation can be either AM, FM, or PM. For example, the source may be amplitude modulated onto the subcarrier, and the latter may be frequency modulated onto the RF carrier. This pair of modulation operations is referred to as AM|FM. One can similarly describe other types of systems as FM|FM, AM|PM, and so on.

At the receiving end, after RF and IF processing, the receiver must invert the two modulation operations and pass the IF signal through two stages of demodulation (Figure 5.3c). The first recovers the baseband subcarrier, and the second recovers the source from the subcarrier. We can analyze the cascade demodulation by applying successively the results of Chapter 4. In the following discussion we examine some specific examples of this type of modulation format.

AM|FM Systems.   In AM|FM systems the source waveform is amplitude modulated onto the subcarrier and the baseband is frequency modulated onto the RF carrier. Denote the source waveform as $d(t)$ and assume it to occupy a bandwidth of B_d. We again assume $d(t) \leq 1$ to prevent over-modulation. The output of the carrier modulator in Figure 5.3 is the baseband signal $m(t)$ and has the normalized form

$$m(t) = [1 + d(t)] \cos(\omega_s t + \theta_s) \qquad (5.3.1)$$

where ω_s and θ_s are the subcarrier frequency and phase angle. The baseband therefore has the spectrum shown in Figure 5.3b, with a subcarrier bandwidth about ω_s of $2B_d$ Hz. The upper frequency of $m(t)$, which defines its overall bandwidth, is then

$$B_m = \frac{\omega_s}{2\pi} + B_d \qquad \text{Hz} \qquad (5.3.2)$$

The baseband signal, in addition, has power

$$P_m = \tfrac{1}{2}(1 + P_d) \qquad (5.3.3)$$

where P_d is the power of the source waveform $d(t)$ ($P_d \leq 1$). The baseband frequency modulates the RF carrier producing at the receiver IF output the carrier waveform

$$c(t) = A \cos\left(\omega_c t + \Delta_\omega \int m(t)\, dt\right) \qquad (5.3.4)$$

where Δ_ω is the frequency deviation of the RF carrier. This carrier occupies an IF bandwidth of approximately

$$B_{IF} = 2\left[\frac{\Delta_\omega \sqrt{P_m}}{2\pi B_m} + 1\right]B_m$$

$$= 2\left[\frac{\Delta_\omega}{2\pi B_m}\left(\frac{1+P_d}{2}\right)^{1/2} + 1\right]B_m \qquad \text{Hz} \qquad (5.3.5)$$

where we have used the rms value of the modulation as the effective frequency deviation of the carrier. If the carrier is received with suitable power to exceed the threshold condition of the first (FM) demodulation,

$$\text{CNR}_{IF} = \frac{A^2/2}{N_0 B_{IF}} \geq Y \qquad (5.3.6)$$

then the output of the demodulator has the spectrum shown in Figure 5.4. The demodulation has recovered the baseband with the addition of the quadratic frequency noise, as shown. This frequency noise has the spectrum $S_{nb}(\omega)$ in (4.9.1) for the FM case. The subcarrier bandpass filter centered at ω_s extracts the signal content only over the bandwidth occupied by the subcarrier. The signal power of the subcarrier at the filter is then

$$P_{sc} = \frac{\Delta_\omega^2}{2}(1 + P_d) \qquad (5.3.7)$$

The noise power at the filter output is obtained by integrating over the filtered spectrum of the noise. For an ideal subcarrier filter over the subcarrier bandwidth, the filter output noise power is

$$P_{ns} = \frac{2}{2\pi}\int_{\omega_s - \pi B_{sc}}^{\omega_s + \pi B_{sc}} \left(\frac{N_0 \omega^2}{A^2}\right) d\omega$$

$$= \frac{N_0}{A^2 3\pi}\left[(\omega_s + 2\pi B_d)^3 - (\omega_s - 2\pi B_d)^3\right] \qquad (5.3.8)$$

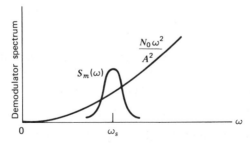

Figure 5.4. Demodulated subcarrier spectrum with frequency noise.

Using the expansion

$$(a \pm b)^3 = a^3 \pm 3a^2 b + 3ab^2 \pm b^3$$

we reduce (5.3.8) to

$$P_{ns} = \frac{N_0}{3\pi A^2}[6\omega_s^2(2\pi B_d) + 2(2\pi B_d)^3] \qquad (5.3.9)$$

If we further assume that the subcarrier frequency is much larger than the source bandwidth $\omega_s \gg 2\pi B_d$, then (5.3.9) is approximately

$$P_{ns} \cong 2\left(\frac{N_0 \omega_s^2}{A^2}\right) 2B_d \qquad (5.3.10)$$

Since $2B_d$ is the subcarrier bandwidth, we see that (5.3.10) is equivalent to assuming that the FM demodulated noise spectrum is flat over the subcarrier bandwidth, with equivalent spectral level $N_0 \omega_s^2 / A^2$, two-sided. The input to the subcarrier demodulator therefore has a CNR, for $P_d = 1$, given by

$$CNR_s = \frac{P_{sc}}{P_{ns}}$$

$$= \left(\frac{\Delta_\omega}{\omega_s}\right)^2 \left[\frac{A^2/2}{N_0 2B_d}\right] \qquad (5.3.11)$$

This CNR therefore represents that available for subcarrier (AM) demodulation. Since the bracket is the IF CNR in the subcarrier bandwidth, we see that CNR_s is increased over that available in the IF. The improvement is by the square of the factor Δ_ω / ω_s. Since Δ_ω represents the RF carrier deviation by the subcarrier, Δ_ω / ω_s plays the role of a subcarrier modulation index. Again we see the availability of trading IF bandwidth for improved CNR_s, simply by increasing the deviation Δ_ω. We also note that, for a fixed deviation Δ_ω, CNR_s decreases with ω_s. This shows the disadvantage of using too large a subcarrier frequency and is simply caused by the increased frequency noise at the higher subcarrier frequencies, as seen from Figure 5.4.

The demodulated subcarrier signal must now pass into the second (AM) demodulator to recover the source signal. To achieve this, it is first necessary that the input CNR_s satisfy the AM demodulation threshold, as discussed in Sections 4.2 and 4.3. Thus we require

$$CNR_s \geq Y_{AM} \qquad (5.3.12)$$

where Y_{AM} is the required AM threshold. This requirement allows us to determine the parameters in (5.3.11). After AM demodulation, the source signal will be recovered with a demodulated SNR given by (4.2.13),

$$SNR_d = CNR_s \qquad (5.3.13)$$

The parameter SNR_d is an indication of the overall fidelity of our communication link and is generally a parameter specified as a design constraint. That is, for a desired value of SNR_d we can use (5.3.11)–(5.3.13) to aid in the selection of the remaining system parameters.

Although the use of a subcarrier gives us control over baseband frequency location, an obvious question is whether we have paid a penalty for this advantage, as opposed to direct FM of the source waveform onto the RF carrier with no subcarrier. Recall from our FM analysis that the latter system, operating with the same peak deviation of Δ_ω, will produce a demodulated SNR of $SNR_d = 3(\Delta_\omega/2\pi B_d)^2(A^2/2N_0B_d)$. Comparing this to (5.3.13) and (5.3.11), we see that the use of the subcarrier has produced an SNR_d that is lower by a factor proportional to $(\omega_s/B_d)^2$. We have therefore achieved a poorer source SNR_d by using the subcarrier to favorably control spectral location. As stated, this is simply due to the fact that the subcarrier forces us to operate at higher frequencies of the baseband where the demodulated baseband FM noise is greater. Trade-offs with other system formats can also be made (Problem 5.6).

FM|FM Systems. Let us extend the discussion to FM|FM systems by repeating the analysis of Section 5.3.1. In this case the source waveform is frequency modulated onto the subcarrier, producing

$$m(t) = \cos\left(\omega_s t + \Delta_s \int d(t)\, dt\right) \qquad (5.3.14)$$

where Δ_s is the frequency deviation of the source onto the subcarrier. The RF carrier again appears as in (5.3.4). The baseband, subcarrier, and RF bandwidths follow as

$$B_m = \frac{\omega_s}{2\pi} + \frac{B_{sc}}{2} \qquad \text{Hz}$$

$$B_{sc} = 2\left(\frac{\Delta_s}{2\pi B_d} + 1\right)B_d \qquad \text{Hz} \qquad (5.3.15)$$

$$B_{RF} = 2\left(\frac{\Delta_\omega}{2\pi B_m} + 1\right)B_m \qquad \text{Hz}$$

If (5.3.6) is satisfied, RF demodulation again produces the spectral diagram in Figure 5.4, except that the subcarrier spectrum is that of an FM signal. The filtered subcarrier power is then $\Delta_\omega^2/2$, and (5.3.10) approximates the noise power. Hence

$$\text{CNR}_s = \left(\frac{\Delta_\omega}{\omega_s}\right)^2 \left[\frac{A^2/2}{N_0 B_{\text{sc}}}\right] \tag{5.3.16}$$

The preceding CNR must be large enough to satisfy the subcarrier demodulation FM threshold; that is,

$$\text{CNR}_s \geq Y_{\text{FM}} \tag{5.3.17}$$

Using our FM results of Section 4.4, the demodulated source signal will then be recovered with SNR of

$$\text{SNR}_d = 3\beta_s^2 \frac{\Delta_\omega^2/2}{(N_0 \omega_s^2/A^2)2B_d}$$

$$= 3\beta_s^2(\beta_s + 1)\left(\frac{\Delta_\omega}{\omega_s}\right)^2 \left(\frac{A^2/2}{N_0 B_{\text{sc}}}\right)$$

$$= 3\beta_s^2(\beta_s + 1)\text{CNR}_s \tag{5.3.18}$$

where β_s is the subcarrier modulation index $\Delta_s/2\pi B_d$. We see now that the system receives the advantage of two consecutive stages of FM improvement. However, the RF bandwidth in (5.3.15) is also increasing faster, since both Δ_ω and B_m increase as the system attempts to take advantage of both SNR improvements. Again, a limit will exist on the amount by which β_s can be increased, because of the subcarrier deviation restriction.

FM|PM Systems.   In FM|PM systems, phase modulation of the subcarrier onto the RF carrier is used. The baseband and carrier waveforms are then

$$m(t) = \cos\left(\omega_s t + \Delta_s \int d(t)\,dt\right) \tag{5.3.19a}$$

$$c(t) = A\cos(\omega_c t + \Delta m(t)) \tag{5.3.19b}$$

where Δ is now the phase modulation index of the carrier. The bandwidths are again given by (5.3.15), and the RF threshold condition is identical to (5.3.6) with the PM modulation threshold value inserted. However, the demodulated phase noise is no longer quadratic as in Figure 5.4, but rather has the two-sided flat spectrum N_0/A^2, given in (4.9.1). Thus the subcarrier CNR now becomes

$$\text{CNR}_s = \frac{\Delta^2/2}{2(N_0/A^2)B_{\text{sc}}}$$

$$= \frac{\Delta^2}{2}\left[\frac{A^2/2}{N_0 B_{\text{sc}}}\right] \tag{5.3.20}$$

The result is similar to the FM|FM case in (5.3.16), except the phase index is involved and the subcarrier frequency ω_s does not appear. Hence FM|PM systems are not directly influenced by the choice of subcarrier frequency. If CNR_s is above threshold, the source signal is finally demodulated with SNR of

$$SNR_d = 3\beta_s^2(\beta_s + 1)CNR_s \qquad (5.3.21)$$

Again the improvement caused by two stages of angle demodulation is apparent and the usual trade-off with RF bandwidth is available. The system is principally constrained by the fact that enough carrier power $(A^2/2)$ must be provided to overcome the noise in the total IF carrier bandwidth in satisfying the PM theshold. When carrier power is limited this condition may not be easily satisfied, especially if wideband operation is used. Fortunately, however, PM systems have a secondary advantage that allows the subcarrier information to be recovered with much less carrier power. This is achieved by taking advantage of the spectral distribution of the PM carrier. This we discuss in the next section.

5.4. PM Subcarrier Demodulation with Carrier Tracking

We showed in Section 2.3 in our discussion of PM power spectra that if the carrier is modulated with a subcarrier (modulated sinusoid), the subcarrier spectrum appears at each sideband in the PM spectrum. This suggests that in a system employing phase modulation of a subcarrier, the subcarrier can be recovered by detecting only one of the sidebands rather than attempting to demodulate the entire PM signal. Once the subcarrier sideband is properly extracted, the source signal can be recovered from the subcarrier by standard subcarrier demodulation. To examine such a system, let the baseband be written as

$$m(t) = \sin(\omega_s t + \theta_s(t)) \qquad (5.4.1)$$

corresponding to an arbitrary angle modulated (FM or PM) subcarrier. This baseband is again used to form the receiver PM carrier as

$$c(t) = A \sin[\omega_c t + \Delta m(t) + \psi(t)] \qquad (5.4.2)$$

where we have now purposely included an extraneous carrier phase variation $\psi(t)$ to account for phase shifts, Doppler, oscillator instabilities, and so on. If we proceed simply to PM demodulate (5.4.2) at the receiver, the equations of Section 5.3 must be used, with the necessity of satisfying the threshold condition over the entire carrier modulation bandwidth.

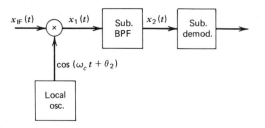

Figure 5.5. PM carrier mixing.

Consider instead the use of a carrier mixer as shown in Figure 5.5, operating on the carrier waveform at either the RF or IF stage. The carrier signal is mixed with a local oscillator signal, $\cos(\omega_c t + \theta_2)$ to form the mixer output signal

$$x_1(t) = K_m A \sin[\Delta m(t) + \psi(t) - \theta_2(t)] + n_m(t) \qquad (5.4.3)$$

where $n_m(t)$ is the multiplier output noise, having baseband spectrum $K_m^2 \tilde{S}_n(\omega)$. If we now introduce the error process

$$\theta_e(t) = \psi(t) - \theta_2(t) \qquad (5.4.4)$$

then (5.4.3) can be expanded as

$$x_1(t) = K_m A \sin[\Delta m(t) + \theta_e(t)] + n_m(t)$$

$$= K_m A \Big\{ (\cos \theta_e) 2 \sum_{k \text{ odd}}^{\infty} J_k(\Delta) \sin[k(\omega_s t + \theta_s(t))]$$

$$+ (\sin \theta_e) \sum_{k \text{ even}}^{\infty} J_k(\Delta) \cos[k(\omega_s t + \theta_s(t))] \Big\} + n_m(t) \qquad (5.4.5)$$

The subcarrier bandpass filter, tuned to ω_s, now passes only the harmonic at $k = 1$, yielding the subcarrier signal

$$x_2(t) = 2K_m A J_1(\Delta) \cos \theta_e \{\sin(\omega_s t + \theta_s(t))\} + n_2(t) \qquad (5.4.6)$$

where $n_2(t)$ is the multiplier noise passing through the subcarrier filter. The curly brackets represent the undistorted modulated subcarrier in (5.4.1). Thus the signal portion of the subcarrier signal appears with the desired subcarrier multiplied by the cosine of the instantaneous error function $\theta_e(t)$. This is now similar to our earlier result with AM coherent demodulation. If $\theta_e(t) = 0$ [i.e., if the local oscillator was phase coherent with the phase variations of $\psi(t)$], then an undistorted subcarrier appears in the subcarrier demodulation channel. Otherwise, the phase error $\theta_e(t)$ modifies the recovered subcarrier. If in fact $\theta_e(t) = 0$, we can write the subcarrier CNR at the subcarrier demodulator input as

$$CNR_s = \frac{2J_1^2(\Delta)A^2}{2N_0B_{sc}}$$

$$= 2J_1^2(\Delta)\left(\frac{A^2/2}{N_0B_{sc}}\right) \tag{5.4.7}$$

Thus the subcarrier can be recovered, with the CNR in (5.4.7), using the mixer system of Figure 5.5, provided that the local oscillator phase is close to the extraneous phase shift variations of the received carrier. This immediately suggests employment of a phase lock tracking technique to track these carrier variations. Since the carrier component of $c(t)$ in (5.4.2) contains the same phase angle $\psi(t)$, [i.e., the $k = 0$ term in (5.4.5)], this trackirg can be obtained by a phase lock loop tracking this component alone. The overall system would then appear as in Figure 5.6. The frequency components in the vicinity of zero frequency are filtered by the loop filter $F(\omega)$ and fed back to control the oscillator phase. The feedback loop, in conjunction with the mixer, now forms a standard phase lock loop, shown dotted in Figure 5.6. The harmonics of the subcarrier frequency should not be accepted by the loop filter, and the loop tracks only the $k = 0$ term in (5.4.5). If we write out this term separately, x_1 appears as

$$x_1(t) = K_m A J_0(\Delta) \sin(\psi - \theta_2) + K_m A (\cos \theta_e) J_1(\Delta) \sin(\omega_s t + \theta_s(t))$$

$$+ [\text{remaining terms for } k \geq 2] + n_m(t) \tag{5.4.8}$$

Comparison of (5.4.8) to our earlier equation (4.6.4) shows that the carrier tracking loop in Figure 5.6 can be represented by the low frequency equivalent system shown in Figure 5.7. Note that the amplitude A is now replaced by the new amplitude parameter

$$A_c = A J_0(\Delta) \tag{5.4.9}$$

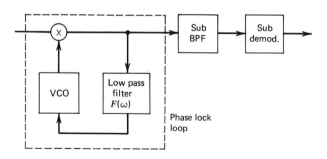

Figure 5.6. Carrier tracking PM subcarrier subsystem.

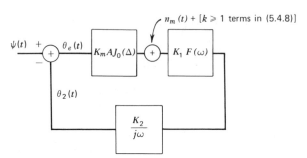

Figure 5.7. Equivalent baseband model of the carrier tracking PM subsystem.

and the input to the loop is the extraneous phase to be tracked (not the entire modulation phase variation as before). This is due to the fact that the loop is tracking only the carrier component at the RF frequency, and not the complete modulation. Note that the subcarrier harmonics appears as an effective "noise" or "interference," as far as carrier tracking is concerned, and enters the loop with the mixer noise $n_m(t)$.

Proceeding analytically as we did earlier, we can write the loop error transform as

$$\Phi_e(\omega) = \Phi_\psi(\omega) - \Phi_2(\omega)$$

$$= [1 - H(\omega)]\Phi_\psi(\omega) - H(\omega)\left(\frac{N_m(\omega)}{K_m A_c} + \frac{\Phi_s(\omega)}{K_m A_c}\right) \quad (5.4.10)$$

where now

$$H(\omega) = \frac{A_c K (F(\omega)/j\omega)}{1 + A_c K (F(\omega)/j\omega)} \quad (5.4.11)$$

Here $H(\omega)$ is the loop gain function of the carrier tracking loop. Note that the loop gain function filters the mixer noise and the subcarrier harmonic spectrum, $\Phi_s(\omega)$. Thus the modulated subcarrier will not affect carrier tracking if the subcarrier is located outside the loop function bandwidth. The loop, however, must be wide enough to track the phase variations $\psi(t)$. This filtering situation is diagrammed in Figure 5.8. We see that for best carrier tracking the subcarrier should be at a frequency much higher than the highest frequency of $\psi(t)$. In this way a loop can be designed to track $\psi(t)$ without tracking the subcarrier.

If we again denote

$$\tilde{B}_L = \frac{1}{2\pi} \int_0^\infty |H(\omega)|^2 |\tilde{H}_{\mathrm{IF}}(\omega)|^2 \, d\omega \quad (5.4.12)$$

as the loop noise bandwidth, then the loop noise $n_m(t)$, contributes a mean

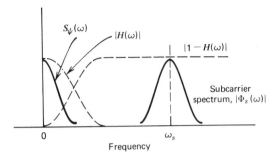

Figure 5.8. Carrier tracking loop filtering.

square error of

$$\sigma_n^2 = \frac{2N_0\tilde{B}_L}{A_c^2} = \frac{1}{\text{CNR}_L} \qquad (5.4.13)$$

where now

$$\text{CNR}_L = J_0^2(\Delta)\left(\frac{A^2/2}{N_0\tilde{B}_L}\right) \qquad (5.4.14)$$

Note that when compared to the CNR_L in (4.7.9) the factor $J_0^2(\Delta)$ appears as a carrier power suppression factor. This suppression increases with phase index Δ, constraining the amount of phase deviation by which the subcarrier can be placed on the RF carrier. It should also be noted, however, that the subcarrier CNR in (5.4.7) increases with subcarrier deviation Δ, forcing a trade-off in design between CNR_s and the carrier component power for tracking. A plot of the specific functions $J_0^2(\Delta)$ and $2J_1^2(\Delta)$ is shown in Figure 5.9. Note the rapid decrease in carrier tracking power that will occur as the phase modulation index Δ is increased. For this reason, subcarrier PM systems of this type are generally designed with a relatively small index $(0.5 \le \Delta \le 1)$, and such systems are referred to as *low indexed* PM systems. This, of course, is in direct contrast with PM systems using standard phase demodulation in which large index (wideband) operation is preferable.

Ideally, zero tracking error $[\theta_e(t) = 0]$ is required in the tracking loop for (5.4.7) to be valid. In practice, however, this tracking error is not zero because of the noise and the inability to track $\psi(t)$ exactly. In this case (5.4.7) must be modified to take into account the cosine of the phase error. The CNR_L needed to maintain a small tracking error (so that the cosine term is close to one) with a sufficiently high probability can be determined by using equations similar to (4.7.17). The contribution to the phase error from $\psi(t)$ [first term in (5.4.10)] as well as from $n_m(t)$ must be properly accounted for. We investigate this in the next section.

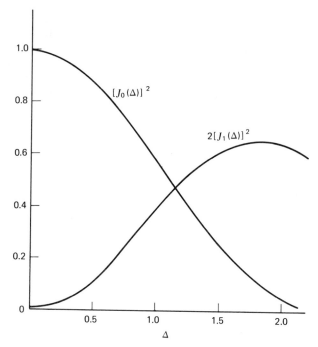

Figure 5.9. Plot of Bessel functions.

It is important to recognize the difference between the role of the PLL when used as a phase tracking demodulator in Figure 4.13 and when used as a carrier tracking loop in Figure 5.6. In the former the loop tracks the complete modulation, and its threshold CNR depends on the total carrier power and noise in the modulation bandwidth. When operating as a carrier tracking loop, it only tracks the extraneous phase shifts on the carrier component. The threshold CNR depends only on the power in the carrier component, but only combats the noise in the bandwidth of $\psi(t)$ and not the complete modulation bandwidth. Since $\psi(t)$ generally varies much more slowly than typical modulation processes, the carrier tracking bandwidth will be much narrower than the modulation tracking bandwidths. Hence carrier tracking can be maintained at an extremely low power level, compared to modulation tracking. In practice, typical carrier tracking loop bandwidths are in the range of 1–100 Hz.

5.5. The Design of Carrier Tracking Loops

In Section 4.8 we considered the design of a modulation tracking phase lock loop. Let us now examine the corresponding design of a carrier tracking PLL, as used in Figure 5.6. We assume the subcarrier frequency is far enough removed from the loop bandwidth so as not to affect tracking performance. The only effect on loop tracking is that due to noise and the inability to track $\psi(t)$ exactly. The loop error transform is given then by (5.4.10) as

$$\Phi_e(\omega) = [1 - H(\omega)]\Phi_\psi(\omega) - H(\omega)\left(\frac{N_m(\omega)}{K_m A_c}\right) \qquad (5.5.1)$$

We see this is identical in form to Equation (4.7.3), defining loop eror in a modulation tracking demodulator, except that the carrier amplitude A is replaced by the carrier component amplitude A_c in (5.4.9), and that the extraneous phase variation $\psi(t)$ is involved rather than the phase modulation. Nevertheless, the same design comments apply, and we again can consider (5.5.1) as a trade-off of signal tracking and noise rejection. Modeling $\psi(t)$ as a random process (corresponding, for example, to oscillator noise) allows us to proceed with design by again resorting to Wiener filter theory just as in Section 4.8.

In practice, the extraneous carrier phase variations of most concern are those due to offset carrier phase shifts, Doppler, and Doppler rates. These tend to be purely deterministic effects, and usually quite predictable (or at least boundable). Such deterministic phase effects contribute a deterministic function to the total loop error in (5.5.1), given by the inverse Fourier transform of the first term. Formal inversion of $[1 - H(\omega)]\Phi_\psi(\omega)$ produces the exact error time response due to $\psi(t)$. Such responses involve a transient behavior, followed by a steady state behavior. Since the steady state is reached in a relatively short time, an adequate indication of the loop response to $\psi(t)$ can be achieved by examining only the steady state value. The steady state value of this deterministic error time response (which occurs after loop transients have diminished) can be obtained by appealing to the final value theorem of transform theory (Section A.2)

$$\begin{bmatrix} \text{steady state values of the time} \\ \text{response of the loop error} \end{bmatrix} = \lim_{j\omega \to 0} j\omega[1 - H(\omega)]\Phi_\psi(\omega)$$

$$= \lim_{j\omega \to 0} \left[\frac{(j\omega)^2 \Phi_\psi(\omega)}{j\omega + A_c KF(\omega)}\right] \qquad (5.5.2)$$

The response caused by deterministic extraneous phase shifting therefore depends on the form of $\psi(t)$ and the type of loop filtering $F(\omega)$. For the important types of extraneous phase shifts mentioned, the resulting steady state error in (5.5.2) is listed in Table 5.1, for the various loop filter functions shown. The loop therefore tracks the stated extraneous effects with zero, constant, or infinite steady state loop errors. The zero error

Table 5.1. Steady State Tracking Errors

			Steady State Error	
$\psi(t)$	Cause	$\Phi_\psi(\omega)$	$F(\omega)=1$	$F(\omega)=\dfrac{j\omega+b}{j\omega}$
ψ_0	Constant phase offset	$\dfrac{\psi_0}{j\omega}$	0	0
$\omega_d t$	Constant Doppler shift	$\dfrac{\omega_d}{(j\omega)^2}$	$\dfrac{\omega_d}{A_cK}$	0
$\dfrac{\alpha t^2}{2}$	Increasing Doppler shift	$\dfrac{\alpha}{(j\omega)^3}$	∞	$\dfrac{\alpha}{A_cK}$

means the loop VCO will eventually lock onto the carrier phase shift. A constant error means the loop VCO will track with the correct frequency, but have a constant offset phase error. An infinite steady state error means that the loop cannot track the frequency shift and the input carrier will eventually "pull away" from the loop VCO phase with time, approaching an infinite phase error in the steady state (i.e., as $t \to \infty$). Note that the loop with the more complicated filter yields better phase tracking capabilities. The ability to track higher order polynomials of the extraneous phase can be related to the number of integrations within the loop (recall that an inherent phase integration is already provided by the loop oscillator). We see, for example, that we need a loop with a total of two integrations to track an increasingly Doppler shift with constant phase error. Since the VCO inherently provides one integration this added integration must be provided by the loop filter.

Consider specifically the case of a Doppler offset ω_d. This means the received carrier RF or IF frequency is offset from the loop VCO center frequency by ω_d rps. The results in Table 5.1 indicate that if a first order loop ($F(\omega)=1$) is used, the VCO frequency is driven to that of the carrier, but an offset phase error of ω_d/A_cK occurs. This phase error can be written in terms of the loop noise bandwidth in Table 4.1, since $B_L = A_cK/4$. Thus the Doppler tracking phase error is small if $\omega_d/4B_L \ll 1$, or if $B_L \gg \omega_d/4$ (i.e., if the loop noise bandwidth is large enough to encompass the Doppler offset). If an integrating filter is used, the carrier tracking VCO is driven to the exact Doppler frequency and phase of the carrier.

An immediate consequence of using higher order and wider bandwidth loops to achieve better Doppler tracking is, of course, that the error contribution from the noise [second term in (5.5.1)] increases. This again suggests a design trade-off, perhaps from a mean squared tracking error point of view. For example, we can determine the carrier loop gain function $H(\omega)$ that minimizes the mean squared tracking error caused by noise, while constraining the total integrated deterministic squared error caused by $\psi(t)$. The latter is the energy in the deterministic time function given by the integral of the squared magnitude of the first term in (5.5.1). By application of the calculus of variations we are then seeking the minimization of a term similar to (5.2.2), and solution follows as in (5.2.3) with proper substitution. The optimal loop function is then the realizable form of

$$H(\omega) = \frac{|\Phi_\psi(\omega)|^2}{|\Phi_\psi(\omega)|^2 + (N_0/A_c^2)} = 1 - \frac{(N_0/A_c^2)}{|\Phi_\psi(\omega)|^2 + (N_0/A_c^2)} \qquad (5.5.3)$$

where $\Phi_\psi(\omega)$ is the transform of $\psi(t)$. For the case of a constant phase offset, $\Phi_\psi(\omega) = \psi_0/j\omega$, the loop function in (5.5.3) becomes

$$H(\omega) = \frac{\psi_0^2}{\psi_0^2 + (N_0/A_c^2)\omega^2} \qquad (5.5.4)$$

The realizable function (upper half ω plane poles) is then

$$H(\omega) = \frac{\psi_0 A_c/\sqrt{N_0}}{(\psi_0 A_c/\sqrt{N_0}) + j\omega} \qquad (5.5.5)$$

The optimal loop for tracking carrier phase offsets in white noise is therefore a first order loop. Note again the dependence of the optimal loop parameters on the specific values of the phase offset and the noise.

For Doppler tracking, $\Phi_\psi(\omega) = \omega_d/(j\omega)^2$, and the optimal realizable loop function follows as (Problem 5.12)

$$H(\omega) = \frac{1 + (0.75/B_L)j\omega}{1 + (0.75/B_L)j\omega - (\omega^2/(32B_L^2/9))} \qquad (5.5.6)$$

where

$$B_L = \frac{0.75\sqrt{A_c}\,\omega_d}{N_0^{1/4}} \qquad (5.5.7)$$

The optimal loop filter function is then

$$F(\omega) = \frac{j\omega + (0.75/B_L)}{j\omega + (9N_0^{1/2}/8A_cB_L^2)} \qquad (5.5.8)$$

The optimal loop for tracking a Doppler offset is a second order loop with a single zero-pole filter function. This result is interesting in that the optimal loop does not include the second integration required from Table 5.1 to drive the Doppler phase error to zero, implying that noise effects here can be more significant than steady state tracking errors. (In the limit as the noise level goes to zero, we note that the filter does indeed approach an integrating filter.) We also see that the optimal design parameters, such as the filter zero and pole locations, depend on the actual Doppler offset value. This requires the system designer to know these values, or at least restrict the range over which they may occur. For example, in tracking Doppler the maximum expected Doppler shift must be determined, since the optimal loop noise bandwidth B_L depends directly on this frequency shift. Additional studies of carrier tracking loop design by these procedures can be found in References 1–4. The topic is considered again in Chapter 9.

As stated earlier, a bandpass limiter is often inserted in the IF stage prior to the carrier tracking loop in order to stabilize the IF carrier amplitude A. By using a hard limiter adjusted to the voltage level V_L, as in Figure 3.21, the IF carrier is limited to a fixed amplitude $A = 4V_L/\pi$. In the noiseless case this amplitude becomes the fixed amplitude parameter of the tracking loop, and loop design can be based on this fixed value. When noise is present at the limiter input, however, only the carrier portion of the limiter output is available for phase tracking. In this case it is convenient to write the limiter output as in (3.9.18), where α_s and α_n account for limiter suppression. The equivalent loop of Figure 5.7 then has effective amplitude $\alpha_s A J_0(\Delta)$ within the loop, and an effective loop input noise power, in the limiter output IF bandwidth B_{IF}, of $\alpha_n^2 R_n(0)$. Treating the limiter output noise as being spectrally flat over the noise bandwidth B_{IF}, this loop noise will have equivalent spectral level $\alpha_n^2 R_n(0)/B_{IF}$. The CNR within the loop noise bandwidth in (5.4.14) is now replaced by

$$\mathrm{CNR}_L = \frac{\alpha_s^2 A^2 J_0^2(\Delta)/2}{(\alpha_n^2 R_n(0)/B_{IF})B_L}$$

$$= \left(\frac{\alpha_s}{\alpha_n}\right)^2\left(\frac{A^2/2}{R_n(0)}\right)J_0^2(\Delta)\left(\frac{B_{IF}}{B_L}\right) \qquad (5.5.9)$$

Using (3.9.19), we now rewrite this as

$$\mathrm{CNR}_L = \mathrm{CNR}_{BL}\left(\frac{B_{IF}}{B_L}\right)J_0^2(\Delta) \qquad (5.5.10)$$

Hence the loop CNR_L is the limiter output CNR in the limiter IF bandwidth multiplied by the ratio of the bandwidths and suppression factor.

This shows why a relatively small limiter output CNR can still be used for adequate carrier tracking if the bandwidth ratio is high enough.

5.6. Selection of Phase Modulation Indices for FM|PM Carrier Tracking Systems

We have seen that an FM|PM system that employs carrier tracking phase demodulation of the RF carrier is basically governed by two key design equations. The first, given by (5.4.7), indicates the subcarrier CNR that will be generated at the subcarrier demodulation input if proper carrier tracking is performed. The second, given by (5.4.14), is the loop CNR_L required to guarantee a locked tracking loop. Both of these equations depend on both the received IF carrier power $A^2/2$ and the carrier phase index Δ, which divides the power among the carrier and subcarrier components. This implies that a trade-off between these parameters must be made. Since carrier power is the most critical, the usual design technique is to utilize phase indices that allow minimal power reception while satisfying the necessary constraints. Since subcarrier demodulation and simultaneous carrier tracking require thresholding conditions, the design problem is the simultaneous satisfaction of the two threshold conditions:

$$\text{CNR}_L = J_0^2(\Delta)\left(\frac{A^2/2}{N_0\tilde{B}_L}\right) \geq Y_c \qquad (5.6.1a)$$

$$\text{CNR}_s = 2J_1^2(\Delta)\left(\frac{A^2/2}{N_0 B_{sc}}\right) \geq Y_s \qquad (5.6.1b)$$

where Y_c and Y_s represent the tracking and subcarrier demodulation thresholds, respectively. The tracking threshold depends on the type of loop and expected steady state errors and is selected to achieve a prescribed mean squared loop tracking error. This can be obtained from a desired probability of keeping the loop in lock (keeping the phase in the linear range). The subcarrier threshold Y_s is that value of CNR_s needed to satisfy the threshold condition of the subcarrier demodulation. Equation (5.6.1) represents a set of simultaneous inequalities in the two unknowns, Δ and $P_c \triangleq A^2/2$. We wish to select these so as to satisfy (5.6.1) with the smallest value of P_c, given the system thresholds, bandwidths, and noise levels.

Suppose we first solve (5.6.1) as equalities, obtaining the solution pair,

$$\frac{2J_1^2(\Delta)}{J_0^2(\Delta)} = \frac{Y_s B_{sc}}{Y_L \tilde{B}_L} \qquad (5.6.2a)$$

$$P_c = \frac{Y_c N_0 \tilde{B}_L}{J_0^2(\Delta)} \qquad (5.6.2b)$$

Equation (5.6.2a) can be solved graphically for Δ by determining the value of Δ where the function on the left equals the value on the right. A plot of the function $2J_1^2(\Delta)/J_0^2(\Delta)$ is shown in Figure 5.10 as a function of the parameter Δ. By entering the ordinate at the value on the right of (5.6.2a), we can read off the required value of Δ from the abscissa. Evaluation of (5.6.2b) at this Δ produces the necessary P_c. When this procedure is carried out as a function of the parameter $(Y_s B_{sc}/Y_c B_L)$ we generate the curve labeled I in Figure 5.11, indicating the resulting P_c needed. As we move along the curve, Δ is changing values, given by the solution to (5.6.2a). The curve shows a rapid increase in P_c beyond a certain point, because as Δ increases toward the value 2.4 [first zero of $J_0(\Delta)$], the rate of falloff of $J_0(\Delta)$ increases sharply, and P_c must be correspondingly increased to maintain the condition of (5.6.2b). Physically, increasing Δ places more power in the higher order Bessel function terms, so that a smaller percentage of the total IF carrier power is available for the carrier component. The total power must therefore be increased to compensate.

Solving (5.6.1) with equalities is not necessarily the best solution, however. An alternative solution is simply to set Δ equal to a specific value,

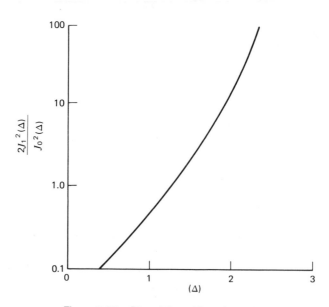

Figure 5.10. Plot of Bessel function ratio.

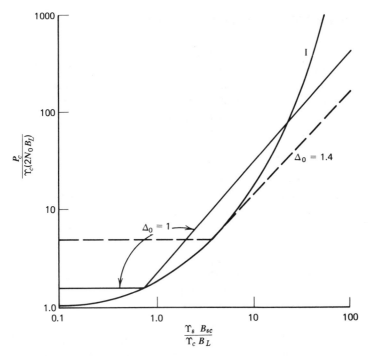

Figure 5.11. Required power in carrier tracking demodulation.

Δ_0 ($\Delta_0 \leq 2.4$), and instead use

$$P_c = \max \left[\frac{Y_s B_{sc} N_0}{2J_1^2(\Delta_0)}, \frac{Y_c \tilde{B}_L N_0}{J_0^2(\Delta_0)} \right] \qquad (5.6.3)$$

That is, we simply provide enough IF carrier power to satisfy both threshold inequalities at that Δ_0. For the index value $\Delta_0 = 1$ and 1.4 we generate the curves in Figure 5.11, each plotted as a function of the same normalized bandwidth B_{sc}. We see that for $B_{sc} \leq 2Y_c J_1^2(\Delta_0)/Y_s J_0^2(\Delta_0)$ the required power by this method exceeds that obtained by the equality solution (curve I). This implies that more power is being used in the region than is necessary, and more efficient use could be obtained by a better selection of Δ for proper distribution between tracking and subcarrier power levels. Beyond this point, however, minimal power occurs if the index is held fixed and the IF carrier power P_c is linearly increased as B_{sc} increases. As Figure 5.11 indicates, minimal power is required if Δ is varied up until a value of $\Delta \cong 1.4$ is reached and then held constant beyond this point.

5.7. Source Multiplexing (FDM Baseband Systems)

When more than one source is involved, and the information of all is to be sent simultaneously, the sources must be multiplexed (combined into a single baseband signal). This multiplexing must be such that the individual source waveforms can be separated at the receiver. A common way to achieve this is to modulate each source onto a separate subcarrier, each located at a different region of the frequency spectrum. The subcarriers can then be summed to form the multiplexed baseband signal, as shown in Figure 5.12*a*. A system of this type is referred to as a *frequency division multiplexed* (FDM) baseband system, since the source waveforms are effectively separated in frequency by the subcarriers. The modulation onto the subcarriers and the RF carrier can be done by any of the standard

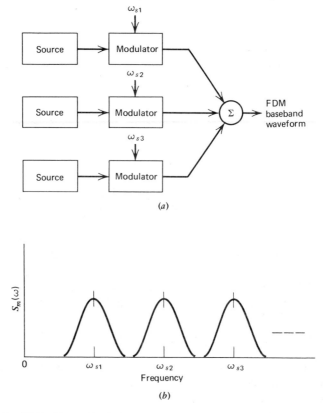

Figure 5.12. FDM subsystem. (*a*) Multiplexing block diagram, (*b*) multiplexed spectrum.

methods. Multiplexed systems of this type are denoted, for example, as FDM|AM|FM, indicating FDM multiplexing is used, with AM on the subcarriers, and FM of the multiplexed baseband on the RF carrier.

A typical FDM baseband spectrum is given in Figure 5.12b, showing the multiplexed sources side by side, each at their respective subcarrier frequencies. The form of the spectrum about each subcarrier frequency depends on the modulation used. Often guard spacing is inserted between the spectra to obtain sufficient spectral separation. In some cases one source may be placed at zero frequency (no subcarrier used), and the remaining sources subcarrier modulated to a specific frequency range. In other cases all sources are placed on subcarriers. Theoretically, the lower end of the lower subcarrier channel can begin at zero frequency, but in practice it is usually located away from zero to avoid low frequency interference.

FDM baseband formats are common in our everyday commercial stereo radio, television, and telephone systems. Each television channel spectrum of Figure 1.10 is basically a form of FDM multiplexing of the video and audio information, with the voice frequency modulated onto the subcarrier at 5.25 MHz and placed above the video spectrum. Figure 5.13a shows the format for a commercial stereo radio transmission channel. The information from the left and right microphones is summed to form a source signal for direct low pass transmission. A second source signal, composed of the difference of the two microphone signals, is amplitude modulated onto a subcarrier of 38 kHz. The resulting baseband, which occupies a spectrum of about 53 kHz, is then transmitted via an FM carrier as a single stereo channel. FDM formats are also used in satellite baseband communications. Figure 5.13b shows a typical satellite repeater FDM telephone channel, containing 12 multiplexed voice channels. The baseband is part of a FDM|SSB-AM|FM system, in which single sideband AM is used to modulate each 4 kHz voice source onto the subcarrier frequencies shown. The resultant baseband spectrum is 48 kHz wide and is designated as a standardized Telephone Group [5].

At the receiver, following RF demodulation, the general multiplexed spectrum of Figure 5.12a is recovered. Source separation is achieved by merely filtering off the individual subcarrier channels and subcarrier demodulating (Figure 5.14). Adequate guard spacing between channels eases this subcarrier separation by the filters. If the information of several sources is desired, a parallel filter-demodulator channel must be available for each subcarrier. For example, in the stereo system of Figure 5.13a a stereo receiver demodulates both the 38 kHz subcarrier and the low pass spectra, which produces microphone sum and difference waveforms. These can then be separated into left and right channels. A receiver without

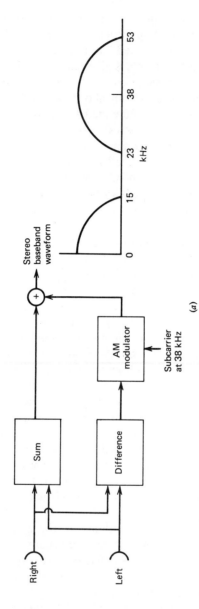

Figure 5.13. Examples of FDM baseband formats. (*a*) (*above*) Commercial stereo, (*b*) (*on page 215*) satellite telephone channels.

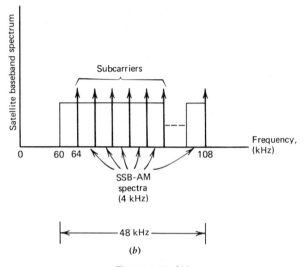

Figure 5.13. (*b*)

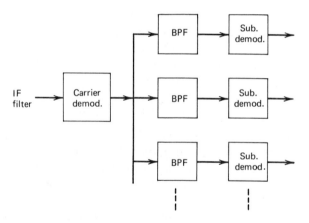

Figure 5.14. FDM receiver demultiplexing.

stereo reception capability (i.e., without a subcarrier demodulator) receives only the sum signal for monoaural performance. Thus a receiver in an FDM system demultiplexes the baseband waveform by simply bandpass filtering the desired subcarrier channel. FDM systems have this prime advantage of extreme simplicity in both multiplexing at the transmitter end and source separation at the receiving end. We emphasize that this frequency multiplexing is accomplished at the baseband level, and only a single RF carrier is being used.

FDM|AM|FM Systems. Let us examine a FDM baseband system in which AM is used on the subcarriers and FM on the RF carrier. The analysis parallels that of Section 5.3, with the added extension to more than one source. Consider N separate source waveforms $d_i(t)$, having power P_{di} and bandwidth B_{di}, respectively. The AM subcarrier multiplexing produces the baseband signal

$$m(t) = \sum_{i=1}^{N} C_i[1 + d_i(t)] \sin (\omega_{si}t + \psi_i) \qquad (5.7.1)$$

where ω_{si} are the subcarrier frequencies and ψ_i their phase angles. The baseband occupies an upper frequency bandwidth of

$$B_m = \frac{\omega_{sN}}{2\pi} + B_{dN} \qquad (5.7.2)$$

and a power level of

$$P_m = \sum_{i=1}^{N} \frac{C_i^2}{2} (1 + P_{di}) \qquad (5.7.3)$$

The RF carrier is formed again as in (5.3.4), so that C_i in (5.7.1) represents the actual frequency deviation of the ith subcarrier signal. Using an effective FM index β defined in (2.2.17), the wideband RF carrier therefore occupies a bandwidth of approximately

$$B_{RF} = 2\left[\frac{\sqrt{P_m}}{2\pi B_m} + 1\right]B_m \approx 2\left(\frac{\sqrt{P_m}}{2\pi}\right) \qquad (5.7.4)$$

The fact that B_{RF} depends on the baseband power level P_m in (5.7.3) is an important point here. Since specification of RF bandwidth is often a critical part of system design, accurate evaluation of P_m is required. Unfortunately, this may not always be an easy task. First of all, direct substitution into (5.7.3) requires knowledge of the values of all P_{di}. Often these tend to be dependent on the type of usage of the source. For example, in telephone channels P_{di} is the power content of a speaker's voice signal, which varies considerably from one person to another. A second point is that even if typical values of P_{di} can be projected, use of (5.7.3) generally leads to an overdesigned B_{RF} since, in most instances, not all sources may be active. This necessitates establishing a source activity, or *loading factor*, for the particular system involved. The problem is further complicated by the fact that activity behavior may be dependent on time of day, location of the source, and so on. In telephone analysis there has been considerable effort to collect empirical data that allow accurate evaluation of typical multiplexed telephone modulation power levels [5, 6].

After RF demodulation the baseband spectrum is recovered in the presence of frequency demodulation noise. For a specific subcarrier channel the analysis of Section 5.3 can be directly applied. Hence the ith subcarrier filter generates a CNR given by

$$\text{CNR}_{si} = \left(\frac{C_i}{\omega_{si}}\right)^2 \left[\frac{A^2/2}{N_0 2 B_{di}}\right] \qquad (5.7.5)$$

assuming $P_{di} = 1$. Each CNR therefore depends on the subcarrier frequency, deviation, and bandwidth. If we assume that all sources have the same bandwidth, $B_{di} = B_d$ and require the same CNR, then (5.7.5) indicates that we need

$$C_i = C\omega_{si} \qquad (5.7.6)$$

where C is a proportionality constant, in order to make CNR independent of i. Equation (5.7.6) implies that the frequency deviations of subcarriers on the RF carrier should be proportional to their subcarrier deviation. This operation of selecting transmitter frequency deviations in this way is another form of transmitter preemphasis. In effect, we purposely emphasize the higher subcarriers at the transmitter since they will be at a higher noise level at the receiver. Equation (5.7.6) corresponds to linear preemphasis, and the parameter C determines the amount of preemphasis applied. The resulting subcarrier CNR in (5.7.5) is then

$$\text{CNR}_{si} = C^2 \left[\frac{A^2/2}{N_0 2 B_d}\right] \qquad (5.7.7)$$

for all channels. Subsequent AM demodulation in each channel then produces a demodulated SNR of the same value. Thus performance is a direct function of the preemphasis parameter C. However, we also have

$$B_{RF} \cong \frac{2}{2\pi}\left[\sum_{i=1}^{N} C_i^2\right]^{1/2} = \frac{C}{\pi}\left[\sum_{i=1}^{N} \omega_{si}^2\right]^{1/2} \qquad (5.7.8)$$

and the RF bandwidth increases with C also. Equations (5.7.7) and (5.7.8) are the key design equations for FDM|AM|FM systems. The latter relates C to the allowable RF bandwidth for a given set of subcarrier frequencies, and the former determines the recovered channel SNR. For smallest RF bandwidth and largest possible SNR, the subcarrier frequency set should be selected as low as possible, which implies placing the subcarrier channels as close as possible to each other, starting at the lowest possible frequency. Unfortunately, practical filter requirements generally dictate guard spacing between channels, and low frequency interference prevents use of extremely low subcarrier frequencies.

FDM|FM|FM Systems. In FDM|FM|FM multiplexed systems the source waveforms are frequency modulated onto the subcarriers prior to FM on the RF carrier. The baseband signal is now

$$m(t) = \sum_{i=1}^{N} C_i \sin\left(\omega_{si}t + \Delta_{si}\!\int d_i(t)\, dt\right) \tag{5.7.9}$$

where C_i is again the carrier deviation caused by the subcarrier and Δ_{si} is the subcarrier frequency deviation caused by the source. The power of $m(t)$ is then

$$P_m = \sum_{i=1}^{N} \frac{C_i^2}{2} \tag{5.7.10}$$

and the resultant RF, subcarrier, and basebandwidths follow as

$$B_{\mathrm{RF}} = 2\left[\frac{\sqrt{P_m}}{2\pi B_m} + 1\right] B_m \approx \frac{1}{\pi}\left[\sum_{i=1}^{N} \frac{C_i^2}{2}\right]^{1/2} \tag{5.7.11a}$$

$$B_m = \frac{\omega_{sN}}{2\pi} + \frac{B_{sN}}{2} \tag{5.7.11b}$$

$$B_{si} = 2\left[\frac{\Delta_{si}}{2\pi B_{di}} + 1\right] B_{di} \tag{5.7.11c}$$

After RF transmission and demodulation, the ith subcarrier is filtered with the CNR obtained from (5.7.5), yielding

$$\mathrm{CNR}_{si} = \left(\frac{C_i}{\omega_{si}}\right)^2\left[\frac{A^2/2}{N_0 B_{si}}\right] \tag{5.7.12}$$

So long as CNR_{si} is above the threshold, the ith FM subcarrier is demodulated with SNR,

$$\mathrm{SNR}_{di} = 3\beta_i^2\left(\frac{B_{si}}{B_{di}}\right)\mathrm{CNR}_{si}$$

$$= 3\beta_i^2\left(\frac{C_i}{\omega_{si}}\right)^2\left[\frac{A^2/2}{N_0 B_{di}}\right] \tag{5.7.13}$$

where $\beta_i = \Delta_{si}/2\pi B_{di}$ is the modulation index of the ith subcarrier. If we assume that all subcarrier channels use the same index and that each channel is to have the same demodulated SNR, then theoretically we can proceed as with AM subcarriers by linearly preemphasizing with $C_i = C\omega_{si}$. Under an equal subcarrier bandwidth condition, all channels would then operate with the same SNR. From a practical point of view, however, the assumption that all subcarrier channels will have the same bandwidth B_{si}

may be difficult to satisfy. This is due to the deviation restriction on the Δ_{si} of the individual subcarrier oscillators. Clearly, if the subcarrier deviations are all equal in (5.7.11c), either the lower subcarriers will be overdeviated or the upper subcarriers will be underdeviated. For this reason, FDM|FM|FM are generally designed with larger deviations and bandwidths for the higher subcarrier channels to utilize always the full allowable frequency deviation. This means that B_{si} in (5.7.12) is also proportional to ω_{si} and that (5.7.12) will produce the same CNR in each channel only if

$$C_i = C\omega_{si}^{3/2} \qquad (5.7.14)$$

This indicates a more rapid nonlinear preemphasis for transmitter subcarriers with FM subcarriers than with AM. Subcarrier channels selected to have bandwidths proportional to their subcarrier frequency, so as to take full advantage of the deviation limitation, are called *proportional* subcarrier bands, as opposed to the constant subcarrier bands of the AM case. On the other hand, if all channels are to have the same modulation index β_i, the source bandwidths B_{di} that can be accommodated with each Δ_{si} must also be proportional to ω_{si}. This means that small bandwidth sources should be placed on the lower subcarriers and larger bandwidth sources should be assigned to the corresponding higher subcarriers, for most efficient design.

5.8. FDM|FM|PM with Carrier Tracking Demodulation

A carrier tracking demodulator for an FDM|FM|PM system is shown in Figure 5.15. Again the objective of the tracking loop is to track out the extraneous phase modulations of the IF carrier while mixing the IF spectrum to zero frequency. The individual subcarrier channels filter off the desired subcarrier spectrum in the shifted IF spectrum, and proceed to subcarrier demodulate for source recovery. If the baseband $m(t)$ has the form in (5.7.9), the IF spectrum of $c(t)$ can be expanded as

$$c(t) = A \sum_{k_N = -\infty}^{\infty} \cdots \sum_{k_1 = -\infty}^{\infty} \left\{ \prod_{i=1}^{N} J_{k_i}(\Delta_i) \right\} \cdot$$

$$\cos \left[\omega_{\mathrm{IF}} t + \sum_{i=1}^{K} k_i \left(\omega_{si} t + \Delta_{si} \int d(t)\, dt \right) + \psi \right] \qquad (5.8.1)$$

where Δ_i is the phase deviation of the ith subcarrier, corresponding to our discussion in Section 2.4. The carrier tracking loop phase locks to the carrier component and its mixer output shifts the entire spectrum to zero frequency, while removing the extraneous phase variation $\psi(t)$ (assuming

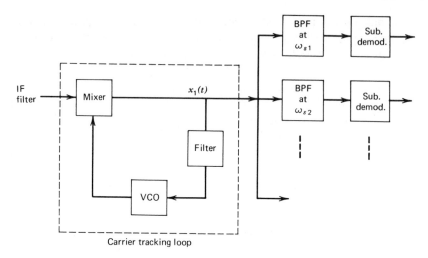

Figure 5.15. FDM demultiplexing with PM carrier tracking.

perfect tracking). The power in the carrier component used for loop tracking is then the power of the frequency component at ω_{IF}, corresponding to the $k_1 = k_2 = k_3 \ldots k_N = 0$ term. Hence

$$P_c = \frac{A^2}{2}\left[\prod_{i=1}^{N} J_0^2(\Delta_i)\right] \qquad (5.8.2)$$

The tracking loop mixer output then has the shifted version of the spectrum of (5.8.1). Hence

$$x_1(t) = A \sum_{k_N=-\infty}^{\infty} \cdots \sum_{k_1=-\infty}^{\infty} \left\{\prod_{i=1}^{N} J_{k_i}(\Delta_i)\right\} \cdot$$

$$\sin\left[\sum_{i=1}^{N} k_i\left(\omega_i t + \Delta_{si} \int d(t)\right)\right] + n_m(t) \qquad (5.8.3)$$

which is the multiplexed equivalent of our earlier result in (5.4.5) when the tracking error is zero. The harmonic extracted from this shifted IF signal by the ith subcarrier filter corresponds to the spectral component at the subcarrier frequency ω_{si}. This has an amplitude given by the term with $k_i = \pm 1$ and $k_j = 0$, $j \neq i$. The power in this harmonic is therefore

$$P_{si} = \frac{A^2}{2}\left[2J^2(\Delta_i)\right]\prod_{\substack{j=1 \\ j \neq i}}^{N} J_0^2(\Delta_j) \qquad (5.8.4)$$

The noise power in the ith subcarrier channel is given by the power of the mixer noise $n_m(t)$ in (5.8.3) passing through the ith subcarrier filter, and therefore has value $N_0 B_{si}$, where B_{si} is the filter noise bandwidth. The resulting CNR of the ith demodulator input is then

$$
\text{CNR}_{si} = \frac{P_{si}}{N_0 B_{si}}
$$

$$
= \left[2J_1^2(\Delta_i) \prod_{\substack{j=1 \\ j \neq i}}^{N} J_0^2(\Delta_j) \right] \left[\frac{A^2/2}{N_0 B_{si}} \right] \qquad (5.8.5)
$$

The second bracket is the total available IF carrier to noise ratio in the subcarrier bandwidth. The first bracket therefore represents the subcarrier suppression term that indicates the portion of the IF CNR available for the ith channel. Again we have the multiplexed equivalent of our earlier result of (5.4.7). Note that the CNR of each subcarrier is affected by the RF phase indices Δ_i of all other subcarriers. That is, as we alter the subcarrier phase indices at the transmitter during RF modulation, we redistribute the carrier power over all the channels, according to (5.8.5). We see that the contribution to this suppression term resulting from another subcarrier actually behaves as $J_0^2(\Delta_j)$, showing that CNR is reduced in all other channels if a particular subcarrier is given an increased index. We also see that since the mixer noise is basically a flat noise spectrum, CNR does not depend on the subcarrier frequencies, so that subcarrier preemphasis is not necessary with PM carrier tracking.

As long as (5.8.5) satisfies the FM subcarrier demodulator threshold, the source waveform will be demodulated with the associated subcarrier FM improvement. For the ith channel this becomes

$$
\text{SNR}_{di} = 3\beta_i^2(\beta_i + 1)\text{CNR}_{si} \qquad (5.8.6)
$$

just as in (5.7.13), where β_i is the FM index of the source on the subcarrier. Being able to use large indices β_i (to within that allowable by the subcarrier deviation constraint) to obtain an improved source SNR is an intrinsic advantage of FM\PM systems.

In typical FDM design, threshold requirements are placed on the CNR in the tracking loop and on the subcarrier CNR of each channel. For fixed bandwidths and noise levels this corresponds to a set of threshold conditions on the power terms in (5.8.2) and (5.8.4). Hence we now have the set of design requirements

$$
P_c \geq \Upsilon_c(2N_0 \tilde{B}_L) \qquad (5.8.7a)
$$

$$
P_{sj} \geq \Upsilon_j(N_0 B_{sj}), \qquad j = 1, 2, \ldots, N \qquad (5.8.7b)
$$

where Υ is the associated threshold values. Equations (5.8.7) represent a set of $N+1$ simultaneous inequalities that must be satisfied for system design. These equations contain the N phase indices (Δ_i) and the IF carrier power $P_c = A^2/2$ that are to be specified, similar to our procedure in Section 5.5. To solve, we again divide (5.8.7b) by P_c, and rewrite the set as

$$\frac{P_{sj}}{P_c} = \frac{2J_1^2(\Delta_j)}{J_0^2(\Delta_j)} \geq \frac{Y_j B_{sj}}{Y_c \tilde{B}_L}, \qquad j = 1, 2, \ldots, N \qquad (5.8.8a)$$

$$P_c = \frac{Y_c N_0 2 B_L}{\prod_{i=1}^{N} J_0^2(\Delta_i)} \qquad (5.8.8b)$$

The jth term in (5.8.8a) depends only on Δ_j and is identical to the form of (5.6.2a). Figure 5.10 can be used successfully to determine each required Δ_j, and carrier power P_c can be determined from (5.8.8b). A simplified example of this computation is shown in Figure 5.16, for the case of equal subcarrier bandwidths and thresholds, with $N = 2$ and 3. The earlier result in Figure 5.11 is superimposed for comparison. Again we see the rapid increase in IF carrier power P_c as subcarrier bandwidth is increased, with an increased power required as the multiplexing load (N) increases. This is

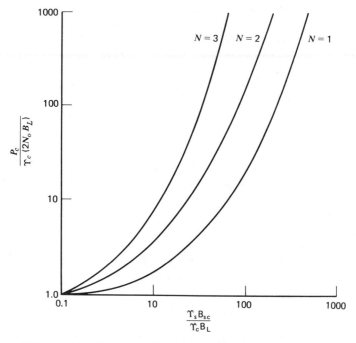

Figure 5.16. Required power in FDM carrier tracking demodulation (N = number of sources, assumes $B_{s1} = B_{s2} = B_{s3} = B_{sc}$ and $Y_j = Y_s$).

due to the added suppression produced in all power values as more sub-carriers are multiplexed.

In concluding our discussion of multiplexing in this chapter we should point out that although we have concentrated on frequency multiplexing at the baseband, the discussion of FDM can be extended to the RF carrier level as well. Individual RF carriers, each with its own baseband and modulation format, can be placed side by side to form an FDM RF spectrum. Such schemes are particularly important in satellite communications where multiple RF carriers are to be relayed simultaneously by a common satellite transponder. Each RF carrier is assigned its own RF band for accessing through the satellite. Such multiple carrier RF spectra are referred to as *frequency division multiple access* (FDMA) formats.

References

1. Lindsey, W. *Synchronization Theory in Control and Communications*, Prentice-Hall, Englewood Cliffs, N.J., 1971, Chap. 4.
2. Lindsey, W. and Simon, M. *Telecommunication Systems Engineering*, Prentice-Hall, Englewood Cliffs, N.J., 1973.
3. Stiffler, J. *Theory of Synchronous Communications*, Prentice-Hall, Englewood Cliffs, N.J., 1971, Chap. 5.
4. Van Trees, H. *Detection, Estimation, and Modulation Theory*, Part II, Wiley, New York, 1971.
5. Freeman, R. *Telecommunication Transmission Handbook*, Wiley, New York, 1975, Chap. 3.
6. *Transmission Systems for Communications*, edited by staff of the Bell Telephone Laboratories, prepared by Western Electric Co., Winston-Salem, N.C., copyrighted by Bell Telephone Laboratories Inc., 1970, Chap. 9.

Problems

1. (5.2) A baseband subsystem has a source producing a waveform with power spectrum $S_d(\omega)$ that is bandlimited white noise spectrum, with bandwidth B_d Hz and spectral level S_0 W/Hz. The carrier system effectively filters the baseband with channel filter $H_{bc}(\omega) = 1/(1 + j\tau\omega)$, and adds in a white noise interference of level N_{b0} W/Hz.

(a) Determine the optimal baseband filter at the receiver for maximizing SDR, assuming the source waveform is used directly as the baseband.

(b) Determine the optimal filters at the transmitter and receiver for maximizing SNR for this system, assuming the transmitter power is limited to P_m.

2. (5.2) Show that the maximal value of SNR in (5.2.8), under the constraint of (5.2.7) is given by

$$\text{SNR}_{\text{max}} = \frac{P_n}{P_m}\left(\frac{1}{2\pi}\int_{-\infty}^{\infty}\frac{S_d(\omega)\,d\omega}{|H_{bc}(\omega)|^2}\right)$$

where P_n is the noise power over the frequency extent of $S_d(\omega)$.

3. (5.2) Formally carry out the minimization of (5.2.9), and derive the results in (5.2.10).

4. (5.2) Apply the results of Section 5.2 and determine the optimal baseband filter that should be used to maximize SDR in (5.2.1) when the source signal is a bandpass waveform over $(\omega_s - 2\pi B_d, \omega_s + 2\pi B_d)$.

5. (5.2) Assume $H_t(\omega)$ and $H_b(\omega)$ are given in Figure 5.2. Determine the optimal channel filter $H_{bc}(\omega)$ that maximizes SDR in (5.2.1).

6. (5.3) There is a voice source with $P_d = 1$ and a bandwidth of 4 kHz. An RF carrier with an allowable RF bandwidth of 1 MHz is available. The RF carrier power just satisfies a threshold value of 15 dB over the noise in its RF bandwidth. Determine the demodulated source SNR_d that will occur for the following formats: (a) standard FM, (b) standard AM, (c) AM|FM, with a subcarrier of 100 kHz, (d) FM|FM with the same subcarrier and index $\beta = 3$.

7. (5.3) A bandlimited source of spectral level S_0 and bandwidth B_d Hz is to be transmitted over a baseband subsystem using an AM|FM format. The channel noise is white with level $N_0/2$. Channel filtering effects are negligible. (a) Using the results of Section 5.2, design the optimal receiver filter that should be placed after the FM demodulator so as to maximize the subcarrier to distortion rate prior to the AM demodulation. (b) Repeat for a transmitter-receiver pair, and comment on the required source preemphasis.

8. (5.4) A carrier signal (amplitude A, frequency ω_c) is phase modulated with a sine wave at frequency ω_m and modulation index Δ. The modulated signal is filtered in a narrow bandpass filter about ω_c (bandwidth $\ll 2\omega_m$). The filtered output then has white noise (level N_0 one-sided) added to it. The combined output, filtered signal plus noise, is then used as the input to a linear carrier tracking phase lock loop having a noice bandwidth B_L Hz. Determine the mean square value of the tracking phase error in the loop.

9. (5.5) A first order PLL is designed with a loop noise bandwidth of 50 Hz. It is to track a carrier in noise with a Doppler shift of 20 rps. Using linear loop theory determine the received carrier power needed to guarantee that the steady state tracking error (signal plus noise component) has a total mean square phase error less than 0.02 rad. (Assume $N_0 = 10^{-9}$.)

10. (5.5) Derive the tabulation of Table 5.1 by formally applying the final value theorem of Section A.2.

11. (5.5) Determine the loop noise bandwidth for the optimal carrier tracking loop in (5.5.5). Rewrite $H(\omega)$ in terms of B_L.

12. (5.5) Derive (5.5.6), beginning with (5.5.3). (*Hint*: Use the fact that the realizable portion of the polynomial

$$\left[a\left(1 + \frac{b}{\omega^2} + \frac{c^2}{\omega^4} \right) \right]$$

is given by

$$\left[\sqrt{a}\left(\frac{c + (b + 2c)j\omega - \omega^2}{\omega^2} \right) \right]$$

13. (5.5) For the optimal Doppler tracking loop of (5.5.6) and (5.5.8), (a) identify the loop damping factor ζ and natural frequency ω_n (see Table 4.1), (b) design an RC filter for the required loop filter of (5.5.8).

14. (5.6) Consider a subcarrier PM system with carrier tracking (Section 5.4). Of the total carrier power $A^2/2$, $J_0^2(\Delta)$ of it is used by the carrier loop and $2J_1^2(\Delta)$ of it is used by the subcarrier. Sketch the system efficiency as a function of Δ, where efficiency is the ratio of carrier power used to total carrier power.

15. (5.6) Assume a carrier is phase modulated directly with a binary square wave having derivation Δ and bandwidth B_r (instead of a subcarrier). Show that the pair of design equations in (5.6.1) are replaced by

$$(\cos^2 \Delta)\left(\frac{P_c}{N_0 \tilde{B}_L} \right) \geq Y_c$$

$$(\sin^2 \Delta)\left(\frac{P_c}{N_0 B_r} \right) \geq Y_r$$

16. (5.7) An AM|FM multiplexed system uses subcarrier frequencies at ω_1 and $\omega_2(\omega_2 > \omega_1)$ with amplitudes k_1 and k_2, respectively. Assume satisfactory threshold operation, white noise interference, subcarrier bandwidths much less than subcarrier frequencies, and ideal (flat filter) functions. If the transmitter uses a preemphasis rule $C_i = C\omega_i^3$, determine: (a) the ratio of subcarrier SNR for the two channels ($\text{SNR}_{sc_1}/\text{SNR}_{sc_2}$), (b) the

RF Carson rule bandwidth used (assume the rms carrier deviation). (c) Generalize the result of (a) to an arbitrary preemphasis rule $C_i = C\omega_i^n$.

17. (5.7) Apply the results of Section 5.2 to the FDM|AM|FM system and determine the optimal preemphasis that should be used. (Do not assume the FM noise is flat over the AM spectra.)

18. (5.7) An FDM|AM|FM system uses subcarrier frequencies at 100 kHz, 300 kHz, and 600 kHz. An RF bandwidth of 4 MHz is available. The RF carrier power is just sufficient to provide a 15 dB SNR in the subcarrier bandwidths (all assumed identical). What is the maximum demodulated source SNR attainable?

19. (5.7) A group of source signals $d_1(t), d_2(t), \ldots, d_N(t)$, each having bandwidth B_d Hz, is multiplexed in an FDM|AM|PM system. Calculate the output SNR_d following subcarrier demodulation for each channel. Assume that all demodulators operate above threshold, ideal flat filters are used, and the receiver carrier noise is white. Identify all parameters used in your equations.

20. (5.7) Using the following parameters, design a set of four preselected FM subcarrier channels (center frequency and bandwidths) for source multiplexing:

$$\text{Lowest usable frequency, 370 Hz}$$

$$\Delta_{si} = 0.075 \, \omega_{si}$$

$$B_{sc} = 2\Delta_{si}$$

Determine what source bandwidth B_d can be frequency modulated onto each channel, if a β_i of 5 is desired in all channels.

21. (5.8) Consider an FDM system of (5.8.1). Assume, in addition, a binary square wave of deviation Δ_r and bandwidth B_r is also phase modulated onto the carrier. Rederive the set of equations in (5.8.7) that must be solved for system design.

CHAPTER 6

DIGITAL SOURCES AND ENCODING

Whereas analog sources produce continuous waveforms, digital sources produce sequences of binary symbols, or *bits*, as their output. These symbols could be produced from (1) command sources, in which the bits form binary words to be used for receiver instruction; (2) computer sources, in which the bit words correspond to the computer language; or (3) conversion of an analog source waveform into a binary sequence. In the latter case the binary sequence is then sent to represent the source waveform. This operation of converting a continuous waveform into a digital sequence is called *analog to digital* (A-D) conversion. At the transmitter the digital source bits, whether representing commands or A-D conversion symbols, are then encoded (transformed) into a baseband signal for transmission over the carrier link (Figure 6.1). After carrier demodulation the recovered baseband signal is then converted back (decoded) into the binary symbols. These recovered bits can then be used for command recognition and execution, for computer data reduction, or, in the case of A-D conversion, for reconstruction of the original source waveform. Reconstruction of an analog waveform from its corresponding digital sequence is called *digital to analog* (D-A) conversion.

With analog sources the primary concern of the communication link was the preservation of the actual source waveform during transmission. We measured the ability of a system to achieve this by computing the SNR of the recovered source signal. With digital sources, the role of the communication link is primarily for the transmission of the digital symbols. When viewed in this way the objective of the system is the accurate

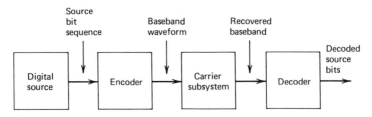

Figure 6.1. The digital baseband subsystem.

reproduction of the symbol sequence at the receiving end. A question then arises as to the best method for assessing link performance. Typically, strong emphasis is placed on the probability of the link making errors in the transmission of each symbol. This probability is referred to as the *bit error probability*, or *bit error rate*, of the system. We then must be able to determine what values of error probability correspond to "good" and to "poor" systems. In the case of command or computer transmission, bit error probability can be easily converted to a percentage of correct command execution, from which evaluation of system performance is fairly obvious. When dealing with A-D systems, however, performance assessment merely in terms of error probability may be misleading. Since the accurate reconstruction of our original source waveform is always of ultimate concern, it appears natural to base performance on the degree of degradation of this waveform caused by the total system. However, to evaluate the degradation, and to design the interfacing communication link, it is first necessary to understand the actual way in which the binary symbols are generated, what they represent, the mechanism of errors, and a qualitative measure of the effect of a bit error. This will then allow us to interpret the meaning of a satisfactory system bit error probability. As we shall see, the operation of A-D and D-A conversion inherently introduces degradation to the source waveform. Careful assessment of these intrinsic conversion errors is necessary in order to prevent overdesign of the symbol transmission portion of the system. These points, and other properties of digital source systems, are investigated in this and the next two chapters.

6.1. A-D Conversion

A digital source employing A-D conversion has the basic block diagram in Figure 6.2. The A-D converter has the task of converting the analog source waveform into an equivalent bit stream. We often say that the A-D converter "*digitizes*" the source. Basically, A-D conversion is composed of

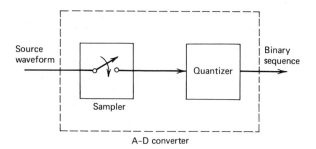

Figure 6.2. A–D converter block diagram.

two operations—sampling and quantizing. By sampling the source waveform periodically we obtain a continual sequence of voltage samples. These samples correspond to the voltage values of the waveform at the sampling times. These voltage samples are then mapped, or quantized, into digital words described by binary symbols. Transmission of these symbols in sequence is then used instead of sending the original source waveform. In some cases, additional processing of the voltage samples or bit words (called *source encoding*) is introduced prior to transmission. Advantages of this additional processing are discussed in Section 6.5.

Both sampling and quantizing inherently introduce discrepancies into the eventual waveform reconstruction. These discrepancies are usually assessed by determining the total mean squared waveform error (MSE) that occurs after waveform reconstruction. The sampling operation produces a mean squared waveform *aliasing* error (MSAE), whereas the quantizing introduces a mean squared quantization error (MSQE). In addition, the physical transmission of the bits from the encoder to the receiver may further increase this effective quantization error. The total MSE is then the sum of the contributions from all these components. Hence we generally write

$$\text{MSE} = \text{MSAE} + \text{MSQE} \qquad (6.1.1)$$

as the total reconstructed mean squared error of the digitized waveform.

Often MSE is normalized by the power of the reconstructed source waveform, P_d, and the ratio $(\text{MSE}/P_d)^{1/2}$ is often interpreted as the rms *accuracy* of the digital system. In addition, it is common practice to refer to the square of its reciprocal as an effective signal to distortion ratio (SDR) for the reconstructed signal [see Equation (1.3.2)]. Thus

$$\text{SDR} = \frac{P_d}{\text{MSE}} \qquad (6.1.2)$$

and is often stated in decibels. Although it is common practice to treat the parameter SDR as if it were the same as an effective SNR, it should be noted that SDR refers to no specific noise addition. Instead, it simply accounts for the inaccuracies in the overall waveform symbol transmission and reconstruction procedures as a mean squared variation from the true waveform. In this sense SDR does retain meaning as a measure of signal fidelity. In the following we examine the specific contributors to MSE in (6.1.1).

Sampling. The key parameter governing the waveform sampling operation in Figure 6.2 is the rate at which the samples are taken. This sampling rate is dictated by the so-called sampling theorem, which serves as a guideline for rate selection. The sampling theorem is important not only for its mathematical implication, but also for its proof, which yields insight to the required construction procedure (i.e., the D-A conversion) at the receiver. Simply stated, the sampling theorem states that any time function having its transform bandlimited to B Hz can be uniquely represented by its voltage values occurring every $1/2B$ sec apart. That is, if a time function has a Fourier transform with no components outside $(-B, B)$ Hz, then we need only sample the time function every $1/2B$ sec to reconstruct completely (identically) the entire time function. When applied to source waveforms, this means we need only accurately transmit these time samples to the receiver in order to communicate the source waveform. This extremely vital theorem is proven as follows.

Let $D(\omega)$ be the Fourier transform of the waveform $d(t)$, and let

$$D(\omega) = 0, \qquad |\omega| > 2\pi B \qquad (6.1.3)$$

Let $\bar{D}(\omega)$ be the periodic extension of $D(\omega)$, as shown in Figure 6.3. We can then write $\bar{D}(\omega)$ as

$$\bar{D}(\omega) = \sum_{k=-\infty}^{\infty} D(\omega + k4\pi B) \qquad (6.1.4)$$

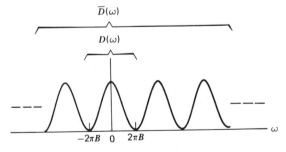

Figure 6.3. Periodic version $[\bar{D}(\omega)]$ of $D(\omega)$.

Since $\bar{D}(\omega)$ is periodic, we can also expand it in a unique Fourier series in the variable ω as

$$\bar{D}(\omega) = \sum_{n=-\infty}^{\infty} d_n \, e^{j(2\pi n/4\pi B)\omega} \qquad (6.1.5)$$

Thus $\bar{D}(\omega)$ in (6.1.4) is completely specified mathematically by its Fourier coefficients $\{d_n\}$. That is, given $\{d_n\}$, $\bar{D}(\omega)$ in (6.1.5) can always be exactly constructed. Now from Fourier series (Problem 1.2) we have

$$d_n = \frac{1}{4\pi B} \int_{-2\pi B}^{2\pi B} \bar{D}(\omega) e^{-j(n/2B)\omega} \, d\omega$$

$$= \left(\frac{1}{2B}\right) \frac{1}{2\pi} \int_{-\infty}^{\infty} D(\omega) e^{-j(n/2B)\omega} \, d\omega$$

$$= \frac{1}{2B} d\left(\frac{-n}{2B}\right) \qquad (6.1.6)$$

Thus the values of $d(t)$ at the points $t = n/2B$, $-\infty < n < \infty$, identify the sequence $\{d_n\}$, which in turn specifies $\bar{D}(\omega)$. From the latter, $D(\omega)$ can be determined exactly, and by use of the inverse Fourier transform, $d(t)$ is uniquely specified at all t. Hence the time samples of $d(t)$, taken at rate $2B$ samples/sec, can be used to produce $d(t)$ itself. This completes the underlying proof of the sampling theorem.

The preceding proof, in addition to establishing the sufficiency of the time samples, also suggests the manner in which the D-A reconstruction can occur. That is, $d(t)$ is simply the Fourier transform of $\bar{D}(\omega)$ limited to the range $-2\pi B \leq \omega \leq 2\pi B$. Therefore

$$d(t) = \frac{1}{2\pi} \int_{-2\pi B}^{2\pi B} \left[\frac{1}{2B} \sum_{n=-\infty}^{\infty} d\left(\frac{-n}{2B}\right) e^{jn\omega/2B}\right] e^{j\omega t} \, d\omega$$

$$= \frac{1}{2B} \sum_{n=-\infty}^{\infty} d\left(\frac{n}{2B}\right) \left[\frac{1}{2\pi} \int_{-2\pi B}^{2\pi B} e^{j(t-n/2B)\omega} \, d\omega\right] \qquad (6.1.7)$$

The bracket evaluates by using

$$\frac{1}{2\pi} \int_{-2\pi B}^{2\pi B} e^{jx\omega} \, d\omega = \frac{\sin(x 2\pi B)}{\pi x} \qquad (6.1.8)$$

Hence (6.1.7) becomes

$$d(t) = \sum_{n=-\infty}^{\infty} d\left(\frac{n}{2B}\right) \frac{\sin(2\pi Bt - n\pi)}{(2\pi Bt - n\pi)} \qquad (6.1.9)$$

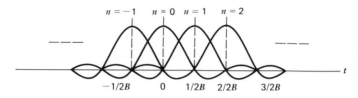

Figure 6.4. Sinc reconstruction functions (sinc($2\pi Bt$) $\triangleq$ sin ($2\pi Bt$)/$2\pi Bt$).

Thus the waveform $d(t)$ can be reconstructed by using the sample values $d(n/2B)$ as coefficients of the corresponding functions shown. These functions are simply shifted versions of the sinc $(2\pi Bt) \triangleq$ sin $(2\pi Bt)/2\pi Bt$ function, the nth one centered as $t = n/2B$ (Figure 6.4). The summation of the contribution from all such functions, with the proper sample value coefficients, at any t will exactly produce $d(t)$. Note that a specific reconstruction function is required by the sampling theorem. Other techniques, such as straightline connections between sampling points, only serve as an approximation to $d(t)$, and generally produce an effective low pass filtering of the desired waveform (Problem 6.2). We also see that an infinite number of samples [terms in (6.1.9)] must be used with each one affecting $d(t)$ at all values of t. If $d(t)$ is assumed to be time limited to a T_0 sec time interval and zero elsewhere, then only $T_0/(1/2B) = 2BT_0$ time samples are needed to represent $d(t)$. However, simultaneous time limiting and bandlimiting of a waveform is contradictory, although such assumptions are often made for analytical convenience, and can be somewhat justified [1].

The reconstruction of $d(t)$ from its time samples, using (6.1.9), although exact, requires generation of the sequence of sinc time functions. From a system design point of view, it would be convenient to have a simple interpretation to the right side of (6.1.9). To examine this, consider using the sequence of samples $d(nT)$, $T = 1/2B$, to adjust the amplitudes of a train of narrow, periodic pulses, as shown in Figure 6.5. If we denote the

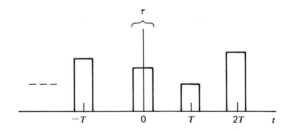

Figure 6.5. Amplitude adjusted pulse waveform.

narrow pulses by

$$w(t) = \begin{cases} \dfrac{1}{\tau}, & -\dfrac{\tau}{2} \le t \le \dfrac{\tau}{2} \\ 0, & \text{elsewhere} \end{cases} \qquad (6.1.10)$$

and let its repetition period be T, then the amplitude adjusted pulse train can be written as

$$q(t) = \sum_{n=-\infty}^{\infty} d(nT) w(t-nT) \qquad (6.1.11)$$

This pulse train has Fourier transform

$$Q(\omega) = \int_{-\infty}^{\infty} q(t) e^{-j\omega t} \, dt$$

$$= \sum_{n=-\infty}^{\infty} d(nT) \int_{-\infty}^{\infty} w(t-nT) e^{-j\omega t} \, dt$$

$$= \sum_{n=-\infty}^{\infty} d(nT) e^{-jn\omega T} \int_{-\infty}^{\infty} w(t-nT) e^{-j\omega(t-nT)} \, dt \qquad (6.1.12)$$

The integral is simply the Fourier transform of $w(t)$ in (6.1.10) and does not depend on n. Hence by use of (6.1.4) and (1.3.7),

$$Q(\omega) = \left[\sum_{n=-\infty}^{\infty} D\left(\omega + \frac{2\pi n}{T}\right) \right] \left[\frac{\sin(\omega \tau/2)}{(\omega \tau/2)} \right] \qquad (6.1.13)$$

Since $T = 1/2B$, the transform $Q(\omega)$ is that of a low pass filtered version of the desired periodic spectrum $\bar{D}(\omega)$, as shown in Figure 6.6. If $\tau \to 0$ (the pulse were infinitesimally narrow and infinitely high, approaching delta functions), then $Q(\omega) \to \bar{D}(\omega)$. Thus the periodic function $\bar{D}(\omega)$, used to prove the sampling theorem, can in fact be generated by using the time samples $d(nT)$ to amplitude modulate a train of delta functions. In this

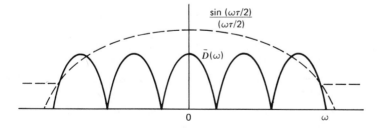

Figure 6.6. Equivalent filtering of desired transform.

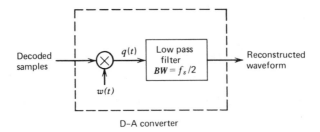

Figure 6.7. *D–A* converter block diagram.

case (6.1.11) becomes

$$q(t) = \sum_{n=-\infty}^{\infty} d(nT)\, \delta(t - nT). \qquad (6.1.14)$$

and its transform $Q(\omega)$ is identical to $\bar{D}(\omega)$. The spectrum $D(\omega)$ is then obtained by filtering off the portion of the spectrum between $(-2\pi B, 2\pi B)$, that is, by low pass filtering. The output of this low pass filter is then $d(t)$. This physically forms the basis of D-A conversion at a receiver (Figure 6.7). The sample values decoded at the receiver are used to produce amplitude modulated pulse trains that are then low pass filtered to produce the reconstructed source waveform $d(t)$. This is equivalent to assuming that the pulse train $q(t)$ in (6.1.11) or (6.1.14) was actually transmitted from the source sampler to the D-A low pass filter. (Recall that only the binary equivalent of the voltage sample values are in fact sent.) For this reason, $q(t)$ in (6.1.11) is often referred to as the *sampler signal* of the A-D converter. The waveform in (6.1.14) using delta functions is called the *idealized sampler* signal, and its transform is given by (6.1.13) with $\tau = 0$. To summarize, then, sampling a source waveform every T sec, sending the sample values, and reconstructing by the low pass filter method described earlier, is equivalent (as far as reconstruction is concerned) of low pass filtering the sampler signal $q(t)$ directly.

From (6.1.13) we see that the pulse train $q(t)$ need not have true delta functions, since little distortion occurs to $D(\omega)$ so long as $\tau \ll T$. On the other hand, the low pass reconstruction filter must be perfectly flat over $(-2\pi B, 2\pi B)$ to avoid distorting $D(\omega)$. Hence the low pass filter must be an ideal low pass filter. This, of course, is evident from (6.1.9), which can also be interpreted as the output produced by an input $q(t)$ of weighted delta functions for a filter whose impulse response is $\sin(2\pi Bt)/2\pi Bt$, that is, an ideal low pass filter. These ideal filter requirements can be weakened by allowing a separation between $D(\omega)$ and the nearest interfering spectrum of $\bar{D}(\omega)$. This can be achieved by interpreting the upper end of the frequency transform $D(\omega)$ in (6.1.3) as being $(B + \delta)$ instead of B. (The function will be zero over the upper δ Hz.) Sampling at the rate $2(B + \delta)$ produces a periodic frequency function $\bar{D}(\omega)$ with a spacing of 2δ between the adjacent spectra. Thus sampling faster than required simplifies the D-A filter design.

When the spectrum is truly bandlimited, sampling at the proper rate or faster allows undistorted waveform reconstruction. However, if we do not sample fast enough or if the spectrum is not truly bandlimited, an error in the reconstruction occurs as a result of sampling. This is called the aliasing error, and its mean squared value can be determined as follows. Let $d(t)$ be the waveform to be sampled and let $D(\omega)$ again be its transform, not necessarily bandlimited. Let us sample at rate f_s samples per second and reconstruct from these samples. We can again interpret the resulting reconstruction as an ideal filtering of the idealized sampler signal $q(t)$ in (6.1.14). This signal has the transform [from (6.1.13) with $\tau = 0$] given by

$$Q(\omega) = \sum_{n=-\infty}^{\infty} D(\omega + n 2\pi f_s) \qquad (6.1.15)$$

This corresponds to shifts of $D(\omega)$ by all multiples of the sampling frequency f_s. The resulting spectrum is illustrated in Figure 6.8. Reconstruction by ideally filtered $Q(\omega)$ from $(-2\pi f_s/2, 2\pi f_s/2)$ no longer yields an undistorted version of the original $D(\omega)$. Instead, portions of the shifted spectrum are folded back into the desired spectrum. Stated in other words, the sampling rate f_s is only capable of reconstructing frequency components of $d(t)$ up to $f_s/2$, and all higher frequencies are lost, producing an effective error in the reconstructed waveform. The mean squared value of this signal error [difference between the true $d(t)$ and that part that will be reconstructed] is therefore given by the area under the waveform power or energy spectrum beyond $f_s/2$ Hz. This portion of the spectrum is identical to that part folded back into the range $(-2\pi f_s/2, 2\pi f_s/2)$ in $Q(\omega)$. The mean squared aliasing error (MSAE) is given by this folded integrated

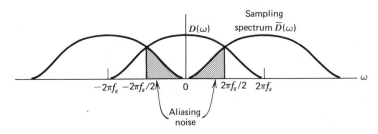

Figure 6.8. Aliasing noise diagram.

area. Hence*

$$\text{MSAE} = \frac{2}{2\pi} \int_{2\pi f_s/2}^{\infty} |D(\omega)|^2 \, d\omega \qquad (6.1.16)$$

Since the energy in the original waveform is the integral of $|D(\omega)|^2$ over all frequencies, the resulting SDR caused by aliasing becomes

$$\text{SDR} = \frac{\int_0^{\infty} |D(\omega)|^2 \, d\omega}{\int_{\pi f_s}^{\infty} |D(\omega)|^2 \, d\omega} \qquad (6.1.17)$$

The preceding can be evaluated for any specific waveform energy spectrum $|D(\omega)|^2$. As a typical example, let

$$|D(\omega)|^2 = \frac{1}{1 + (\omega/2\pi f_d)^{2\mu}} \qquad (6.1.18)$$

where f_d is the 3 dB frequency and μ determines the spectrum roll-off. Substituting into (6.1.17) yields

$$\text{SDR} = \frac{\int_0^{\infty} (dx/1 + x^{2\mu})}{\int_{f_s/2f_d}^{\infty} (dx/1 + x^{2\mu})} \qquad (6.1.19)$$

The result is plotted in Figure 6.9, as a function of the normalized sampling rate f_s/f_d, for various fall-off parameter values μ. Note that sampling rates greatly in excess of twice the 3 dB bandwidth of the waveform are required to maintain acceptable SDR values, say, 30–40 dB (0.5% accuracy). This illustrates the danger in simply interpreting the 3 dB bandwidth as the bandwidth extent of $d(t)$ and in blindly applying the sampling theorem. As μ increases, the sampling rate required to produce a fixed value of SDR also decreases, approaching the theoretical minimum rate. Since spectrum fall-off can be increased by additional filtering of the waveform, the results indicate the advantage of presampling filtering of the source waveform, as far as aliasing error is concerned.

In certain situations we may wish to digitize a bandpass signal. This occurs if the source produces a bandpass waveform or if the source signal is placed on a subcarrier prior to sampling. Reconstruction of the bandpass signal would be obtained by ideal bandpass filtering over the original bandwidth of the bandpass signal. The previous sampling discussion extends to bandpass sampling as well. Suppose $D(\omega)$ has the bandpass spectrum shown in Figure 6.10a. If we sample the waveform $d(t)$ at rate f_s samples per second and produce the idealized sampler signal in (6.1.14)

---

* We use energy spectra here since the $d(t)$ we have been considering are deterministic waveforms. The resulting SDR is then a ratio of energy values. If $d(t)$ is a random process, we can equivalently deal with power spectra and compute SDR as a power ratio.

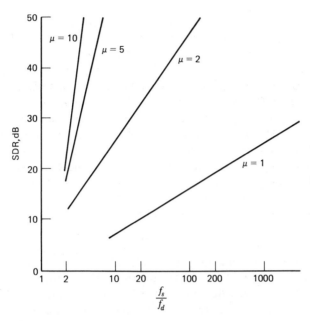

Figure 6.9. SDR due to aliasing for Butterworth shaped spectra (f_d = spectrum 3 dB bandwidth, μ is the spectral fall-off rate).

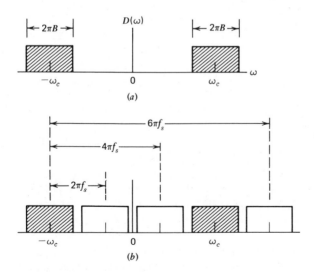

Figure 6.10. Bandpass sampling. (*a*) Source spectrum, (*b*) sampled spectrum.

with these samples, its transform is again given by (6.1.15). This sampled spectrum is sketched in Figure 6.10b, and again corresponds to the periodic shifting of $D(\omega)$ every $2\pi f_s$ rps. The actual value of $D(\omega)$ at any ω depends on the relation between the sampling frequency f_s and the original bandpass spectrum $D(\omega)$. If reconstruction is obtained by ideal bandpass filtering of $\bar{D}(\omega)$ over the bandwidth of $D(\omega)$ (i.e., replacing the low pass filter in Figure 6.7 by a bandpass filter), a distortionless spectrum appears only if shifts of $D(\omega)$ do not overlap $D(\omega)$ anywhere. This requires that if the negative frequency spectrum of $D(\omega)$ is shifted just to the left of the positive frequency spectrum for some integer n, the $(n+1)$ integer shift must move it completely to the right to avoid overlap. This requires that the shifting interval $2\pi f_s$ be at least as large as twice the positive frequency bandwidth $2\pi B$ of $D(\omega)$. Thus we require

$$f_s \geq 2B \qquad (6.1.20)$$

and the bandpass spectrum must be sampled at a rate equal to at least twice the bandpass bandwidth to avoid overlap. The actual required rate depends on the relative values of B and its spectral location (Problem 6.5).

Let us determine the condition necessary on the spectral location of $D(\omega)$ in order to allow a sampling rate of exactly $2B$ (the minimal possible in (6.1.20)). Using Figure 6.10b, it is evident that the bandpass spectrum should be located at a center frequency $\pm f_c$ such that when sampling at $f_s = 2B$ an integer number of positive shifts slides the right-hand edge of the negative frequency spectrum just up to the left-hand side of the positive spectrum. For this to occur it is necessary that $nf_s = 2[f_c - (B/2)]$, or equivalently,

$$f_c = nB + \frac{B}{2} = \left(\frac{n}{2} + \frac{1}{4}\right) f_s \qquad (6.1.21)$$

for some integer n. A bandpass spectrum of bandwidth B should therefore be located at any of the center frequencies given in (6.1.21) to avoid aliasing errors (spectrum overlap) when sampling with rate $f_s = 2B$. This aids in the selection of subcarrier frequencies relative to sampling frequencies for use in bandpass sampling.

Quantization. After the waveform is sampled the A-D subsystem must transmit the sample values to the receiver for the reconstruction. However, the sample value can take on any of a continuum of values over the voltage range, and the A-D operation can only transmit a finite number of sample values at each sampling point. It accomplishes this by mapping, or *quantizing*, the complete sample value voltage range into a finite number of selected values. This quantization operation is shown as Figure 6.11. The

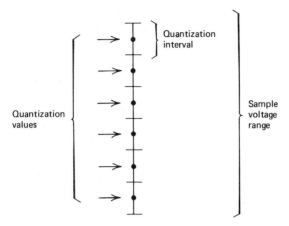

Figure 6.11. Quantization diagram.

sample voltage range is divided into preselected intervals, and each interval is given a prescribed binary code word. All sample values occurring in a specific interval are transmitted with that interval binary word. At the receiver each binary code word is converted to a specific voltage value in the corresponding interval. This reconstructed voltage value is called the *quantization value* of the interval. Hence A-D quantization is effectively the same as converting all voltage values in an interval to the quantization value of that interval. Thus there is an inherent waveform error introduced by the quantization operation even if perfect D-A conversion followed.

The selection of the binary code for the intervals is arbitrary, but obviously a distinct code word must be available for each quantization interval. If there are L intervals, then clearly $\log_2 L$ (or the nearest largest integer) bits are needed to represent all intervals uniquely. A common technique is to use *natural* coding, in which the intervals are numbered consecutively from zero to $L-1$, and the binary representation of each number is used as the code word for that interval. Thus $\log_2 L$ bits are produced at each sample time, and if the sampler operates at f_s samples per second, the A-D subsystem generates bits at the rate

$$\mathcal{R} = f_s \log_2 L \qquad \text{bits/sec} \qquad (6.1.22)$$

Quantizers can take on a variety of forms, each describable by the number, size, and location of their intervals, in addition to their quantization values. Quantizers are generally constructed from arrays of diode voltage gates that excite binary flip-flop circuits, producing the necessary binary word for each voltage sample. The simplest type of quantizer is a *uniform* quantizer, in which all intervals are selected to have equal width over a prescribed voltage range. If a sample voltage value exceeds the quantizer range, the quantizer is said to be *saturated*. Saturating samples are generally associated with the end intervals (which is identical to assuming the end intervals extend to infinity). A quantizer with unequal intervals is called a *nonuniform* quantizer.

As stated previously, quantization introduces an intrinsic waveform error even if the remaining part of the system is error-free. The mean squared quantization error (MSQE) that will be introduced at each sample value can be determined by treating the voltage sample as a random variable and averaging the mean squares sample error over all intervals. If we let $MSQE_i$ be the mean squared quantization error when the random voltage sample is in the ith quantization interval, and if we let P_i be the probability that the sample is in the ith interval, then

$$MSQE = \sum_{i=1}^{L} (MSQE_i) P_i \qquad (6.1.23)$$

The mean squared error in an interval can be determined by computing the squared error from a sample point in the interval to the quantization value that will be reconstructed for that interval, and averaging over the probability density of the sample point. Hence if the voltage sample has the probability density $p_d(x)$, and if ξ_i is the quantization value of the ith interval, then

$$MSQE_i = \int_{I_i} (x - \xi_i)^2 p(x|i) \, dx \qquad (6.1.24)$$

where I_i represents the ith interval range and $p(x|i)$ is the probability density of the sample value when in I_i. This is given by

$$p(x|i) = p_d(x)/P_i, \qquad \text{for} \quad x \text{ in } I_i \qquad (6.1.25)$$

where now

$$P_i = \int_{I_i} p_d(x) \, dx \qquad (6.1.26)$$

Equation (6.1.24) can be computed for each quantizer interval and used in (6.1.23). Thus MSQE depends on the form of the quantizer and the probability density of the sample voltage value. It is again convenient to normalize by the mean squared value of the sample voltage value

$$P_d = \int_{-\infty}^{\infty} x^2 p_d(x) \, dx \qquad (6.1.27)$$

and compute an effective SDR due to quantization, just as in (6.1.17) for the aliasing error.

Example 6.1. Assume a uniform quantizer with L intervals operating over a voltage range $(-V, V)$ volts, so that each interval has width $\varepsilon = 2V/L$. The center point of each interval is used as the quantization value. Let each voltage sample have a uniform distribution over $(-A, A)$ volts. Consider first the case of no saturation, $A \le V$. Equation (6.1.24) becomes

$$\text{MSQE}_i = \int_{-\varepsilon/2}^{\varepsilon/2} x^2 \left(\frac{1}{\varepsilon}\right) dx = \frac{\varepsilon^2}{12} \qquad (6.1.28)$$

for each quantization interval falling within $(-A, A)$. (We take A to be an integer multiple of ε.) Since the result is the same for all such intervals and all intervals are equally likely to be occupied (because of the uniform density on the sample value), Equation (6.1.23) becomes

$$\text{MSQE} = \frac{\varepsilon^2}{12} \qquad (6.1.29)$$

Similarly, (6.1.27) evaluates to $A^2/3$. The resulting SDR due to quantization errors alone is then

$$SDR = \frac{A^2/3}{\varepsilon^2/12} = \frac{L^2}{(V/A)^2} \qquad (6.1.30)$$

We see that the quantization SDR varies directly with the number of intervals, but decreases by the square of ratio of the quantizer range to the signal sample range. Hence it is advantageous to insure that the quantizer "match" the expected sample range and not be any larger. Note that increasing L to improve SDR will require a subsequent increase in the A-D bit rate in (6.1.22).

Consider now $A > V$, so that saturation may occur. We assume a voltage sample outside the quantizer range will be associated with the end intervals. The mean squared interval error in (6.1.29) occurs only if the sample value falls within $(-V, V)$. Outside this range (i.e., during saturation), the mean squared saturation error becomes

$$\text{MSSE} \triangleq 2 \int_{V}^{A} \left[x - \left(V - \frac{\varepsilon}{2}\right)\right]^2 \left(\frac{1}{A-V}\right) dx$$
$$= \frac{(A - V + (\varepsilon/2))^3}{3(A-V)} - \frac{(\varepsilon/2)^3}{3(A-V)} \qquad (6.1.31)$$

The total MSQE in (6.1.23) is then

$$\text{MSQE} = \left(\frac{V}{A}\right)\left(\frac{\varepsilon^2}{12}\right) + \left(\frac{A-V}{A}\right)\text{MSSE} \qquad (6.1.32)$$

The desired signal power is again $P_d = A^2/3$. The SDR due to quantization with saturation is therefore

$$SDR = \frac{L^2}{\left(\frac{V}{A}\right)^3 + \left(1 - \frac{V}{A}\right)^3 \left[\left(1 + \frac{V}{L(A-V)}\right)^3 - \left(\frac{V}{L(A-V)}\right)^3\right]L^2} \quad (6.1.33)$$

Combining (6.1.30) and (6.1.33) gives the general expression for a uniform quantization with L intervals and range $(-V, V)$ opperating on a uniform voltage sample of range $(-A, A)$ as

$$SDR = \begin{cases} L^2/u^2, & u \geq 1 \\ \dfrac{L^2}{1 + \left[\left(\dfrac{1-u}{u} + \dfrac{1}{L}\right)^3 - \left(\dfrac{1}{L}\right)^3\right]L^2}, & u < 1 \end{cases} \quad (6.1.34)$$

where $u = V/A$. The result is plotted in Figure 6.12 as a function of A/V for $L = 16$ and 256, corresponding to 4 and 8 bit quantizer words. The maximum value at $A = V$ is L^2 and only occurs if the quantizer range is identical to the signal range. Note the degradation that occurs as the quantizer is mismatched, being especially severe as saturation occurs. This

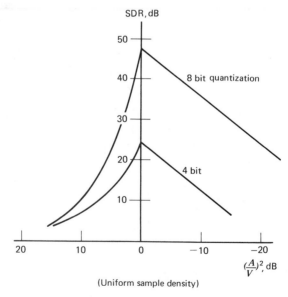

Figure 6.12. Quantizer SDR for uniform voltage sample density $(-A, A)$ (quantizer range $= (-V, V)$.

is due to the fact that mean squared errors caused by saturating voltages are more severe than those due to interval quantization. For this reason, quantizers are usually designed to encompass the signal range, not to limit it. This desired matching of the quantizers and signal range becomes difficult if the signal amplitude is unknown or perhaps changing with time. For this reason automatic amplitude control circuitry is often inserted prior to quantizing to overcome this problem.

Example 6.2.   Consider again the uniform quantizer of Example 6.1 and let the sample value have a Gaussian distribution of zero mean and variance σ^2. Because of the Gaussian symmetry, the MSQE becomes

$$\text{MSQE} = \begin{cases} \dfrac{2\sigma^2}{2\pi} \sum\limits_{i=1}^{L} \int\limits_{ih-h/2}^{ih+h/2} \left(x - ih + \dfrac{h}{2}\right)^2 e^{-x^2/2}\, dx, & L \text{ even} \\[2em] \dfrac{2\sigma^2}{2\pi} \sum\limits_{i=0}^{L-1/2} \int\limits_{ih-h/2}^{ih+h/2} (x - ih)^2 e^{-x^2/2}\, dx, & L \text{ odd} \end{cases} \qquad (6.1.35)$$

where $h = 2V/L\sigma$ and the upper end of the last interval is taken as infinity. Normalizing by the signal power σ^2 and inverting produces the SDR shown in Figure 6.13, plotted as a function of the parameter V/σ for $L = 16$ and 256. We again note the presence of a peak value, and a degradation as the quantizer range V is offset from this optimal value. Note the optimal quantizer value depends on L, and again the most rapid degradation occurs

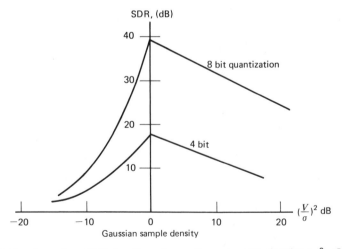

Figure 6.13. Quantizer SDR for Gaussian voltage sample density ($\sigma^2 =$ Gaussian variance, quantizer range $= (-V, V)$.

as the probability of saturation increases; i.e., the signal power σ^2 greatly exceeds the quantizer range. Again we see the importance of matching the quantizer design to the power level of the signal being sampled.

6.2. Optimal Sample Quantization

We have seen that knowledge of the quantizer structure and sampled voltage probability density leads directly to computation of the mean squared quantization error. We may therefore inquire if a quantizer can be found for a particular sample probability density that will minimize the resulting MSQE. Let us assume a general quantizer having intervals defined by the sequences of end points $l_1, l_2, \ldots, l_{L+1}$, as shown in Figure 6.14. We let ξ_i be the quantization value of the ith interval defined between $(l_i + l_{i+1})$. We can therefore write the general form of (6.1.23) as

$$\text{MSQE} = \sum_{i=1}^{L} \int_{l_i}^{l_{i+1}} (x - \xi_i)^2 p_d(x)\, dx \tag{6.2.1}$$

For a specified $p_d(x)$, we see that MSQE is a function only of the $\{l_i\}$ and $\{\xi_i\}$. We therefore would like to find their values such that MSQE is minimized. Proceeding by formal differentiation, we have

$$\frac{\partial \text{MSQE}}{\partial \xi_i} = 2 \int_{l_i}^{l_{i+1}} (x - \xi_i) p_d(x)\, dx = 0, \qquad i = 1, 2, \ldots, L \tag{6.2.2a}$$

$$\frac{\partial \text{MSQE}}{\partial l_1} = -(l_1 - \xi_1)^2 p_d(l_1) = 0 \tag{6.2.2b}$$

$$\frac{\partial \text{MSQE}}{\partial l_{L+1}} = (l_{L+1} - \xi_L)^2 p_d(l_{L+1}) = 0 \tag{6.2.2c}$$

$$\frac{\partial \text{MSQE}}{\partial l_i} = [(l_i - \xi_{i-1})^2 - (l_i - \xi_i)^2] p_d(l_i) = 0, \qquad i = 2, 3, \ldots, L \tag{6.2.2d}$$

Equation (6.2.2) represents a set of $2L + 1$ coupled simultaneous equations that must be solved for the $2L + 1$ unknowns $\{l_i\}$ and $\{\xi_i\}$ that minimize MSQE. Note that (6.2.2a) always has the solution

$$P_i \xi_i = \int_{l_i}^{l_{i+1}} x p_d(x)\, dx \tag{6.2.3}$$

or

$$\xi_i = \frac{1}{P_i} \int_{l_i}^{l_{i+1}} x p_d(x)\, dx \tag{6.2.4}$$

Figure 6.14. Generalized quantizer (ξ_i = quantization values).

Thus the quantization value of an interval should always be selected as the mean value of the sample probability density in that interval no matter where the interval is located. That is, we should always quantize to the conditional mean sample value of each interval.

An exact solution to (6.2.2) for the case when $p_d(x)$ is a zero mean Gaussian variable, for several values of L, was obtained by Max [2]. The solution is listed in Table 6.1 for $L = 2$, 4, and 8, and shows the optimal interval locations and quantization values for minimal MSQE. Note that

Table 6.1. Optimal Quantizer for Gaussian Sample Value

	$L = 2$	$L = 4$	$L = 8$	
Interval locations	$l_3 = \infty$ $l_2 = 0$ $l_1 = -\infty$	$l_5 = \infty$ $l_4 = 0.98\sigma$ $l_3 = 0$ $l_2 = -0.98\sigma$ $l_1 = -\infty$	$l_9 = +\infty$ $l_8 = 1.76\sigma$ $l_7 = 1.05\sigma$ $l_6 = 0.5\sigma$ $l_5 = 0$	$l_4 = -0.5\sigma$ $l_3 = -1.05\sigma$ $l_2 = -1.76\sigma$ $l_1 = -\infty$
Quantization values	$\xi_2 = 0.79\sigma$ $\xi_1 = -0.79\sigma$	$\xi_4 = 1.56\sigma$ $\xi_3 = 0.45\sigma$ $\xi_2 = -0.45\sigma$ $\xi_1 = -1.56\sigma$	$\xi_8 = 2.15\sigma$ $\xi_7 = 1.34\sigma$ $\xi_6 = 0.75\sigma$ $\xi_5 = 0.24\sigma$	$\xi_4 = -0.24\sigma$ $\xi_3 = -0.75\sigma$ $\xi_2 = -1.34\sigma$ $\xi_1 = -2.15\sigma$
$\dfrac{\sqrt{\text{MSQE}}}{\sigma}$	0.364	0.1175	0.0345	

the optimal Gaussian quantizer is nonuniform, concentrating its intervals over the region where the Gaussian density is highest. The end intervals are infinite, avoiding the saturation effect. For the case $L = 2$, the quantizer contains only two intervals and makes a simple sign decision on the sample value, but quantizes to a specific voltage value. The resulting MSQE for each value of L is also listed in Table 6.1 and shows an obvious improvement as larger values of L are used. Note again that the optimal quantization intervals depend on the source signal power σ^2 and the latter parameter must be known, or measured, in order to construct the optimal quantizer.

6.3. Effect of Transmission Errors on Quantization

In discussing MSQE in the previous sections, we have assumed that the quantized code word will be received correctly at the receiver D-A subsystem. This means the waveform reconstruction error is only due to the errors inherent in the quantization operation itself. However, if interval code word errors occur during transmission, an additional error must be accounted for because of the incorrect quantization value used for reconstruction. That is, if the sample value falls in the ith interval and we transmit the corresponding code word in order to quantize to value q_i, transmission errors may cause the receiver code word to be interpreted as a word associated with an incorrect interval. We analyze this effect as follows. Let ξ_i again be the quantization value of the ith interval but let $\hat{\xi}_i$ be the value actually reconstructed at the receiver when the original sample is in the ith interval. If no transmission error is made, $\hat{\xi}_i = \xi_i$. If an error is made, then $\hat{\xi}_i$ is the quantization value of some other interval. Thus the mean squared error of the sample value is increased if a transmission error is made. The resulting mean squared quantization error, when we allow for transmission word errors, is then

$$\overline{\text{MSQE}} = \sum_{i=1}^{L} \int_{I_i} (x - \hat{\xi}_1)^2 p_d(x)\, dx$$

$$\triangleq \mathscr{E}[(x - \hat{\xi}_i)^2] \tag{6.3.1}$$

If we now write $(x - \hat{\xi}_i) = (x - \xi_i + \xi_i - \hat{\xi}_i)$, we can then expand (6.3.1) as

$$\overline{\text{MSQE}} = \mathscr{E}[(e_q + e_t)^2]$$

$$= \mathscr{E}[e_q^2] + \mathscr{E}[e_t^2] + 2\mathscr{E}[e_q e_t] \tag{6.3.2}$$

where $e_q = x - \xi_i$ and $e_t = \xi_i - \hat{\xi}_i$. Thus e_q is the error in the interval quantization itself and e_t is the error introduced in the transmitted and reconstructed quantization value. The $\overline{\text{MSQE}}$ therefore is composed of the sum of three terms. The first is due purely to quantization alone, and its mean square value is precisely MSQE defined earlier. The second term is an effective mean squared transmission error term, and the last is the average cross-product of the two error types. We can assume e_q and e_t are statistically uncorrelated errors (e_q is due to the sample statistics and e_t is related to the transmission system interference) so that

$$\mathscr{E}[e_q e_t] = \mathscr{E}[e_q]\mathscr{E}[e_t] \qquad (6.3.3)$$

Now

$$\mathscr{E}[e_q] = \mathscr{E}[x - \xi_i] = \int_{I_i} x p_d(x)\, dx - \xi_i P_i \qquad (6.3.4)$$

and will be zero if the optimal quantizer values in (6.2.4) are used. Under this condition (6.3.2) reduces to

$$\overline{\text{MSQE}} = \mathscr{E}(e_q^2] + \mathscr{E}[e_t^2]$$
$$= \text{MSQE} + \mathscr{E}[e_t^2] \qquad (6.3.5)$$

and the $\overline{\text{MSQE}}$ uncouples into a separate computation of quantizer error, MSQE, and mean squared transmission error. The latter can be determined by averaging over all possible squared errors, remembering that the reconstructed sample value will always be one of the incorrect quantization values. Let P_{ij} be the probability that the transmission system causes the ith interval code to be recovered as the jth interval code word. We then have the mean squared error contributed by the transmission system as

$$\text{MSTE} \triangleq \mathscr{E}[e_t^2] = \sum_{i=1}^{L} \sum_{j=1}^{L} (\xi_i - \xi_j)^2 P_{ij} P_i \qquad (6.3.6)$$

The value of MSTE depends on the set of quantization values $\{\xi_i\}$, the interval probabilities of the sample values (P_i), and the transmission word error probabilities P_{ij}. The latter depend on the manner in which the transmission system is implemented and the way in which the bits are actually sent (a topic to be discussed subsequently). Note that transmission errors always increase the mean squared error, but their effect can be reduced by lowering the P_{ij} (transmitting the words more accurately). Equation (6.3.6) shows how the A-D and D-A operations must interface with the actual communication system design in evaluating the overall performance of a digital system. Unfortunately, (6.3.6) is not always easy to compute, even if the P_{ij} have been specified, and closed form expressions will occur only in special cases.

Example 6.3. Consider the quantizer problem in Example 6.1 with $A = V$. Assume the transmission system is such that at most one bit will be in error during any word transmission, and this bit error occurs with probability PE. If L quantization intervals are used with natural coding, then it is seen that an error in the last bit of the interval code word causes a quantization value difference of $\pm\varepsilon$ volts. (Reconstructed value will be one interval away from the correct value.) Similarly, an error in the second to last bit produces an error of $\pm2\varepsilon$ volts. Thus an error in the nth (from end) bit produces an error of $\pm2^{n-1}\varepsilon$. In addition, $P_{ij} = PE$ for all $i \neq j$, and (6.3.6) takes the form

$$\mathscr{E}[e_t^2] = \sum_{i=1}^{L} \left(\frac{1}{L}\right) \sum_{n=1}^{\log_2 L} (2^{n-1}\varepsilon)^2 PE$$

$$= \varepsilon^2 (PE) \sum_{n=1}^{\log_2 L} 4^{n-1}$$

$$= \frac{\varepsilon^2 PE(L^2 - 1)}{3} \tag{6.3.7}$$

Therefore, using the results of Example 6.1,

$$SDR = \frac{P_d}{\overline{MSQE}} = \frac{P_d}{MSQE + MSTE}$$

$$= \frac{A^2/3}{(\varepsilon^2/12) + \varepsilon^2 PE(L^2-1)/3}$$

$$= \frac{L^2}{1 + 4(L^2-1)PE} \tag{6.3.8}$$

This example shows quantitatively how transmission bit errors degrade the system SDR. If $PE \to 0$, $SDR \to L^2$ and the system is limited only by its quantization error. As $L \to \infty$, $SDR \to \frac{1}{4}PE$ and the actual transmission bit error probability determines performance. Note that a performance crossover point occurs where $4(L^2-1)PE \approx 1$ or where error probability is approximately equal to $\frac{1}{4}(L^2-1)$. If the system PE is larger than this value, there is no advantage in further increasing L, since the system is essentially transmission limited. On the other hand, there is no need to improve PE much beyond this value, as far as SDR is concerned, since the system is quantization limited. This example manifests the close interdependence of source (A-D) and link (PE) design that must be properly balanced for successful engineering.

6.4. Digitizing Noisy Waveforms

We have been assuming that the source waveform to be sampled represents the desired waveform that is to be reconstructed at the receiver. Often we are involved in a situation where the source waveform to be A-D converted is contaminated with additive noise prior to the sampling. This means that the effective system errors introduced by the A-D and D-A operations must be considered relative to the intrinsic noise already present in the source waveform. There is no need to strive to construct a digital system with an extremely high SDR value, if the source waveform is already degraded to a large extent by the noise. To aid in understanding this trade-off, consider the case where the source waveform being digitized represents the summation of a desired signal and an additive noise signal. The reconstructed SDR in (6.1.2), which takes into account both aliasing and quantization errors, refers only to that of the total sampled waveform. This SDR must be modified to obtain the true desired signal to distortion ratio after reconstruction, as was discussed in Section 1.7.

Let P_d be the power of the source waveform and let

$$P_d = P_s + P_n \qquad (6.4.1)$$

where P_s and P_n are the source signal and noise powers, respectively, prior to sampling. The reconstructed signal to noise ratio for the desired source signal is the ratio of the signal power P_s to the total discrepancy in the reconstructed waveform. The latter is due both to the noise originally present in the sampled signal and to the errors of the digitizing operations. The output SNR is then

$$\text{SNR} = \frac{P_s}{P_n + \text{MSE}} \qquad (6.4.2)$$

where MSE is the total mean squared error introduced by the A-D and D-A operations. From (6.1.2), this can be written as

$$\text{SNR} = \frac{P_s}{P_n + (P_d/\text{SDR})} \qquad (6.4.3)$$

Using (6.4.1) this becomes

$$\text{SNR} = \frac{(P_s/P_n)(\text{SDR})}{(\text{SDR}) + (P_s/P_n) + 1} \qquad (6.4.4)$$

The reconstructed signal SNR will therefore be related to the source SNR (P_s/P_n) and to the SDR of the digitizing system by (6.4.4). If SDR $\to \infty$, the signal SNR is limited by that of the source itself. If $P_s/P_n \to \infty$, the system is limited by SDR. Note that the form of (6.4.4) shows that the reconstructed

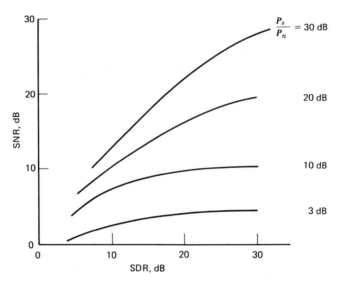

Figure 6.15. SNR versus SDR for noisy A–D conversion.

SNR is always less than the smaller of either the (P_s/P_n) or SDR ratios. Equation (6.4.4) is plotted in Figure 6.15 for several values of source SNR. Plots of this type are convenient for assessing the effect of the A-D and D-A operations on the overall waveform distortion.

6.5. Generalized A-D Conversion (Source Encoding)

So far we have considered only the most straightforward method of A-D conversion. Each source sample, whether representing a time waveform voltage sample or a command word, was quantized one at a time, in a sequential manner, to form the source sequence. In forming this sequence we are constrained to operate over the communication link at the source rate given in (6.1.22). In the design of any practical digital system there is invariably a requirement to reduce, as much as possible, the operating source bit rate. With the standard A-D conversion we have considered, this rate reduction can be achieved only by lowering the sampling rate f_s or by reducing the number of quantization levels per sample, each of which increases the MSE of the reconstructed signal. For this reason there is interest in developing modifications of the standard A-D conversion that will achieve some degree of rate reduction without sacrificing significant reconstruction fidelity. This operation of converting source samples to bit sequences so as to minimize the resulting bit rate is called *source encoding*.

Over the past decade, analytical approaches to generalized source encoding have been developed [3–5], and many elegant results have been derived that predict the theoretical limitations and capabilities of source encoding. Although such theoretical results are usually confined to specific system models, and unfortunately do not always yield practical design solutions, they do serve as a useful guideline for system implementations. A vigorous development of source encoding is well beyond our scope here, but we may digress momentarily to indicate some ways in which standard A-D conversion may be modified to achieve some degree of source rate reduction.

One simple way to source encode is to utilize a more sophisticated quantization encoding, other than natural encoding. Consider the example illustrated in Table 6.2. A four level quantizer (or perhaps a four level command generator) can produce four possible quantization or sample values, which we simply number, as shown in column 1, but each occurs with the probability shown in column 2. Natural encoding would use the quantization words of column 3 for representations, ignoring the relative frequency of occurrence, and would therefore continually operate at the encode rate of two bits per sample. As a modification for source encoding, we might consider using instead quantization words of different lengths, using the shortest words to code the quantization values occurring most often, and using longer words for those least used. An example of such encoding is shown in column 4. The average number of source bits used per sample will be given by the sum of the number of bits per word times the probability of each occurring. Hence, from columns 1 and 4,

$$\text{Average bits per sample} = \tfrac{1}{2}(1) + \tfrac{1}{4}(2) + \tfrac{1}{4}(3)$$

$$= 1.75 \qquad (6.5.1)$$

We have therefore reduced the average rate of bit transmission over that of natural encoding by 12%. The required receiver reconstruction operation

Table 6.2. Source Encoding Example

Quantization or Sample Values	Probability of Occurrence	Natural Encoding	Unequal Word Length Encoding
0	$\frac{1}{8}$	00	001
1	$\frac{1}{2}$	01	1
2	$\frac{1}{4}$	10	01
3	$\frac{1}{8}$	11	000

is now more complicated, however, since the words of different lengths must be recognized. This requires that the reconstruction subsystem be exactly in step with the received words, so that individual words can be separated. (Note, in this example, the appearance of either a one or three consecutive zeros signifies the end of a quantization word, provided that the examination is started at the beginning of each word.) Hence rate reduction is obtained at the expense of more stringent reconstruction processing.

The average number of bits per sample in (6.5.1) is in fact the minimum possible for the example in Table 6.2. The minimum value of the average number of bits per sample needed to encode a specific set of quantization values with known a priori probabilities is called the *entropy* of the set. If the samples are independent,* this entropy can be determined from the occurrence probabilities P_i of each quantization value by

$$\mathcal{H} = - \sum_{i=1}^{L} P_i \log_2 P_i \qquad (6.5.2)$$

where the probabilities P_i are obtained from (6.1.26). For a given quantizer, the entropy is therefore directly dependent on the source density $p_d(x)$. In some cases (e.g., Table 6.2) we can encode each sample so as to operate at the average bits per sample given by the quantizer entropy. A coding procedure, called *Huffman* encoding [6], generates the required binary word that should be used for each quantization value so as to achieve the smallest average number of bits per sample for the given a priori probabilities. This coding technique is diagrammed in Figure 6.16 and is explained as follows. The original quantization values to be encoded are first ranked in descending order of their probabilities. The last two values are combined into a single sample value and labeled with the sum probability of the two. This combined sample is then grouped with the remaining $L-2$ samples to form a new set of $L-1$ sample values that are then ranked again according to their probabilities. The process is repreated for the new sample set, again adjoining the two lowest into a combined sample to be reranked with the remainders, and forming a new set of $L-2$ samples. The process is continually repeated, keeping track of the original samples and the way they are combined (Figure 6.16), until a final set of only two samples remain. Encoding is then achieved by starting with the final two samples and assigning zeros and ones as we traverse back through the diagram. Initially we assign a zero and one to each of the last two

---

* If the samples are not independent (i.e., correlation exists from one sample in time to the next) then average entropy over a sequence of samples must be determined, using joint sample probabilities.

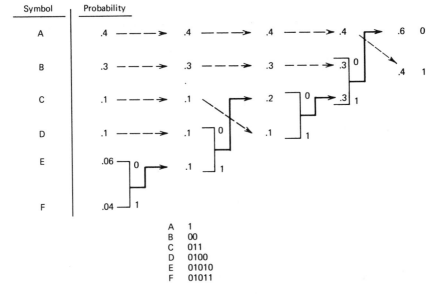

A   1
B   00
C   011
D   0100
E   01010
F   01011

Figure 6.16. Huffman encoding example.

samples. As we move back one step we append bits to our earlier bit assignments, adding a one and zero to each member of a combined sample. Whenever we label an original sample, it is removed from the procedure with the code sequence up to that point. Continuing back in this manner, we eventually code all original samples with recognizable codes. It has been shown that this method achieves the minimum average number of bits per sample [7, 8]. Note that in Figure 6.16 we do not achieve the entropy of the sample set, even though we have minimized the average number of bits to be used per sample.

Another way to achieve bit rate reduction is by the generalized A-D system shown in Figure 6.17. Rather than quantize directly the message

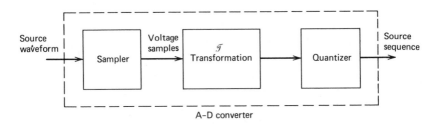

Figure 6.17. Generalized A–D converter.

samples as they occur, we perform a transformation $\mathcal{T}$ on the sample values prior to quantizing. Thus we instead quantize a function of a sequence of samples rather than the samples themselves. Such generalized A-D conversion systems have been referred to as *block quantizers* [9, 10] and form the basis for *redundancy removal* systems and *orthogonal transformation* systems. In redundancy removal systems, the basic role of the transformation $\mathcal{T}$ is to take advantage of correlation that may exist from sample to sample. This means that knowledge of n samples provides some information about the $(n + 1)$ sample, and the latter can be quantized differently from the way it would if we were quantizing it alone. Since some information about the $(n + 1)$ voltage value can be obtained from the previous n samples, we require less bits to quantize it with the same accuracy, and the overall source bit rate can be reduced. At the receiver each sample value is reconstructed from its quantization word and the values of the previous samples. Roughly speaking, the previous samples are used to project an estimate of the present sample, and the quantization word is used to refine that estimate. We shall not derive here the internal structure of the transformation estimator $\mathcal{T}$, except to comment that its optimal design requires exact knowledge of the source sample statistics. It is common to model sources as being some form of nth order *Markov sources*, defined as a source whose present voltage sample value is statistically dependent on n of its past sample values. Obviously, the complexity of the transformation depends on the order n of the Markovian model. Thus redundancy removal systems, constructed as in Figure 6.17 with a fixed preselected transformation, are extremely model dependent in their design and performance.

When source sample statistics are not known and cannot be adequately modeled, the transformation operation can be used to perform sample to sample comparison, so that adjacent samples need not be sent unless sufficiently different. In essence, the transformation $\mathcal{T}$ is being used to perform an "instantaneous correlation" measurement. Such approaches have led to the implementation of various types of *differential quantizers*, in which only sample differences are quantized instead of actual sample values [11, 12]. Hence if a long sequence of almost identical samples occurred, a significant reduction in the bit transmission rate can be made during their occurrence. This allows a form of adaptive adjustment of the instantaneous transmisson rate, according to the actual time variations of the sample differences. However, such variations of the A-D output rate require that some type of storage (*buffers*) be provided if the transmission over the link is desired at a constant bit rate.

There has been particular interest in considering generalized A-D conversion where the transformation $\mathcal{T}$ corresponds to a class of linear orthogonal transforms. Such transformations can be regarded as a matrix multiplication of the source samples into a set of new coordinates to represent the source message. These new coordinates are then quantized and transmitted in lieu of the original samples. At the receiver, the inverse orthogonal transform is used to convert the recovered coordinates back to the original source samples for reconstruction. The orthogonal transform causes the coordinate energy to be redistributed, and a transmission advantage occurs if the new coordinates can be sent with less bits (or more accuracy). For example, if N successive voltage samples $\{d_i\}$ had equal variance, each sample would have to be quantized with the same number of bits. If an orthogonal transformation of the N samples could be found that will produce N new coordinates with an unequal variance distribution, a smaller MSQE will be produced at the same bit rate by variable coordinate quantization. That is, the same number of total bits that were available for quantizing the original N samples can now be distributed to each new coordinate in proportion to its variance. The coordinates with the higher variance (energy) are therefore quantized more accurately, resulting in a smaller MSQE. (The proof of this can be found in Reference 17). Conversely, we can transmit the coordinates with the same accuracy at a smaller bit rate. The larger the differences in the coordinate variance distribution, the greater will be the degree of improvement. Hence orthogonal transformations that redistribute the sample energy in the most nonuniform manner are the most advantageous. We point out, however, that physical implementation of the quantizer requires unequal numbers of quantization intervals for each coordinate, varying according to the variance distribution. In addition, performance improves with N, the number of samples being transformed at one time. However, the orthogonal transform requires an N by N matrix multiplication and is therefore limited by computational complexity. (In general, on the order of N^2 computations must be performed on the original source samples.) For this reason there has been interest in finding orthogonal transforms that allow simplicity and ease in computation. Classes of Haar [13], Walsh [14], Hadamard [15], and discrete Fourier transforms [16] have been found useful in this regard. More recently, computer algorithms (called *fast transforms*) have been developed that allow versions of these transforms to be performed with considerable reduction in computation. These fast transforms require on the order of only $(N \log N)$ computations to carry out the matrix transformation. It must also be remembered that for any transform, its inverse must also be determined for use at the reconstruction end for coordinate inversion. Thus practical source encoding by means of orthogonal coordinate transforms reduces to primarily a search for orthogonal matrixes that are computational simplified, are easy to invert, and yield acceptable variance distributions for quantizing. The previously stated orthogonal transforms have been reported as being successively applied to voice and image transmission [17–20].

6.6. Time Division Multiplexing of Digital Sources

In Chapter 5 we considered the simultaneous transmission of waveforms from a set of analog sources by multiplexing the source waveforms into a single waveform. With digital sources there is often a similar requirement for simultaneous transmission of the output bit sequences of a set of digital sources in the same way. For example, there may be several digital sources operating simultaneously in a particular system. The sources produce bits at various rates, and it is desired to transmit these symbols simultaneously to the receiver. One method is again to use a FDM format in which specific parts of the frequency spectrum are assigned to a particular source in forming the baseband. (This is discussed in detail after discussion of the possible signaling formats that can be used for bit transmission of a single source.) A second method is to attempt to interleave the bits from each source prior to encoding and to produce a single bit sequence for transmission. This represents a form of *time division multiplexing* (TDM) of the sources. The operation of combining the digital sequences is sometimes referred to as *digital multiplexing,* or *parallel-serial conversion.* After reception of the multiplexed bit sequence at the receiver, a demultiplexing, or deinterleaving, operation must be performed to separate the individual bit streams. After separation these bit streams can be simultaneously used in separate channels for D-A conversion or command execution.

Interleaving bits is fairly straightforward when all sources produce bits at the same rate. We need only commutate over the source bit set with a high speed rotating switch, reading out one bit at a time from each source in sequence and continually recycling (Figure 6.18a). When the source rates are unequal, however, a slightly more complicated procedure must be used involving a parallel-serial converter (PSC). A PSC is composed of unequal *bit shift registers** placed in parallel with a single commutation switch (Figure 6.18b). A register must be available for each different source. The different bit streams are simultaneously loaded into the appropriate registers, and the entire register contents read out in sequence at fixed time intervals to accomplish the multiplexing.

---

* A shift register is a cascade of binary flip-flop circuits. At prescribed time intervals (called the register clock rate) each stage of the register can take on the binary value of the previous stage, with the first stage taking on a new value and the last stage dropping its bit. Thus input bit sequences can be made to shift through the register at a prescribed clock rate.

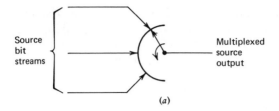

(a)

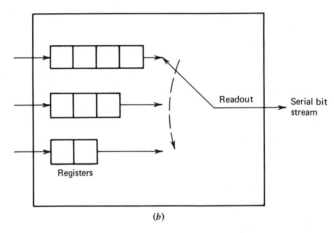

(b)

Figure 6.18. Parallel-serial converter. (*a*) Equal bit rates with commutating switch, (*b*) unequal bit rates with shift registers.

Assume there are N sources, having rates $\mathscr{R}_i$ bits/sec, $i = 1, 2, \ldots, N$, fed into a specific register of a PSC at its own rate. The register lengths (number of register stages) must be selected such that after some interval in time, all the registers will fill simultaneously. If h_i is the register length of the ith such register of a particular PSC, this loading condition requires that a time τ exists such that the number of bits of rate $\mathscr{R}_i$ occurring in τ sec is exactly h_i. Hence τ must be such

$$\mathscr{R}_i \tau = h_i \tag{6.6.1}$$

for all inputs to the PSC. This immediately implies that for any two input rates, say $\mathscr{R}_i$ and $\mathscr{R}_j$, the corresponding register lengths must be related by

$$\frac{h_i}{h_j} = \frac{\mathscr{R}_i}{\mathscr{R}_j} \tag{6.6.2}$$

For smallest register lengths (simplest PSC) it is necessary that the (h_i) have their smallest possible integer value. Thus each h_i should be the smallest factor of the set of input rates $(\mathcal{R}_i)$. Equivalently, the set of input rates should be related to the set of register lengths by

$$\mathcal{R}_i = \alpha h_i \tag{6.6.3}$$

where α is the largest common multiple of all rates. We see that a set of PSC input rates specifies the required register lengths, whereas a set of registers specifies the ratios of input rates. The minimal register size will occur if the h_i ar the smallest prime factors, and the rates $\mathcal{R}_i$ are selected to be multiples of these prime numbers. As an example, suppose we wish to multiplex three digital sources having the rates $\mathcal{R}_1 = 72$ bps, $\mathcal{R}_2 = 48$ bps, and $\mathcal{R}_3 = 60$ bps. The smallest integer factors of these rates are then $h_1 = 6$, $h_2 = 4$, and $h_3 = 5$, and the common factor is 12 (i.e., $\mathcal{R}_i = 12h_i$). A PSC as in Figure 6.18b would therefore require three parallel registers with 6, 4, and 5 stages, respectively.

After the registers of the PSC are filled, the bits can be read out from each register in sequence at a constant readout rate. That is, the commutating switch in Figure 6.18b moves to the first register, reads out all stored bits in sequence, then moves to the second register and reads out all bits, and so on. One complete cycle of register readout must occur in τ sec, the time it takes to fill each, with each register refilling immediately after emptying. Hence the output serial bit rate of the PSC is

$$\mathcal{R}_o = \frac{1}{\tau} \sum_{i=1}^{N} h_i = \sum_{i=1}^{N} \mathcal{R}_i \text{ bits/sec} \tag{6.6.4}$$

The multiplexed output rate is therefore always equal to the sum of the individual source rates.

Whenever several data sources are multiplexed, it is generally necessary to insert synchronization bits periodically to aid in the receiver demultiplexing. If K_s sync bits are inserted after every K_b data bits at the output of each PSC, then the data rate of the PSC output bit stream is increased by the factor

$$\mathcal{R}_s = \frac{K_s}{(K_b/\mathcal{R}_o)} = \mathcal{R}_o\left(\frac{K_s}{K_b}\right) \tag{6.6.5}$$

The total effective PSC multiplexed output bit rate is then

$$\mathcal{R}_o + \mathcal{R}_s = \mathcal{R}_o\left(1 + \frac{K_s}{K_b}\right) \tag{6.6.6}$$

Deinterleaving of the multiplexed bit stream is accomplished by a commutator operating in synchronism with the arriving bit stream. The synchronized commutator taps off the group of bits of each source in sequence, feeding them to the proper destination. The sync bits, inserted at the PSC, maintain the required commutation synchronism for this deinterleaving.

6.7. Symbol Transmission and Channel Encoding

We have been discussing digital systems only in terms of the generation of the source symbols and the manner in which the recovered symbols are used for reconstruction. In Section 6.3 we acknowledged the possibility of a particular source symbol being recovered erroneously because of the transmission system and evaluated its effect on the mean squared reconstruction error. However, we have essentially neglected the basic details of the link that transmits these symbols from source to the reconstruction point. This transmission is typically accomplished by converting the source symbols into baseband waveforms for RF transmission, as was shown in Figure 6.1. Following RF demodulation the recovered baseband is inverted back to the symbol stream from which reconstruction, deinterleaving, or command recognition can take place. The operation of converting the source bit stream into baseband waveforms is referred to as *channel encoding*. The operation of recovering the source symbols from the demodulated baseband is called *channel decoding*. Since the effect of transmitting the baseband waveform over the RF links and demodulating has been studied previously, our earlier results are applicable here and there remains only the investigation of the actual channel encoding and decoding operations (i.e., the formulation of the baseband waveform and the associated decoder processing of the recovered baseband). These operations will eventually determine the bit and word error probabilities that we assumed in our earlier discussions and therefore will greatly impact system performance and related design.

Source symbols are transmitted by one of three basic channel encoding methods—*binary*, *block*, or *convolutional* encoding. In binary encoding, the source sequence of zeros and ones is transmitted bit by bit, encoding each one at a time, in turn, into a known baseband signal for channel transmission. Thus there must be two available baseband waveforms, one for a binary one and the other for a binary zero, to allow subsequent bit transmissions, each limited to a finite time interval. The bits are therefore converted, in sequence, to the appropriate baseband signal. This binary encoding operation can be represented symbolically as

$$1 \rightarrow s_1(t), \qquad 0 \le t \le T$$
$$0 \rightarrow s_0(t), \qquad 0 \le t \le T \tag{6.7.1}$$

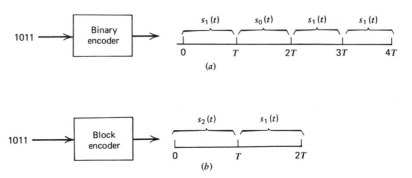

Figure 6.19. Encoding model. (*a*) Binary, (*b*) block.

That is, a binary one is converted to the waveform $s_1(t)$ whenever it occurs in the binary sequence, and a binary zero is converted to $s_0(t)$. Each signal is T sec long, where T is now the time to transmit a data bit. As an example, the binary sequence 1011 would be binary encoded into the baseband waveform shown in Figure 6.19*a*. This waveform will then be transmitted over the carrier link described earlier.

The receiver decoder attempts to reconstruct the source sequence by determining whether a zero or one is transmitted during each T sec bit period. This is equivalent to deciding whether the known waveforms, $s_1(t)$ or $s_0(t)$, are being received during each T sec interval. A decoder decision-ing error will therefore produce a bit error in the decoded sequence. Note the decoder requirement is not one of recovering the baseband waveforms, but simply of determining which one is being transmitted. For this reason channel decoding theory can be encouched within a framework of decision theory and hypothesis testing. Although we choose not to develop this approach here, the reader may be interested in pursuing this theoretical notion in the texts of Middleton [21], Van Trees [1], and others [22, 23].

In block coding, the source sequence is transmitted in blocks of bits, rather than one at a time. The encoder converts each possible block of (say) k bits into a known baseband signal for transmission. Since a block of k binary symbols can take $M = 2^k$ different forms, there must be M distinct signals available to the encoder. For example, the block encoding opera-tion for $k = 2$ would be represented by

$$11 \rightarrow s_1(t),$$
$$10 \rightarrow s_2(t),$$
$$01 \rightarrow s_3(t), \qquad 0 \le t \le T \qquad (6.7.2)$$
$$00 \rightarrow s_4(t),$$

The baseband waveform for block encoding the sequence 1011 using (6.7.2) is shown in Figure 6.19*b*. A decoder decisioning error will again produce bit errors in the decoded sequence. The parameter k is called the block *length*, and we note that the parameter T now denotes the time to transmit k bits of data. The receiver decoder now must decide which of $M = 2^k$ possible waveforms is being received during each T sec interval, and a correct decision now represents the correct decoding of k consecutive source bits.

In convolutional encoding, the transmitted waveforms again correspond to blocks of symbols, but the symbols do not correspond directly to sequential blocks of source bits. Instead the transmitted waveforms are generated from symbol blocks that effectively slide along the source sequence, rather than jumping to consecutive, nonoverlapping block segments. This produces a form of interleaved block encoding that is difficult to describe by simple waveform allocation, as in (6.7.2). Convolutional encoding will be described in more detail in Chapter 8.

An implicit requirement in all decoding operations is that the T sec intervals be known exactly at the receiver (i.e., when they start and when they end). Designation of these intervals is referred to as receiver *timing* and is provided by a timing subsystem that operates in conjunction with the decoder. This timing subsystem therefore becomes part of the synchronization block depicted back in Figure 1.2. Timing subsystems must make use of timing information provided by the transmitter encoder. The methods by which this information is provided, and the subsequent timing subsystem design, are discussed in Chapter 9.

The task of selecting waveforms for the channel encoding can be generalized somewhat by introducing a general form for the waveform representation. This can be accomplished by use of orthonormal expansions to represent each signal. Let $\Phi_j(t), j = 1, 2, \ldots, \nu$ be a set of orthonormal functions over $(0, T)$. That is, let

$$\int_0^T \Phi_i(t)\Phi_j(t)\, dt = 1, \qquad i = j$$

$$= 0, \qquad i \neq j \qquad (6.7.3)$$

The waveforms to be used for symbol transmission can then be represented by an orthonormal expansion with these functions. Thus we write

$$s_i(t) = \sum_{j=1}^{\nu} s_{ij}\Phi_j(t), \qquad i = 1, 2, \ldots, M \qquad (6.7.4)$$

where M is the number of signals needed in the block encoding set and s_{ij} are the signal coordinates with respect to each orthonormal function. For a given orthonormal set $\{\Phi_j(t)\}$ we see that each signal $s_i(t)$ is determined only by its coordinates $\{s_{ij}\}$. We can therefore denote the signal vector $\mathbf{s}_i$ corresponding to the waveform $s_i(t)$ by the vector of its coordinates

$$\mathbf{s}_i = (s_{i1}, s_{i2} \ldots s_{iv}), \qquad i = 1, 2, \ldots, M \qquad (6.7.5)$$

Thus a set of signals can be equivalently represented by the vectors set $\mathbf{s}_i$. Given each signal vector $\mathbf{s}_i$, the waveform of the signal can be generated by summing (6.7.4), using the vector coordinates and the set of orthonormal functions. Since the orthonormal set is arbitrary, there may be many different waveforms that can be associated with a particular signal vector. Likewise, there may be many different waveform sets that can be associated with a particular vector set. The selection of the orthonormal set of time functions to be used in the signal generation generally is based on engineering design considerations. This is explored in Chapters 7 and 8.

Once the orthonormal function set has been selected, the block encoding operation of converting blocks of bits to waveforms is therefore identical to converting the block to signal vectors, since the vectors specify the waveforms. Thus the basic block encoding mapping, as shown, for example, in (6.7.2), can equivalently be viewed as a mapping from binary blocks to a set of signal vectors. Note that by this equivalence we see that the basic channel encoding operation is now divided into separate and distinct tasks: (1) finding the appropriate vector set for the encoding and (2) finding an orthonormal function set that generates waveforms suitable for signal transmission. The former involves finding vector sets with an inherent encoding advantage, whereas the latter is influenced by engineering considerations (ease of generation, bandwidth utilization, processing convenience, etc.). This dichotomy allows us to separate the channel encoding operation of Figure 6.19 into a cascade of two separate operations, as shown in Figure 6.20. In the first operation the block of bits is converted to the appropriate vector, and in the second the vector is converted to its corresponding waveform. Sometimes, in the literature, the first block is considered to perform the channel encoding and the second block is referred to as an encoding *modulator*, in which the encoding vector is modulated onto

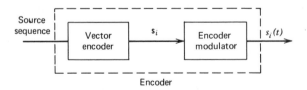

Figure 6.20. Encoder equivalent model.

the orthonormal functions. The actual demarcation between channel encoding and modulation need not concern us, since the overall result is the same; that is, an encoded baseband waveform is formed from the source sequence.

The basic objective in the selection of the vector sets, and the mapping of the source blocks onto the vectors, is of course to achieve maximum decoding performance. It is difficult at this point to discuss the vector encoding procedure, since the structure of the decoder has not yet been developed. We must therefore postpone temporarily a rigorous development of channel encoding until the various decoding alternatives are presented. However, we may digress here to point out one application that demonstrates the usefulness of a channel encoding operation. Let us restrict attention to the special case where the signal vector components themselves are binary symbols. That is, each s_{ij} in (6.7.5) can take on only one of two values $(0, 1)$, and the signal vectors s_i are themselves binary sequences. The channel encoding operation is therefore equivalent to encoding blocks of source bits into binary sequences. In other words, for this example, channel encoding converts one binary word into another binary word. Thus included in the framework of channel encoding is the operation of simply converting source block words into newer words. The selection of the newer encoded words is to achieve some transmission advantage. Consider the example summarized in Table 6.3. The four possible forms of a two bit source word are shown in column 1. A given source word will be transmitted in error if either bit of the word is incorrectly decoded, no matter what transmission method is used. Suppose instead the source words are channel encoded into the new words in column 2 prior to transmission. These new words are formed by simply adding a bit (called a *check* bit or *parity* bit) to the end of each source word such that the total number of ones in each word is always even. We now binary transmit, one symbol at a time, the new digital word instead of the original two bits. Introducing this check symbol allows the receiver to

Table 6.3. Encoding Example

Source Word	Three Bit Encoded Word	Five Bit Encoded Word
00	000	00000
01	011	00111
10	101	11011
11	110	11100

recognize immediately when a single bit error of the new word has been made during decoding, since the parity check on the ones will not be satisfied. Such erroneous words can therefore be rejected rather than used incorrectly. A true word error occurs only if two or three symbols of the new word are incorrectly decoded. Thus the check symbol has improved the transmission capability, since it requires more than one new word symbol error to make a source word error. However, the new words use three symbols per sample, and the required channel transmission bit rate with this encoding is increased over the source bit rate in (6.1.1). Word protection is therefore achieved by this form of channel encoding, at the expense of a higher required transmission rate.

The result can be extended to the new words in column 3. Here five symbols are used to transmit the original two bits. Each of the five symbols are binary encoded and transmitted individually to represent the word. Decoding is accomplished by comparing the five decoded symbols (which may have errors) with each of the four new words, and selecting the word that differs in the fewest symbols. We see that one or two symbol errors in decoding the new word will always be recognized as a word different from the four, and three symbol errors are necessary before a word error is made. The word error probability is further improved, but again the rate is increased. Note that this word set has the property that a single symbol decoding error in the new words will actually be corrected, since the subsequent word comparison will always conclude the correct word was transmitted. Two symbol errors will produce a word that may differ in two places from two different words of the original set, and therefore cannot always be corrected. Hence the channel encoded words in column 3 have the ability to correct single errors while at least recognizing the presence of two symbol errors (in which case the erroneous word can be rejected).

This idea of encoding source blocks into new words with bit error correction and bit error recognition properties can be unified into a basic *error correction* coding theory [24]. To formulate this briefly, we define a code set as the set of binary vectors used by the channel encoder for its output. The words of columns 2 and 3 constitute such a code set. We define the *Hamming distance* between any two binary vectors as the number of positions where corresponding bits differ. The *code set distance* is the smallest distance among all pairs of codes in the set. We see that the code set in column 2 has distance 2, whereas that of column 3 has distance 3. (Column 1, considered as another code set, would have distance 1.) The code set distance determines its ability to correct and recognize bit transmisssion errors. If the code distance is 1, a single bit error can cause a word error. If the code distance is 2, a single bit error will always be recognized (since by definition it will not appear as any other word). If the code distance is 3, a single bit error will always produce a word that is of distance one from the correct word, and a distance 2 from all others. Hence a channel decoding operation that selects the "nearest" word (in terms of the defined distance) will always decode correctly, effectively correcting the bit error. Two errors, although recognizable as not being a word of the set, may lie equidistant from two different code set words, and therefore are not always correctable. We see therefore why the original bits of column 1 could withstand no errors, the words of column 2 recognize single errors, and the words of column 3 correct single errors. To generalize this notion, a code set with distance d can always correct $(d-1)\frac{1}{2}$ errors and detect $d-1$ errors.

The code set in column 3 is referred to as a linear $(5, 2)$ code set, since it uses words of five binary symbols to encode the original two bits. Column 2 is a $(3, 2)$ code set. The notion of code set distance allows us to generalize error correction encoding to arbitrary (n, k) codes. We see that we will be basically interested in the relation between n, k, and the code set distance. For a given n and k, we would like to find the particular code set with the largest distance. On the other hand, for a specific distance (error correcting capability) we would like to determine the smallest n (lowest transmission rate). In addition, we may wish to impose additional constraints on the code set. For example, we might require the words of a code set to have also the property that no overlap of any two code words forms another code word of the set (*comma-free code sets*) or that the beginning and ending of the code set words are always distinguishable (*synchronizable code sets*). The derivation of such code sets, and the relations among code set word length and distance, is a basic part of error correction theory. The interested reader could pursue these topics in References 4, 8, 9, 24.

In summary, then, we have formulated in this section the basic channel encoding operation of digital system design. We have described channel encoding both in terms of a symbol waveform transmission operation and, equivalently, as a mathematical mapping of source bits into signal vectors. It is important to keep both of these approaches continually in mind. The waveform transmission formulation retains us in the realm of practical engineering design, where hardware considerations play a dominating role. On the other hand, a mathematical model allows us deeper insight into the encoding operation while dictating design directions. We have shown how channel encoding, in the special case of binary vectors and binary transmission, can be used to achieve source word protection through error correction. Unfortunately, we cannot pursue more specific applications of the block encoding operation until the mechanics of receiver decoding have been considered. In the next two chapters we investigate channel decoding systems in detail, devoting Chapter 7 to the binary encoding case and Chapter 8 to block encoding systems. In the latter chapter especially, the basic signal vector model for channel encoding presented in this section is revisited.

6.8. Information Theory and Digital Transmission

Before an engineer delves into the detailed design of a digital transmission system (as we do in the next chapter) it is advantageous to be aware of any inherent limitations to transmitting binary symbols over a communication link. That is, it is helpful to know (apart from device imperfections) the overriding system parameters and the way in which they eventually govern bit transmission capability. In our case, the basic parameters are those of the baseband and carrier subsystems. Initial formulation of this approach to digital transmission began with the work of Hartley [25], was further advanced by Shannon [26], and was subsequently extended by many others. The outgrowth of such studies was the development of an area referred to as *information theory*. Information theory deals with the more profound topic of exactly what constitutes information, and it emphasizes the analytical and theoretical aspect of the transfer of such information from one point to another. Any engineer deeply involved with the transmission of digital information will doubtlessly become involved, in some fashion, with its fundamental principles. Information theory is, of course, a complete topic in itself, as evidenced by the many texts encompassing this field [4, 8, 27–29]. Obviously, a complete treatise is beyond our objective here. However, several basic points, applicable to our discussion, might be summarized for a better overall understanding of our design goals.

The most fundamental result in information theory is the celebrated theorem of Shannon. A version of this result, stated simply without proof in our terminology, is as follows. If the baseband subsystem over which the bit symbols are to be sent as baseband waveforms is limited to a power level of P_m and bandlimited to B_m Hz, then if the baseband interference is composed of white Gaussian noise of spectral level N_{b0}, the maximum rate at which source bits can be sent over the link is given by

$$\mathscr{C} = B_m \log_2\left(1 + \frac{P_m}{N_{b0}B_m}\right) \text{bits/sec} \qquad (6.8.1)$$

The parameter $\mathscr{C}$ is called the *channel capacity* of the link. Shannon showed that if we attempt to send source bits at a rate faster than $\mathscr{C}$ over the link, we are certain to make bit errors, no matter how sophisticated the encoding and decoding procedure. Remarkably, he also showed that if we send source bits at a rate less than $\mathscr{C}$, there exists an encoding and decoding procedure for which no bit errors are ever made. That is, we achieve errorless transmission. Unfortunately, the encoding and decoding required for this errorless transmission cannot always be found, and the price that must be paid for such a perfect system is not always obvious from the theorem itself. Nevertheless, we are guaranteed that there is indeed a way to achieve error-free communication, so long as we transmit source bits at a rate less than $\mathscr{C}$. Equation (6.8.1) is therefore a theoretical upper limit on errorless signaling. This result allows us, at least in a gross sense, to match the source rate and the baseband channel properly. For example, if we had an A-D source producing bits at the rate $\mathscr{R}$, then it is necessary that we construct a baseband subsystem whose capacity $\mathscr{C}$ exceeds $\mathscr{R}$, if we are to have any hope of errorless transmission. Likewise, if we are dealing with a source whose inherent entropy in (6.5.2) predicts a minimal possible source rate exceeding the capacity of the baseband, then we know there is no possible way to transmit the source bits without error. Our only alternative is to find encoding and decoding procedures that produce as little distortion as possible for the error rate that occurs.

In examining the capacity formula in (6.8.1), we note that $\mathscr{C}$ depends on both link bandwidth and link SNR, the latter given by $P_m/N_{b0}B_m$. This immediately points out that the theoretical link capacity can be increased either with bandwidth or with power levels. We emphasize this point, since there is a tendency to believe that transmission bit rates are related only to the available bandwidth—that is, how fast pulses can be forced over the channel. In fact, we see that a capacity much greater than B_m can be achieved if we have a high enough SNR. On the other hand, $\mathscr{C} \to 0$ as $P_m \to 0$, no matter what B_m is. Furthermore, if we allow $B_m \to \infty$ (no bandwidth restrictions) we see

$$\lim_{B_m \to \infty} \mathscr{C} = \lim_{B_m \to \infty} \log_2 \left(1 + \frac{P_m}{N_{b0}B_m}\right)^{B_m}$$

$$= (\log_2 e)\frac{P_m}{N_{b0}} \tag{6.8.2}$$

The limiting capacity is therfore not infinite and is ultimately limited by the effective SNR above. Nevertheless, (6.8.1) does imply an inherent trade-off of baseband bandwidth and power in achieving a desired capacity. We can maintain the same information capacity by decreasing B_m, provided we achieve an adequate increase in P_m. Conversely, baseband power can be reduced if we expand B_m sufficiently. We had noted this type of trade-off before with analog transmission, and we see here that it is an intrinsic property of digital transmission as well. We emphasize that (6.8.1) is valid only for white noise, which is not true for all our baseband subsystem models. In later work we shall reexmine some of the comments made here.

References

1. Van Trees, H. *Detection, Estimation, and Modulation Theory*, Part 1, Wiley, New York, 1968, Chap. 3.
2. Max, J. "Quantizing for Minimum Distortion," *IRE Trans.*, vol. IT-6, March 1960, pp. 7–12.
3. Berger, T. *Rate Distortion Theory*, Prentice-Hall, Englewood Cliffs, N.J., 1971.
4. Gallagher, R. *Information Theory and Reliable Communications*, McGraw-Hill, New York, 1969.
5. Wozencraft, W. and Jacobs, I. *Principles of Communication Engineering*, Wiley, New York, 1965.
6. Huffman, D. "A Method for Constructing Minimum Redundancy Codes," *Proc. IRE*, vol. 40, May 1952, pp. 1098–1101.
7. Abramson, N. *Information Theory and Coding*, McGraw-Hill, New York, 1963.
8. Stiffler, J. *Theory of Synchronous Communication*, Part 3, Prentice-Hall, Englewood Cliffs, N.J., 1971.
9. Huang, J. and Shultheiss, P. "Block Quantization of Correlated Gaussian Random Variables," *IEEE Trans. Comm. Syst.*, vol. CS-11, September 1963, pp. 289–296.
10. Kurtenback, A. and Wintz, P. "Quantizing for Noisy Channels," *IEEE Trans. Comm.*, vol. COM-17, April 1969, pp. 291–302.
11. Bodycomb, J. and Haddad, A. "Some Properties of Predictive Quantizing Systems," *IEEE Trans. Comm.*, vol. COM-17, October 1970, pp. 682–685.
12. Davisson, L., "Theory of Adaptive Data Compression," in *Advances in Communication Systems*, edited by A. Balakrishman, Academic, New York, 1966.
13. Andrews, H. and Caspari, K. "A Generalized Technique for Spectral Analysis," *IEEE Trans. Comput.*, vol. C-19, January 1970, pp. 16–25.
14. Harmuth, H. *Transmission of Information by Orthogonal Functions*, Springer, New York, 1969.
15. Pratt, W. and Kane, J. "Hadamard Transform of Image Coding," *Proc. IEEE*, vol. 57, no. 1, January 1969, pp. 58–65.
16. Oppenheim, A. and Schaefer, R. *Digital Signal Processing*, Prentice-Hall, Englewood Cliffs, N.J., 1975.
17. Campenella, S. and Robinson, G. "A Comparison of Orthogonal Transformations for Digital Speech Processing," *IEEE Trans. Comm.*, vol. COM-19, December 1971, pp. 1645–1649.
18. Crowther, W. and Rader, C. "Efficient Coding of Vocoder Signals Using Linear Transformations," *Proc. IEEE*, vol. 54, November 1966, pp. 1594–1601.
19. Habibi, A. and Wintz, P. "Image Coding by Linear Transformations, *IEEE Trans Comm.*, vol. COM-18, February 1971, pp. 56–62.
20. Pratt, W. "Transform Coding of Color Images," *IEEE Trans. Comm.*, vol. COM-18, February 1971, pp. 980-988.
21. Middleton, D. *Introduction to Statistical Communication Theory*, McGraw-Hill, New York, 1960, Chap. 18.
22. Helstrom, C. *Statistical Theory of Signal Detection*, Pergamon Press, New York, 1960.
23. Hancock, J. and Wintz, P. *Signal Detection Theory*, McGraw-Hill, New York, 1966.

24. Peterson, W. *Error Correcting Codes*, MIT Press, Cambridge, and Wiley, New York, 1961.

25. Hartley, R. "Transmission of Information," *Bell Syst. Tech. J.*, 1928, pp. 535–563.

26. Shannon, C. "A Mathematical Theory of Communications, *Bell Syst. Tech. J.*, vol. 27, July 1948, pp. 379–423.

27. Ash, R. *Information Theory*, Interscience, New York, 1965.

28. Fano, R. *Transmission of Information*, MIT Press, Cambridge, Mass., 1961.

29. Feinstein, A. *Foundations of Information Theory*, McGraw-Hill, New York, 1958.

Problems

1. (6.1) Prove the converse of the sampling theorem: If $d(t)$ is time limited to T sec, then it can always be represented by its frequency samples taken every $1/T$ sec apart.

2. (6.1) A system reconstructs a waveform from its time samples $d(nT)$ by using the general summation

$$\hat{d}(t) = \sum_{n=-\infty}^{\infty} d(nT)h(t-nT)$$

(a) Show that the resulting waveform corresponds to the waveform $d(t)$ filtered by the transform of $h(t)$.

(b) Use the result of (a) to determine the frequency spectrum when $h(t)$ corresponds to a sample-and-hold circuit, that is, $h(t) = 1$, $0 \le t \le T$.

(c) Repeat when a linear extrapolator is used. (A straight line is drawn between sample values.) (*Hint:* First write the reconstructed waveform as in $\hat{d}(t)$ above, using the fact that

$$\hat{d}(t) = d(nT) + [d((n+1)T) - d(nT)](t-nT)/T$$

3. (6.1) Find the distortion effect that occurs with taking nonperfect waveform samples; that is,

$$\hat{d}(nT) = \int_{-\infty}^{\infty} d(t)p(t-nT)\,dt$$

where $p(t)$ is a function over $(-T, T)$, centered at $t = 0$.

4. (6.1) Prove the following identities:

(a) $$\sum_{k=-\infty}^{\infty} \delta(\omega - kp) = \frac{1}{p} \sum_{n=-\infty}^{\infty} e^{-jn2\pi/p}$$

(b) $$\mathscr{F}\left[\sum_{n=-\infty}^{\infty} \delta(t-nT)\right] = \frac{2\pi}{T} \sum_{k=-\infty}^{\infty} \delta\left(\omega - \frac{2\pi k}{T}\right)$$

(c) $\qquad \mathscr{F}\left[\sum_{n=-\infty}^{\infty} x(t-nT)\right] = \frac{2\pi}{T} \sum_{i=-\infty}^{\infty} X\left(\frac{2\pi i}{T}\right)\delta\left(\omega - \frac{2\pi i}{T}\right)$

where $\mathscr{F}$ represents the Fourier transform [Hint: for (a) use Fourier series; for (b) make use of the result in (a); for (c) write the left hand summation as a convolution involving (b).]

5. (6.1) Show that the required rate for ideal bandpass sampling of a bandpass waveform with upper end frequency B_2 and bandwidth B is given by

$$f_s = 2B\left(1 + \frac{k}{M}\right)$$

where M is the largest integer not exceeding B_2/B, and $k = (B_2/B) - M$.

6. (6.1) A source of bandwidth 250 Hz is to be AM onto a subcarrier and then sampled at the rate of 1000 samples per second. Indicate the permitted subcarrier frequencies that may be used for sampling at the minimal rate.

7. (6.1) Bandpass sampling can also be achieved by first quadrature mixing to zero frequency and then sampling. (a) Determine the required sampling rate. (b) By expanding the bandpass sampled waveform, prove this quadrature sampling is identical to bandpass sampling directly.

8. (6.1) Show that transforming a voltage sample that has probability density $p_d(x)$ by the transformation $y = \int_{-\infty}^{x} p_d(u)du$ will generate a uniformly distributed sample over $(0, 1)$.

9. (6.1) Given $p_d(x) = ce^{-cx}$, $x \geq 0$. It is quantized by a two level $(L = 2)$ uniform quantizer over $(0, 2c)$. Determine the MSQE.

10. (6.1) A source signal voltage is uniform over ± 100 volts. Its power spectrum is $S(\omega) = 1.8$ (volt)^2/Hz, $|\omega| \leq 2\pi(1 \text{ kHz})$. The signal is to be digitized and transmitted. It is sampled at rate 1990 samples per second. The quantizer has eight levels and the sampled words are transmitted over a binary channel with bit error probability $PE = 10^{-3}$. Assuming perfect reconstruction and one bit error per sampled word, determine the resulting digital SDR at the receiver.

11. (6.2) Derive the Max Quantizer for the case $L = 2$.

12. (6.2) Determine the optimal (minimum MSQE) quantizer for $L = 2$, for the density in Problem 6.9.

13. (6.2) Prove that a uniform quantizer is optimal for a sample having a uniform probability density.

14. (6.2) From the equation set (6.2.2) show that (a) the end intervals are always selected so that the quantizer spans the range of the density, (b) each interval edge is always at the mean of the density over the two adjacent intervals.

15. (6.3) A quantizer has three equally likely intervals labeled 1, 2, and 3, with corresponding quantization values ξ_1, ξ_2, and ξ_3. The digital transmission system causes the following interval error probabilities

$$P_{ij} = \begin{cases} 0 & \text{for} \quad (i = 1, j = 2,3), (i = 3, j = 1,2) \\ 1 & \text{for} \quad (i = 1, j = 1), (i = 3, j = 3) \\ \frac{1}{3} & \text{for} \quad (i = 2, j = 1,2,3) \end{cases}$$

Determine the MSTE.

16. (6.4) A noisy source has a SNR of 7 dB. The source output is digitized and transmitted over a link that produces an overall SDR of 10 dB. (a) What will be the final reconstructed SNR of the source signal? (b) What must the system SDR be to ensure that the final SNR will be at least 5 dB?

17. (6.5) Determine the entropy of the Max Quantizer for the case $L = 2$ and $L = 4$. How does it compare to the number of bits per sample used?

18. (6.5) *Minimum entropy* encoding occurs when the original sample values are converted to new sample values with minimal entropy. Show that this occurs if the conversion produces a new sample value that is as close to being equally probable as possible.

19. (6.6) Show that the sources in the example in Section 6.6 can be multiplexed with fewer total register stages by first multiplexing two sources in one PSC, and then multiplexing the output with the third source in another PSC.

20. (6.6) Design a parallel-series converter for sources with rates $\mathcal{R}_1 = 100$ kB/sec, $\mathcal{R}_2 = 45$ Kb/sec, $\mathcal{R}_3 = 6$ Kb/sec, and $\mathcal{R}_4 = 3$ Kb/sec. (a) What will be the bit rate at the PSC output? (b) What will the bit rate be if a sync bit is inserted every 200 bits?

21. (6.7) Show that for any (n, k) code the minimum distance must satisfy the inequality

$$2^k \leq 2^n \Big/ \sum_{i=0}^{(d_{min}-1/2)} \binom{n}{i}$$

22. (6.7) A linear (n, k) code set is one formed by starting with k distinct binary words of length n and adding all module two linear combinations to these k to form the entire 2^k set. Show that for any linear code set the minimum Hamming distance of the set is equal to the minimum number of ones occurring in all words of the set (excluding the all zero word).

23. (6.8) A source has sample entropy $\mathcal{H}$, and samples are taken at a rate of f_s per second. Show that if errorless transmission is to occur, we need a SNR of at least

$$\frac{P_m}{N_{b0}B_m} = 2^{(\mathcal{H}f_s/B_m)} - 1$$

24. (6.8) A 4 kHz voice link is digitally transmitted by A-D converting according to the sampling theorem and quantizing to 16 levels. If the information is to be sent over a 16 kHz baseband channel, what SNR is required for errorless transmission to occur?

25. (6.8) Plot a curve of normalized channel capacity $\mathcal{C}/B_m$ as a function of $P_m/N_{b0}B_m \triangleq$ SNR. Determine the SNR needed to achieve $\mathcal{C}/B_m = 0.2, 0.5, 1.2, 10$.

CHAPTER 7

BINARY
DIGITAL SYSTEMS

In digital systems using binary channel encoding, the source bits are transmitted one at a time in sequence to the receiver. This is accomplished by mapping each bit onto one of two possible baseband waveforms for carrier transmission. In this chapter we examine in detail the design and performance of a binary digital system. Channel decoding is accomplished by a linear processing decoding subsystem, initially justifying its design from intuition and later commenting on its generality and optimality. The carrier subsystem interconnecting the channel encoder and decoder is modeled by the additive noise channels derived in Chapter 4. Our primary concern is with the design of the baseband subsystem and the resulting channel decoding performance, the latter measured in terms of an average probability of recovering a bit in error. We are also interested in the way in which baseband power levels, bandwidth, and waveform structure influence this probability. Design techniques for reducing this bit error probability by proper decoder design and signal selection are also presented.

7.1. Binary Encoder and Filter-Sampler Decoding

We consider the binary digital system model of Figure 7.1. The binary channel encoding operation is mathematically described by (6.7.1). The signals $s_1(t)$ and $s_0(t)$, each T sec long, are to represent the baseband waveforms for transmitting the one and zero bit, respectively. We assume

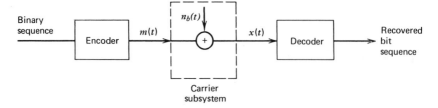

Figure 7.1. The binary digital system baseband model.

the bits generated from the source occur independently, and the received baseband waveforms, $s_1(t)$ and $s_0(t)$, contain equal energy. That is,

$$E = \int_0^T s_i^2(t) \, dt \triangleq P_m T \tag{7.1.1}$$

where P_m is the signal power of the transmitted baseband. The first assumption typifies the output of most digital sources, whereas the second assumes the channel encoder does not favor either binary symbol. The carrier subsystem is modeled as an additive Gaussian noise channel with baseband noise $n_b(t)$ corresponding to the demodulated output noise. This baseband noise has the spectral density $S_{nb}(\omega)$ given in (4.9.1), and its form depends on the type of carrier system. This carrier channel model is in accordance with our carrier subsystem discussion when the latter operates above the demodulator threshold.

The baseband waveform $m(t)$ is composed of sequences of $s_1(t)$ and $s_0(t)$, as was depicted in Figure 6.19, in which the presence of either signal in any bit interval depends on the particular bit. The baseband bandwidth required to transmit such binary encoded waveforms therefore corresponds to the bandwidth occupied by such sequences. Mathematically, we can write the baseband encoded signal as

$$m(t) = \sum_{k=-\infty}^{\infty} [a_k s_1(t - kT + t_0) + (1 - a_k) s_0(t - kT + t_0)] \tag{7.1.2}$$

Here t_o is a random time shift, uniformly distributed over $(0, T)$, to account for the arbitrariness of the beginning of the bit intervals with respect to a fixed time axis, and $\{a_k\}$ is an independent sequence of binary $(0, 1)$ random variables, accounting for the selection of the signal in each bit interval according to the corresponding bit. We assume the probability that $a_k = 1$ is p, the latter equal to the probability that a one is sent in the kth interval. It then follows that

$$\mathscr{E}[a_k] = p$$
$$\mathscr{E}[a_k^2] = p$$
$$\mathscr{E}[a_k a_j] = p^2, \qquad k \neq j \tag{7.1.3}$$

The autocorrelation function of $m(t)$ is then

$$R_m(\tau) = \mathscr{E}[m(t)m(t+\tau)]$$
$$= \sum_{k=-\infty}^{\infty} \sum_{j=-\infty}^{\infty} \mathscr{E}\left[\frac{1}{T}\int_0^T a_k s_1(t-kT+t_o) + (1-a_k)s_0(t-kT+t_o)\right]$$
$$\cdot [a_j s_1(t+\tau-jT+t_o) + (1-a_j)s_0(t+\tau-jT+t_o)] \, dt_o \tag{7.1.4}$$

Changing integration variables, and using (7.1.3), yields

$$R_m(\tau) = \sum_{k=-\infty}^{\infty} \left[\frac{p}{T}\int_{-kT}^{(1-k)T} s_1(u)s_1(u+\tau) \, du\right.$$
$$+ \frac{1-p}{T}\int_{-kT}^{(1-k)T} s_0(u)s_0(u+\tau) \, du\right]$$
$$+ \sum_{\substack{k=-\infty \\ k \neq j}}^{\infty} \sum_{j=-\infty}^{\infty} \int_{-kT}^{(1-k)T} s(u)s[u+\tau+(k-j)T] \, du \tag{7.1.5}$$

where

$$s(t) \triangleq ps_1(t) + (1-p)s_0(t) \tag{7.1.6}$$

The summation over k extends the integrals over the infinite interval, and by rearranging indices, (7.1.5) can be rewritten as

$$R_m(\tau) = \frac{p}{T}r_{s1}(\tau) + \frac{1-p}{T}r_{s0}(\tau) + \sum_{\substack{j=-\infty \\ j \neq 0}}^{\infty} r_s(\tau+jT)$$
$$= \frac{p}{T}r_{s1}(\tau) + \frac{1-p}{T}r_{s0}(\tau) - r_s(\tau) + \sum_{j=-\infty}^{\infty} r_s(\tau+jT) \tag{7.1.7}$$

where now

$$r_{s_i}(\tau) = \int_{-\infty}^{\infty} s_i(t)s_i(t+\tau) \, dt, \qquad i = 0, 1$$
$$r_s(\tau) = \int_{-\infty}^{\infty} s(t)s(t+\tau) \, dt \tag{7.1.8}$$

Thus the correlation function of a binary encoded baseband signal is related in the preceding manner to the correlations of the individual waveforms. The corresponding spectral density of the baseband then follows as the Fourier transform. We now use the fact that if $s_1(t)$ has transform $F_{s1}(\omega)$, then the Fourier transform of $r_{s1}(\tau)$ in (7.1.8) is $|F_{s1}(\omega)|^2$ (Problem 7.1). Making this substitution in (7.1.7), and applying the Poisson summation (Problem 6.4) to the second term, then yields

$$S_m(\omega) = \frac{p(1-p)}{T}[|F_{s1}(\omega) - F_{s0}(\omega)|^2]$$

$$+\frac{2\pi}{T^2}\sum_{i=-\infty}^{\infty}\left|pF_{s1}\left(\frac{2\pi i}{T}\right)+(1-p)F_{s0}\left(\frac{2\pi i}{T}\right)\right|^2\delta\left(\omega-\frac{2\pi i}{T}\right) \quad (7.1.9)$$

as the power spectrum of the binary waveform. The first term relates the encoded spectrum to the spectra of the individual bit waveforms. The second term corresponds to harmonic lines in the spectrum at multiples of the bit frequency $1/T$ Hz, and therefore exhibit any inherent periodicity within the encoded waveform. We see that no such periodic components appear if, and only if,

$$pF_{s1}\left(\frac{2\pi i}{T}\right)+(1-p)F_{s0}\left(\frac{2\pi i}{T}\right)=0, \qquad i=1,2,\ldots \quad (7.1.10)$$

Spectral lines of an encoded spectrum usually represent unusable power, as far as bit transmission is concerned, since they contain no information about the individual bits. However, harmonic lines at the bit frequency multiples can be used advantageously to aid in extracting timing information for bit synchronization, as is discussed in Chapter 9.

After the baseband waveform is encoded and transmitted, the data bits are recovered at the receiver by the decoding operations. The channel decoder has the form shown in Figure 7.2, composed of a linear filter followed by a voltage sampler. This decoder processes the recovered baseband during the bit time, then uses the voltage value at time $t = T$ of the processing to make its bit decision. The fact that the sample is taken at

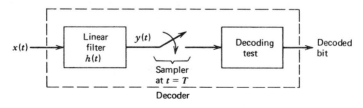

Figure 7.2. Decoder filter-sampler model.

the end of the bit interval appears intuitively plausible, but the assumption is not really constraining, since the effective sampling point can be moved anywhere in the bit interval by inserting delay in the processing filter. We would expect that any advantages of inserting such delay should become apparent in the subsequent analysis. We will also determine the manner in which the decoder voltage sample is used for the final decisioning. Processing derived for one bit interval must be repeated every bit interval, because of an independent bit assumption. Thus the decoding operation over $(0, T)$ is to be repeated periodically every T sec to decode the entire source bit sequence.

For the present we assume the decoder filter has an impulse response $h(t)$ for $0 \leq t \leq T$. We initially consider the decoder to be perfectly timed by the bit timing from the synchronization subsystem so that the bit interval $(0, T)$ is known precisely by the decoder and the sampler is closed at exactly $t = T$ sec in each bit interval. We are interested in the performance of the preceding decoder during the reception of a particular bit. The decoder input in Figure 7.1 is therefore

$$x(t) = m(t) + n_b(t), \qquad 0 \leq t \leq T \qquad (7.1.11)$$

where $m(t)$ is either $s_1(t)$ or $s_0(t)$, the latter to be decided by the decoder. The filter output is

$$y(t) = \int_0^T h(\rho)x(t-\rho)\,d\rho$$

$$= \int_0^T h(\rho)m(t-\rho)\,d\rho + \int_0^T h(\rho)n_b(t-\rho)\,d\rho \qquad (7.1.12)$$

The sampled value at $t = T$ is then

$$y(T) = \int_0^T h(\rho)m(T-\rho)\,d\rho + \int_0^T h(\rho)n_b(T-\rho)\,d\rho \qquad (7.1.13)$$

The second term is due to the noise and therefore represents a random variable. The first term is due to the baseband signal being received and is a constant depending on whether $s_1(t)$ or $s_0(t)$ was sent. Clearly, $y(T)$ is a Gaussian random variable, since $n_b(t)$ is a Gaussian process. Let us denote $y_1(T)$ as the sample value when $s_1(t)$ is transmitted during the bit interval, and denote $y_0(T)$ when $s_0(t)$ is transmitted. Then

$$m_1 \triangleq \text{mean of } y_1(T) = \int_0^T h(\rho)s_1(T-\rho)\,d\rho \qquad (7.1.14a)$$

$$m_0 \triangleq \text{mean of } y_0(T) = \int_0^T h(\rho)s_0(T-\rho)\,d\rho \qquad (7.1.14b)$$

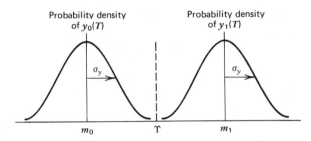

Figure 7.3. Probability densities of decoder sample.

The variance is given by the mean squared value of the filter output due to noise. Hence

$$\sigma_y^2 = \text{variance of } y_1(T)$$

$$= \text{variance of } y_0(T)$$

$$= \frac{1}{2\pi} \int_{-\infty}^{\infty} |H(\omega)|^2 S_{nb}(\omega) \, d\omega \qquad (7.1.15)$$

where $H(\omega)$ is the decoder filter transfer function [Fourier transform of $h(t)$]. The variances are identical, since the random variable (second integral) in (7.1.13) does not depend on which signal was transmitted. The sample value $y(T)$ therefore has one of the two Gaussian probability densities shown in Figure 7.3, where the mean values m_1 and m_0 have been arbitrarily located, and σ_y^2 is the variance in (7.1.15). The decoder need only decide whether the observed sample $y(T)$ comes from the density of $y_1(T)$ or from that of $y_0(T)$. This can be accomplished by determining which bit was most likely to have been sent, based on the sample voltage $y(T)$. Decoding the bit by this type of decision is referred to as *maximum likelihood decoding* and can be shown actually to minimize the average bit error probability (see Problem 7.2). To determine the most likely bit, for an observed $y(T) = y_T$, we must compute $P(i|y_T)$, the probability that bit i was sent given the value y_T, and compare for $i = 0$ and $i = 1$. If $P(1|y_T) > P(0|y_T)$, we conclude a one bit was most probable, and decode a one. If $P(0|y_T) > P(1|y_T)$, we decode a zero bit. When $P(1|y_T) = P(0|y_T)$, each bit is equally likely, and either bit may be selected. (We will be correct half the time no matter which bit we select in this case. We shall hereafter assume we will always decide a one bit when the equality holds.) To carry out this decisioning, we must formally compute $P(i|y_T)$. However, by application of Equation (B.2.7), we can instead write

$$P(i|y_T) = \left[\frac{P(i)}{\frac{1}{2}p_1(y_T) + \frac{1}{2}p_0(y_T)}\right] p_i(y_T) \qquad (7.1.16)$$

where $p_i(y)$ are the densities in Figure 7.3, and $P(i)$ is a priori probability of bit i occurring. If the bits are initially equally likely, $P(i) = \frac{1}{2}$, and the bracket in (7.1.16) does not depend on i. Thus a comparison of $P(1|y_T)$ with $P(0|y_T)$ is identical to a comparison of $p_1(y_T)$ with $p_0(y_T)$. That is, in carrying out the maximum likelihood test, we need only evaluate the two probability densities in Figure 7.3 at the point $y = y_T$ and compare the two. Since these densities are Gaussian densities, we are comparing only $\exp -(y_T - m_1)^2$ to $\exp -(y_T - m_0)^2$. We therefore decide a one bit if

$$(y_T - m_1)^2 \leq (y_T - m_0)^2 \tag{7.1.17}$$

That is, we only need to determine whether the observed decoder sample $y(T) = y_T$ is closer to the fixed value m_1 or m_0, the latter given in (7.1.13). This can be carried out by setting a voltage threshold equal distance between the two mean values, at

$$Y = \frac{m_1 + m_0}{2} \tag{7.1.18}$$

and deciding on the received bit by the following procedure:

decode a one bit if $y(T) \geq Y$

decode a zero bit if $y(T) < Y$ \hfill (7.1.19)

Thus the maximum likelihood decoding procedure, using the decoder system of Figure 7.2 to generate the voltage sample $y(T)$, decodes the data bits by the test in (7.1.19). This type of bit decisioning procedure is often called a *threshold comparison* test, or simply a *threshold* test.

We can now determine the probability that an error in bit decoding will be made using this decoding threshold test. Assume a binary zero is in fact transmitted during $(0, T)$. The decoder is unaware and performs the test in (7.1.19). An error is made if $y(T) \geq Y$ and a correct decision is made if $y(T) < Y$. Thus the probability of an error (PE) is the probability that $y(T)$ observed will equal or exceed Y. Hence

$$PE|\text{zero bit} = \frac{1}{\sqrt{2\pi}\sigma_y} \int_Y^\infty e^{-(y-m_0)^2/2\sigma_y^2} \, dy$$

$$= \frac{1}{\sqrt{\pi}} \int_{(m_1-m_0)/2\sqrt{2}\sigma_y}^\infty e^{-u^2} \, du$$

$$\triangleq \frac{1}{2} \text{Erfc} \left[\frac{m_1 - m_0}{2\sqrt{2}\sigma_y} \right] \tag{7.1.20}$$

where Erfc $(x) \triangleq 1 - \text{Erf}(x)$, and Erf (x) is given in (4.7.14). Similarly, we have

$$\text{PE}|\text{one bit} = \frac{1}{\sqrt{2\pi}\sigma_y} \int_{-\infty}^{Y} e^{-(y-m_1)^2/2\sigma_y^2} \, dy$$

$$= \tfrac{1}{2} \text{Erfc} \left[\frac{m_1 - m_0}{2\sqrt{2}\sigma_y} \right] \qquad (7.1.21)$$

If m_1 had been smaller than m_0 in Figure 7.3, (7.1.20) and (7.1.21) would still be valid with the parameters m_1 and m_0 interchanged. That is, each PE depends only on the separation $|m_1 - m_0|$ no matter where each is located. Thus, for equally likely bits, the average bit error probability is always

PE = [PE|zero bit][Prob. zero bit] + [PE|one bit][Prob. one bit]

$$= \tfrac{1}{2} \text{Erfc} \left[\frac{|m_1 - m_0|}{2\sqrt{2}\sigma_y} \right] \qquad (7.1.22)$$

Equation (7.1.22) is plotted in Figure 7.4. We see that the average bit error probability, when using the filter-sampler decoder, depends only on the parameter $|m_1 - m_0|/2\sigma_y$. Hence best performance, in terms of minimization of bit error probability, would require maximization of this parameter. Since m_1, m_0, and σ_y depend on the decoder filter impulse response $h(t)$, we can find the filter response that maximizes the parameter

$$\Lambda \triangleq \frac{|m_1 - m_0|}{2\sqrt{2}\sigma_y} \qquad (7.1.23)$$

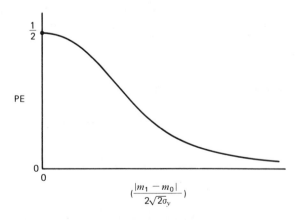

Figure 7.4. Average bit error probability.

We first proceed under the assumption that the baseband additive noise $n_b(t)$ is white and denote its one-sided spectral level as N_{b0}. This would correspond to several cases in (4.9.1) where the filters involved are assumed flat and wider than the baseband bandwidths. The variance in (7.1.15) becomes

$$\sigma_y^2 = \frac{N_{b0}}{2}\left(\frac{1}{2\pi}\int_{-\infty}^{\infty}|H(\omega)|^2\,d\omega\right)$$

$$= \frac{N_{b0}}{2}\int_0^T h^2(t)\,dt \qquad (7.1.24)$$

by Parseval's theorem (Appendix A). The maximization of Λ in (7.1.23) is quite easily obtained, since Λ can be upper bounded and the condition for the bound to occur is readily apparent. Using the Schwartz inequality (Section A.2), we write

$$\Lambda = \frac{\left|\int_0^T h(\rho)[s_1(T-\rho)-s_0(T-\rho)]\,d\rho\right|}{2[N_{b0}\int_0^T h^2(\rho)\,d\rho]^{1/2}} \qquad (7.1.25a)$$

$$\leq \frac{[\int_0^T h^2(\rho)\,d\rho]^{1/2}[\int_0^T [s_1(T-\rho)-s_0(T-\rho)]^2\,d\rho]^{1/2}}{2[N_{b0}\int_0^T h^2(\rho)\,d\rho]^{1/2}} \qquad (7.1.25b)$$

$$= \frac{1}{2}\left[\frac{1}{N_{b0}}\int_0^T [s_1(T-\rho)-s_0(T-\rho)]^2\,d\rho\right]^{1/2} \qquad (7.1.25c)$$

Most important, the equality in (7.1.25b) occurs only if

$$h(\rho) = h_0[s_1(T-\rho)-s_0(T-\rho)], \qquad 0\leq\rho\leq T \qquad (7.1.26)$$

where h_0 is an arbitrary gain constant. The corresponding maximum value of Λ is then given by (7.1.25c). Note that the maximum value does not depend on h_0 (for this reason we now assume $h_0 = 1$). The impulse response in (7.1.26) is therefore that which minimizes bit error probability for the filter-sampler decoder in a binary digital system with white noise. If we expand out (7.1.25c) and define

$$\gamma \triangleq \frac{\int_0^T s_1(t)s_0(t)\,dt}{E} \qquad (7.1.27)$$

as the *correlation coefficient* of the signal pair, then we have

$$\Lambda_{\max} = \left[\frac{2E-2\gamma E}{4N_{b0}}\right]^{1/2} = \left[\frac{E}{N_{b0}}\left(\frac{1-\gamma}{2}\right)\right]^{1/2} \qquad (7.1.28)$$

The error probability obtained using this optimal filter is, from (7.1.22),

$$PE = \tfrac{1}{2}\operatorname{Erfc}\left[\left[\left(\frac{E}{N_{b0}}\right)\left(\frac{1-\gamma}{2}\right)\right]^{1/2}\right] \qquad (7.1.29)$$

Table 7.1. Tabulation of Baseband E/N_{b0} Values

	AM	PM	PM Carrier Tracking
E	$\dfrac{C^2 P_m T}{2}$	$\Delta^2 P_m T$	$\dfrac{A^2}{2}(2J_1^2(\Delta))T$
N_{b0}	$N_0 G_1$	$\dfrac{G_1 N_0}{A^2}$	$\dfrac{G_1 N_0}{A^2}$
$\dfrac{E}{N_{b0}}$	$\dfrac{C^2 P_m T}{2N_0 G_1}$	$\Delta^2 P_m\left(\dfrac{A^2 T}{G_1 N_0}\right)$	$2J_1^2(\Delta)\left(\dfrac{A^2 T}{2N_0}\right)$
	$=\dfrac{P_m}{1+P_m}\left(\dfrac{P_r T}{N_0}\right)$	$=\Delta^2 P_m\dfrac{P_r T}{N_0/2}$	$=2J_1^2(\Delta)\left(\dfrac{P_r T}{N_0}\right)$

$N_0 = \kappa T_{eq}^0$
$G_1 = $ RF, IF gain
$P_r = $ antenna carrier power

$\dfrac{A^2}{2} = P_r G_1$

$P_m = $ modulating waveform power

$\dfrac{C^2}{2}(1+P_m) = P_r G_1$

Thus the minimum PE depends only on the baseband signal energy E, the interfering noise spectral level N_{b0}, and the correlation of the signals γ. The fact that (7.1.29) depends on E shows why the decoder decisioning sample must in fact be taken at the end of the bit interval where it collects the total signal energy, justifying our earlier conjecture. Since E and N_{b0} are the baseband energy and noise level, the value of the ratio E/N_{b0} depends on the specific RF system model. Table 7.1 is a listing of the specific relationship of the baseband ratio E/N_{b0} to the parameters of the RF and IF link, as derived in our earlier discussions. The tabulation is given for each type of RF system that produces white baseband noise.

It is evident that for any signal pair, the ratio E/N_{b0} is a basic parameter in determining decoding performance. This parameter is often written out as

$$\frac{E}{N_{b0}} = \frac{P_m T}{N_{b0}} = \frac{P_m T B_m}{N_{b0} B_m} = TB_m\left(\frac{P_m}{N_{b0}B_m}\right) \qquad (7.1.30)$$

where we have arbitrarily introduced the baseband signal bandwidth B_m.

When written in this way the energy ratio E/N_{b0} can be interpreted as a signal to noise ratio, obtained by multiplying the baseband power ratio $P_m/N_{b0}B_m$ by the factor (TB_m). The latter is often referred to as the *gain* of the decoder, whereas $P_m/N_{b0}B_m$ is the SNR at the decoder input in the signal band B_m. It therefore appears that the decoding operation multiplies up its input SNR by its gain factor to provide an output SNR that determines performance. This interpretation often leads to confusion, since it implies that for best decoding gain, the signal bandwidth B_m should be as large as possible. However, we see clearly in (7.1.30) that B_m does not really affect performance at all, for the latter depends only on the ratio E/N_{b0} for signals of any bandwidth.

The optimized decoder must use a filter with the impulse response prescribed in (7.1.26) over each bit interval. This can be written as

$$h(t) = h_1(t) - h_0(t) \qquad (7.1.31)$$

where

$$h_1(t) \triangleq s_1(T-t), \qquad 0 \leq t \leq T$$
$$h_0(t) \triangleq s_0(T-t), \qquad 0 \leq t \leq T \qquad (7.1.32)$$

and, from (1.6.22), corresponds to the parallel system shown in Figure 7.5. The filters with impulse response given in (7.1.32) are said to be "matched" to the corresponding signals and are therefore called *matched filters*. Note that the filter must present exactly the same impulse response over each bit interval, which may often require capacitor and inductor discharge prior to each bit interval. Roughly speaking, this optimal decoder operates by having each branch of the parallel decoder in Figure 7.5 "look for" one of the signal pairs, and the difference voltage produced determines which one is most likely to have received its signal. It can be shown

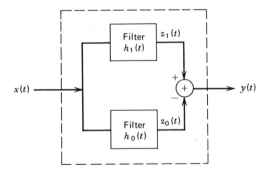

Figure 7.5. Matched filter block diagram.

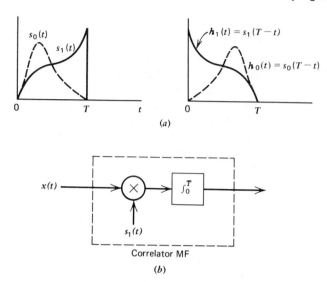

Figure 7.6. Matched filter models. (*a*) Impulse response, (*b*) equivalent correlator design.

[1–4] that this particular decoder is indeed optimal over all possible decoding systems, when the baseband channel is modeled as an additive Gaussian white noise. Furthermore, for the Gaussian channel, the linear filter-sampler decoder is in fact optimal under a wider range of optimality criteria (other than only minimizing PE).

The required matched filter impulse response in (7.1.32) corresponds to inverted versions of the signals involved (Figure 7.6*a*). These are obtained by reflecting the signal into the interval $(-T, 0)$, then shifting the reflected signal to $(0, T)$ to define $h(t)$. This means the complexity of the impulse response will be directly related to the complexity of the signal shape itself. However, there are alternatives to having to construct a filter with the required response. Note that the output of a matched filter, say $h_1(t)$, at time of sampling, is given by

$$\int_0^T h_1(\rho)x(T-\rho)\,d\rho = \int_0^T s_1(T-\rho)x(T-\rho)\,d\rho$$

$$= \int_0^T s_1(t)x(t)\,dt \qquad (7.1.33)$$

This sample output could be produced by simply multiplying the input $x(t)$ by $s_1(t)$ and integrating for T sec over the bit period, as shown in Figure 7.6*b*. Such a system avoids the problem of having to construct a passive filter with the proper impulse response, but it requires the generation of the signal waveform $s_1(t)$ during each bit period. The preceding is called a *correlator matched filter* and can be used in the optimal decoder branches of Figure 7.5. It is important to note that the correlator output is the same as that of the filter $h_1(t)$ only at time $t = T$, which is the time of sampling. Hence Figure 7.6b should be interpreted as a device that produces no output until integration is completed; that is, at $t = T$. The correlator is often called a *synchronous*, or *coherent*, *correlator* since the correlating waveform $s_1(t)$ [or $s_0(t)$] must be actively generated in exact time synchronism with the matched signal if the latter occurred during the bit period.

Equation (7.1.25c) and (7.1.29) assume that the optimal impulse responses are used. Let us consider the effect of using, in (7.1.25a), an incorrect $\hat{h}(t)$, due perhaps to the inability to construct properly the desired $h(t)$ of (7.1.26). We then see that

$$\Lambda = \frac{|\int_0^T \hat{h}(\rho)h(\rho)\,d\rho|}{2[N_{b0}\int_0^T \hat{h}^2(\rho)\,d\rho]^{1/2}} \tag{7.1.34}$$

If we multiply numerator and denominator by

$$\left[\int_0^T h^2(\rho)\,d\rho\right]^{1/2} = [2E(1-\gamma)]^{1/2} \tag{7.1.35}$$

and rewrite, we see that

$$\Lambda = \Lambda_{\max}\left[\left|\int_0^T \left(\frac{\hat{h}(\rho)}{E_{\hat{h}}^{1/2}}\right)\left(\frac{h(\rho)}{E_h^{1/2}}\right) d\rho\right|\right] \tag{7.1.36a}$$

$$= \Lambda_{\max}\left[\left|\int_0^T \left(\frac{\hat{s}(t)}{E_{\hat{s}}^{1/2}}\right)\left(\frac{s(t)}{E_s^{1/2}}\right) dt\right|\right] \tag{7.1.36b}$$

where $\hat{s}(t), s(t)$ are the signals to which the filter functions $\hat{h}(t), h(t)$ are effectively matched, and $E_{\hat{h}}, E_s$ are their energies. That is, $s(t)$ is the difference of the desired signal pair and $\hat{s}(t)$ is the difference of the incorrect signal pair to which $\hat{h}(t)$ is matched, according to (7.1.31) and (7.1.32). Hence the effect of using a mismatched filter $\hat{h}(t)$ instead of $h(t)$, or a correlator with signal $\hat{s}(t)$, instead of $s(t)$, is to multiply the $\Lambda_{\max}$ by the bracketed terms in (7.1.36). We can therefore always account for improperly designed decoders by evaluating one of these expressions.

7.2. Signal Selection for Binary Digital Systems

Up to the present we have assumed a pair of arbitrary signals, $s_1(t)$ and $s_0(t)$, to represent the binary bits. The only condition we placed on the signal waveform is that they have equal energy E. We then determined the optimal linear decoder for this signal pair. We found that when the optimal decoder is used, the bit error probability was given by (7.1.29), where γ was defined in (7.1.27). Thus the actual waveshape of the signals enters only through the parameter γ. (The constraint of E involves only a scale factor on the signals.) For this reason there is little gained by resorting to orthonormal signal expansions, and we can deal with the signals directly. We may first inquire if there is a best pair of signals to use, assuming we can construct their proper matched filter. Clearly, from (7.1.29), with E and N_{b0} fixed, we want the most negative value of γ. From the Schwarz inequality, we note

$$|\gamma| \leq \frac{\left[\int_0^T s_1^2(t)\, dt \int_0^T s_0^2(t)\, dt \right]^{1/2}}{E} = 1 \tag{7.2.1}$$

We see that the most negative possible value is $\gamma = -1$. Furthermore this only occurs when

$$s_0(t) = -s_1(t) \tag{7.2.2}$$

That is, the signal pair are negatives of each other. The preceding signals are called *antipodal* signals and yield the minimum PE possible with linear, matched decoders. For this case,

$$\text{PE} = \tfrac{1}{2} \text{Erfc}\left[\left(\frac{E}{N_{b0}} \right)^{1/2} \right] \tag{7.2.3}$$

The curve is plotted in Figure 7.7 as a function of E/N_{b0}. Note that PE now depends only on the energy of the signals, so that any pair of antipodal signals with the same energy is as good as any other, in terms of the minimal possible PE. (However, there may be other reasons for selecting one signal pair over another, as we shall see.) Also included in Figure 7.7 is the curve for the signal pair such that

$$\int_0^T s_1(t) s_0(t)\, dt = 0 \tag{7.2.4}$$

Such signals are said to be *orthogonal*. For orthogonal signal pairs, $\gamma = 0$ and the resulting PE curve requires twice the value of E to obtain the same error probability as in the antipodal case.

For antipodal signals we see from (7.1.14) that $m_0 = -m_1$ and therefore the threshold in (7.1.18) has $Y = 0$. Furthermore, in Figure 7.5, $z_1(T) = -z_0(T)$, so that the required threshold test comparing $y(T)$ to Y is identical to comparing $y(T) = z_1(T) - z_0(T) = 2z_1(T)$ to zero. This means simply

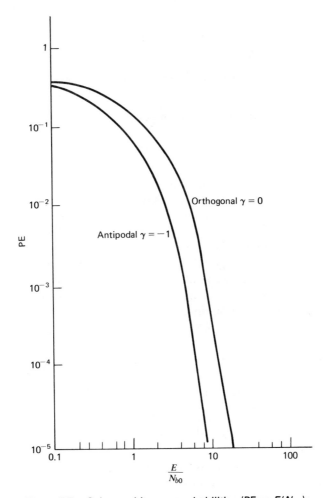

Figure 7.7. Coherent bit error probabilities (PE vs E/N_{b0}).

determining the sign of $z_1(T)$. That is

$$y(T) \lessgtr Y = 0 \qquad \text{if} \quad z_1(T) \lessgtr 0 \qquad (7.2.5)$$

for antipodal signals. Hence the lower matched filter (or correlator) path in Figure 7.5 can be eliminated, and the decoder decisioning need only determine if $z_1(T)$ is positive or negative (Figure 7.8). Antipodal signals therefore have the added advantage of a simplified decoder.

The bandwidth required in the baseband channel when antipodal signaling is used can be determined from (7.1.9). Since $s_1(t) = -s_0(t)$, we have

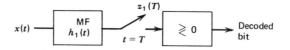

Figure 7.8. Decoder for antipodal signaling.

$F_{s1}(\omega) = -F_{s0}(\omega)$, and, for the case of equally probable source bits ($p = \frac{1}{2}$),

$$S_m(\omega) = \frac{1}{T}|F_{s_1}(\omega)|^2 \qquad (7.2.6)$$

The bandwidth required is that necessary to pass (7.2.6) without appreciable distortion. Note in particular that no harmonic lines appear in the spectrum, since (7.1.10) is satisfied. Although this may be advantageous from a power efficiency point of view, it may be undesirable in terms of synchronization. For this reason it may be necessary to insert the lines purposely when using antipodal signaling. This can be achieved by forcing $p \neq \frac{1}{2}$, by inserting binary ones at prescribed intervals so as to favor their appearance in the bit stream. This operation is often called bit *stuffing*.

Some common types of antipodal signals used in modern binary digital systems are shown in Figure 7.9 and summarized as follows.

PCM Pulses

$$s_1(t) = \left(\frac{E}{T}\right)^{1/2}, \qquad 0 \leq t \leq T$$

$$s_0(t) = -\left(\frac{E}{T}\right)^{1/2}, \qquad 0 \leq t \leq T$$

$$(7.2.7)$$

The preceding signal pair is called a *pulse code modulation* (PCM) signal pair and is sometimes referred to as an NRZ, or *nonreturn to zero*, signal set (since a string of identical binary bits will appear as a single, long, constant signal). With this signal pair the passive matched filters have impulse responses that are themselves pulses, as is evident from Figure 7.6a. Such impulse responses can be well approximated by RC circuits with appropriate time constants. The correlator matched filter of Figure 7.6b can remove the multiplier and simply integrate over each bit interval. Such a decoder is often called an *integrate-and-dump decoder*. The frequency function of PCM signals can be obtained from (7.2.6), and therefore corresponds to the energy spectrum of a single T sec pulse. This was derived earlier in (1.3.7) and is seen to have most of its energy concentrated about zero frequency out to a band of approximately $1/T$ Hz (see Figure 7.9).

Coded PCM Pulses

$$s_1(t) = \begin{cases} \left(\dfrac{E}{T}\right)^{1/2}, & 0 \le t \le \dfrac{T}{2} \\ -\left(\dfrac{E}{T}\right)^{1/2}, & \dfrac{T}{2} \le t \le T \end{cases}$$

$$s_0(t) = \begin{cases} -\left(\dfrac{E}{T}\right)^{1/2}, & 0 \le t \le \dfrac{T}{2} \\ \left(\dfrac{E}{T}\right)^{1/2}, & \dfrac{T}{2} \le t \le T \end{cases}$$

$$(7.2.8)$$

The preceding signals are called *coded PCM* signals and are sometimes referred to as *Manchester coded* signals or RZ (return to zero) signals, since they guarantee a zero crossing during each bit, which can be used to aid synchronization. Passive matched filters are slightly more difficult to construct, but correlators utilizing proper switching, as shown in Figure 7.9, can be used. Coded PCM signals can be interpreted as the amplitude modulation of a square wave of frequency $1/T$ by a PCM bit stream of period T. The spectrum in (7.2.6) of a coded PCM is therefore that of such an amplitude modulated square wave and contains the PCM spectrum shifted to the harmonics of the square wave and summed. The shifting of the spectrum away from zero frequency may be helpful to avoid low frequency noise.

PSK (Phase Shift Keyed) Signals

$$s_1(t) = \left(\frac{2E}{T}\right)^{1/2} \sin(\omega_s t + \theta)$$

$$s_0(t) = -\left(\frac{2E}{T}\right)^{1/2} \sin(\omega_s t + \theta)$$

$$= \left(\frac{2E}{T}\right)^{1/2} \sin(\omega_s t + \theta \pm \pi) \qquad (7.2.9)$$

The signal $s_0(t)$ can be seen to be a phase shifted version (by $\pm\pi$) of $s_1(t)$. Note that the phase shift must be exactly $\pm\pi$ to represent an antipodal set, but the phase angle θ is arbitrary. The decoder, however, must know this arbitrary phase exactly in constructing matched filters or correlators. The operation of determining and maintaining this phase angle at the decoder is called *phase referencing*. The energy in the signals in (7.2.9) is

$$\frac{2E}{T} \int_0^T \sin^2(\omega_s t + \theta) = \frac{2E}{T} \left[\frac{T}{2} - \frac{1}{2} \int_0^T \cos(2\omega_s t + 2\theta)\, dt \right] \qquad (7.2.10)$$

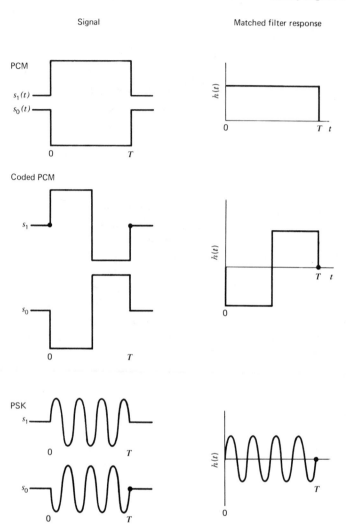

Figure 7.9. Some antipodal binary signaling formats.

and is exactly E if the frequency ω_s is harmonically related to $2\pi/T$, causing the integral to be zero, and is approximately E if $T \gg 2\pi/\omega_s$ (i.e., there are many cycles of ω_s within the bit interval T), causing the integral to be negligibly small. PSK signals are extremely popular because they are relatively easy to generate and involve only the phase modulation of a free-running encoder oscillator by the source bit stream. Note that the spectrum of a PSK signal (Figure 7.9) corresponds to that of sinusoid of

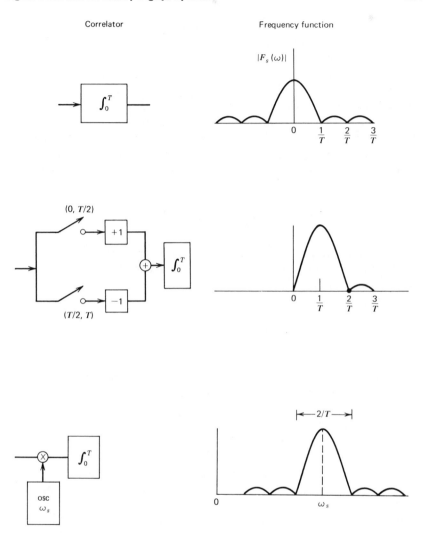

Figure 7.9. (*Continued*).

frequency ω_s amplitude modulated (or phase modulated) by a PCM bit stream of period T, similar to subcarrier modulation. Recall from (2.3.8) that this effectively shifts the spectrum of the PCM signal to the frequency ω_s. Thus the PSK spectrum is approximately $2/T$ Hz wide, centered around the oscillator (subcarrier) frequency $\omega_s/2\pi$. The fact that we can arbitrarily locate the PSK spectrum by control of ω_s is another advantage of PSK systems. This lends itself nicely to an FDM multiplexing format for more than one digital source. By simply assigning different oscillator frequencies to each source at the encoder, we can space the bit modulated waveforms along the frequency axis.

PSK signaling is also compatible with low index, carrier tracking PM systems. In this format the binary encoded PSK signals use an oscillator frequency ω_s sufficiently spaced away from the carrier to permit RF carrier tracking demodulation. The shifted RF spectrum produced by the carrier tracking is then bandpass filtered around ω_s to retrieve the PSK signal for the decoder. Note that in this case the baseband signal energy E is the same as the subcarrier energy (subcarrier power times T) and therefore has the value obtained by using (5.4.7):

$$E = P_s T = 2J_1^2(\Delta)\left(\frac{A^2 T}{2}\right) \tag{7.2.11}$$

where Δ is the phase index of the PSK subcarrier on the RF carrier and $A^2/2$ is the RF carrier power. Under the wideband assumption, the carrier tracking mixer noise is flat with level equal to that of the RF noise level N_0 [see (4.9.1))], and we see that

$$\frac{E}{N_{b0}} = 2J_1^2(\Delta)\left(\frac{A^2 T/2}{N_0}\right) \tag{7.2.12}$$

for carrier tracking demodulation. This result had been included in Table 7.1. Note that again we have the trade-off in design between good carrier tracking (small Δ) and accurate bit decoding (large E/N_{b0}) and much of our analysis from Section 5.6 is applicable here. In this case the subcarrier threshold condition is replaced by a specified value on E/N_{b0}, dictated by the desired bit error probability (see Problem 7.14).

Matched filters for decoding PSK signals would require properly shaped bandpass filters. (The impulse response of a bandpass filter will ring at the filter center frequency.) On the other hand, correlators involve simply multiplying by a phase referenced local oscillator at the frequency ω_s, followed by a bit interval integration (Figure 7.9). Thus the PSK decoder merely mixes the baseband and local oscillator signal continuously in time and integrates repeatedly over T sec intervals. The sequence of integrator output values is then used for bit decisioning. This combination of a mixer and integrate-and-dump filter appears as a cascade of an ideal phase detector followed by an integrator for the resulting PCM signal. We emphasize that this combination is not an ad hoc decoder but is dictated by our matched filter theory.

The basic problem with PSK implementation is the necessity for phase referencing (i.e., knowing exactly the arbitrary phase θ of the PSK subcarrier with respect to the bit period). Consider, for example, a coherent correlator for PSK decoding that uses an inaccurate phase $\hat{\theta}$ in Figure 7.9. We see that we have a mismatched correlator, and (7.1.36b) indicates

$$\Lambda = \Lambda_{max}\left[\int_0^T \frac{\sin(\omega_s t + \theta)}{(T/2)^{1/2}} \cdot \frac{\sin(\omega_s t + \hat{\theta})}{(T/2)^{1/2}} dt\right]$$

$$= \left(\frac{E}{N_{b0}}\right)^{1/2}\left[\frac{2}{T}\int_0^T \frac{\cos(\theta - \hat{\theta})}{2} dt\right] \tag{7.2.13}$$

If the phase referencing error $\theta_e \triangleq \theta - \hat{\theta}$ remains constant during the bit period, the resulting bit error is

$$\text{PE} = \tfrac{1}{2}\text{Erfc}[\Lambda]$$

$$= \tfrac{1}{2}\text{Erfc}\left[\left(\frac{E}{N_{b0}}\right)^{1/2}\cos\theta_e\right] \tag{7.2.14}$$

Hence the bit error probability is increased from its minimal value according to the cosine of the local oscillator phase referencing error. PSK therefore requires extremely accurate phase referencing to operate satisfactorily. If θ_e varies with time during the bit interval, then the integral in (7.2.13) must be evaluated. Unfortunately, the phase reference $\hat{\theta}$ is generated in practical systems by phase tracking loops with inherent noise, so that θ_e evolves as a random process in time and (7.2.13) involves a stochastic integral. Even if it is assumed that θ_e does not change with time, it should in fact be considered a random variable in (7.2.14). This means that PE is really a conditional error probability, conditioned on the random variable θ_e. We shall reexamine (7.2.14) after investigating phase referencing methods in Chapter 9.

We emphasize that the signal sets considered are all antipodal, and therefore all achieve the minimal PE when used with the same energy level and with the optimal decoder. Each signal set has characteristics, advantages, and disadvantages in terms of ease of generation, simplicity in decoding, baseband spectrum, and so on, that may favor one set over another in a particular application. In fact, in practice we may be willing to sacrifice the antipodal signal advantage for simplicity in encoder and decoder design. A common example of this is the following signal set.

Coherent Frequency Shift Keying (FSK)

$$s_1(t) = \left(\frac{2E}{T}\right)^{1/2}\sin(\omega_1 t + \theta)$$

$$s_0(t) = \left(\frac{2E}{T}\right)^{1/2}\sin(\omega_0 t + \theta) \tag{7.2.15}$$

This signal pair corresponds to the frequency shift to ω_1 or ω_0, depending on the bit. Each bit is therefore sent as a T sec burst of a sine wave at one of two possible frequencies. The encoder can therefore be easily implemented by a single free-running oscillator that is frequency switched. This also corresponds to frequency modulating the oscillator with a PCM waveform and a specified frequency deviation, and for this reason it is often referred to as PCM-FM. The signals are not antipodal, however. This can be seen by computing the signal correlation γ under the assumption $(\omega_1 + \omega_0) \gg 2\pi/T$. This produces

$$
\gamma = \frac{1}{E} \int_0^T s_1(t) s_0(t) \, dt
$$
$$
= \frac{\sin\left((\omega_1 - \omega_0)T\right)}{(\omega_1 - \omega_0)T} \qquad (7.2.16)
$$

The coefficient γ is shown in Figure 7.10 as a function of the normalized frequency difference $(\omega_1 - \omega_0)T$. The value γ never reaches minus one (required for the antipodal case) and approaches orthogonal signals as the frequency separation becomes large. However, we see that the most negative correlation value occurs when $(\omega_1 - \omega_0)T \cong 3\pi/2$, or when

$$
\frac{\omega_1 - \omega_0}{2\pi} \cong 0.71\left(\frac{1}{T}\right) \qquad (7.2.17)
$$

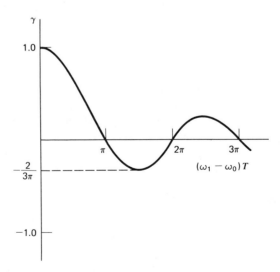

Figure 7.10. FSK signal correlation coefficient versus frequency separation.

When this separation is used, $\gamma = -2/3\pi$, which effectively multiplies E/N_{b0} by the factor $1 - \gamma = 1.21$. Thus the signal pair in (7.2.15), when used with the frequency separation in (7.2.17), is about 1.59 dB better than orthogonal signaling. That is, placing the signal frequencies close together, as dictated by (7.2.17), gives a slight advantage over wide frequency separation. This result occurs for any angle θ, but a pair of coherent correlators (two must be used, one matched to each frequency, since the signals are not antipodal) must be phase synchronized to the angle θ, just as in PSK. For this reason this signaling scheme is called *coherent FSK*.

When $(\omega_1 - \omega_0)T = \pi$, the FSK system operates at the first zero crossing in Figure 7.10, which is the minimal deviation that can be used for orthogonal binary signaling. This format is often called *minimum shift keying* (MSK) [5, 6]. Note that for MSK the frequency deviation between the one and zero bit is $1/2T$ Hz, which is one-half the bit rate frequency. MSK signaling can also be related to quadrature amplitude modulation (Problem 7.16). The bandwidth required for transmitting MSK signals can be obtained directly from (7.1.9) by recognizing that each $F_{s_i}(\omega)$ has the same frequency function as the PSK signal set in (7.2.9), except shifted to the frequency ω_1 or ω_0. The overall spectrum of the MSK signal pair is sketched in Figure 7.11.

In our signal discussion of this section we have dealt with the specific signal waveshapes directly and have found it unnecessary to utilize the orthonormal signal expansions in Section 6.9. This is simply due to the fact that performance depended only on signal cross-correlation, which can best be obtained directly from the signals themselves. Of course, each of the signals considered can in fact be represented by a specific orthonormal

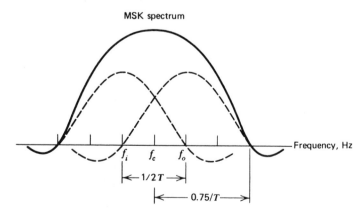

Figure 7.11. MSK signal spectrum.

expansion. By resorting to such expansions we can often derive equivalent forms for the decoder design (Problem 7.7). In dealing with block coded systems in Chapter 8, we find the orthonormal expansions much more useful in deriving system design and performance.

7.3. Effect of Imperfect Bit Timing

We have assumed exclusively the existence of perfect bit timing in all our decoder analyses. This means that the location in time of each bit interval is known precisely at the decoder. In a practical system there are often errors in the decoder's location of the bit intervals, because of imperfections in the timing subsystem. These errors must be accounted for in system analysis.

Consider the situation depicted in Figure 7.12. In (a) is the actual time location of the arriving bit intervals. In (b) is the decoder's interpretation of these interval locations, showing an offset by an amount δ sec. The decoder proceeds to correlate over each bit interval according to its timing schedule, unaware of the δ offset. This produces a correlation that overlaps the adjacent bit interval, and the resulting decoder bit decision is influenced by neighboring source bits. If we assume antipodal signals $\pm s(t)$, with the decoder of Figure 7.8, the situation in Figure 7.12 produces the signal sample value

$$z_1(T, \delta) = \int_0^T m(t)s(t)\, dt$$

$$= \pm \int_0^{T-\delta} s(t+\delta)s(t)\, dt \pm \int_{T-\delta}^T s(t+\delta - T)s(t)\, dt \qquad (7.3.1)$$

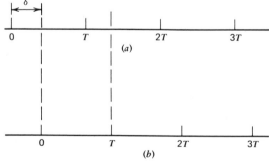

Figure 7.12. Timing error diagram.

where the first $\pm$ depends on the bit in $(0, T)$ and the second $\pm$ depends on the next bit. It is convenient to denote

$$\gamma(\tau) = \frac{1}{T} \int_0^T s(t + \tau)s(t) \, dt \tag{7.3.2}$$

as the *signal correlation function* for the periodic signal $s(t)$. This allows us to write (7.3.1) as

$$z_1(T, \delta) = \pm \int_0^T s(t + \delta)s(t) \, dt = \pm E\left[\frac{\gamma(\delta)}{\gamma(0)}\right] \tag{7.3.3}$$

if the bits are the same, and

$$z_1(T, \delta) = \pm E\left[\frac{\gamma(\delta)}{\gamma(0)} - \frac{2}{E}\int_{T-\delta}^T s(t + \delta - T)s(t) \, dt\right] \tag{7.3.4}$$

if different. Thus bit timing errors causes the signal energy E to be effectively degraded by the bracketed term, depending on the sign of the adjacent bit. The resulting degradation depends on the value of δ and the correlation function of the antipodal signals. If the shift δ in Figure 7.12 had been in the opposite direction, the same result would occur, only the previous bit, instead of the next bit, would be involved. The bit error probability of an imperfectly timed decoder is obtained by averaging the PE over the possibility of the adjacent bits being the same or opposite. Hence the antipodal bit error probability with a timing offset δ is

$$PE(\delta) = \frac{1}{4} \text{Erfc}\left[\left(\frac{E}{N_{b0}}\right)^{1/2}\left(\frac{\gamma(\delta)}{\gamma(0)}\right)\right]$$

$$+ \frac{1}{4} \text{Erfc}\left[\left(\frac{E}{N_{b0}}\right)^{1/2}\left[\frac{\gamma(\delta)}{\gamma(0)} - \frac{2}{E}\int_{T-\delta}^T s(t + \delta - T)s(t) \, dt\right]\right] \tag{7.3.5}$$

Evaluation of (7.3.5) requires specification of the particular antipodal signals involved.

Example 7.1 **(PCM Signals).** Let the antipodal signals be given by the PCM pulsed set in (7.2.8). We immediately see that, for $s(t) = s_1(t)$,

$$\gamma(\tau) = \frac{E}{T}$$

$$\frac{1}{E}\int_{T-\delta}^T s(t + \delta - T)s(t) \, dt = \frac{\delta}{T} \tag{7.3.6}$$

Thus

$$z_1(T, \delta) = \begin{cases} E, & \text{same adjacent bit} \\ E\left(1 - \dfrac{2|\delta|}{T}\right), & \text{opposite adjacent bit} \end{cases} \tag{7.3.7}$$

The signal energy is not degraded for any δ if the adjacent bit is the same and is reduced linearly with $|\delta|$ if the adjacent bit is opposite. The resulting bit error probability is then

$$PE(\delta) = \tfrac{1}{4}\,\mathrm{Erfc}\left[\left(\frac{E}{N_{b0}}\right)^{1/2}\right] + \tfrac{1}{4}\,\mathrm{Erfc}\left[\left[\frac{E}{N_{b0}}\right]^{1/2}\left(1 - \frac{2|\delta|}{T}\right)\right] \tag{7.3.8}$$

It is evident that timing offsets δ should be restricted to a small fraction of the bit interval T in order to prevent significant degradation. Equation (7.3.8) is plotted in Figure 7.13 for several values of δ/T.

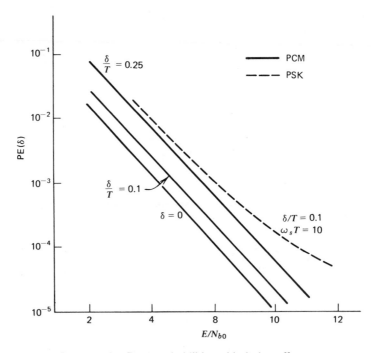

Figure 7.13. Error probabilities with timing offsets.

Example 7.2 (***PSK Signals***). When the PSK signals of (7.2.9) are used with $\omega_s \gg 2\pi/T$, we have

$$\gamma(\tau) = \frac{1}{T}\left(\frac{E}{T}\right)\int_0^T \cos\left(\omega_s(t+\tau)+\theta\right)\cos\left(\omega_s t + \theta\right) dt$$

$$= \frac{E}{2T}\cos\left(\omega_s\tau\right) \tag{7.3.9}$$

and

$$\frac{1}{E}\int_{t-\delta}^T s(t+\delta-T)s(t)\,dt = \frac{1}{2T}\cos\left(\omega_s\delta\right)[T-(T+|\delta|)] \tag{7.3.10}$$

The bit energy degradation is then

$$z_1(T,\delta) = \begin{cases} E\cos\left(\omega_s\delta\right), & \text{same adjacent bit} \\ E\cos\left(\omega_s\delta\right)\left[1 - \frac{2|\delta|}{T}\right], & \text{opposite adjacent bit} \end{cases} \tag{7.3.11}$$

and

$$\text{PE}(\delta) = \tfrac{1}{4}\,\text{Erfc}\left[\left(\frac{E}{N_{b0}}\right)^{1/2}\cos\left(\omega_s\delta\right)\right] + \tfrac{1}{4}\,\text{Erfc}\left[\left(\frac{E}{N_{b0}}\right)^{1/2}\cos\left(\omega_s\delta\right)\left(1 - \frac{2|\delta|}{T}\right)\right]$$
$$\tag{7.3.12}$$

Equation (7.3.12) is included in Figure 7.13. Note that for PSK the effective energy degrades as $\cos\left(\omega_s\delta\right)$ when a timing error of δ occurs. This means that to avoid significant degradation δ must be a small fraction of the oscillator period, rather than only of the bit interval as with PCM signals. This is simply due to the fact that the effective signal energy in imperfectly timed systems falls off according to the signal correlation function in (7.3.2), and PCM signals have a slower correlation roll-off than PSK signals. Since correlation roll-off varies inversely with the signal bandwidth, signals with the higher upper frequency tend to be more susceptible to timing errors. Thus timing error effect may be still another criterion for favoring one set of antipodal signals over another in practical system design.

7.4. Noncoherent Binary Signaling

The fact that the PSK signaling requires only the phase modulation of a subcarrier makes it an attractive scheme for binary digital systems. However, the requirement to maintain phase synchronization over a long stream of bits is often difficult to accomplish. An alternative procedure that retains the encoding and spectral advantage is to use modulated subcarriers but to rely on envelope detection rather than phase detection. In this case the phase referencing is no longer required, but the optimality of true coherent detection is sacrificed. In this section we examine this type of noncoherent binary operation.

Consider a binary digital system using the following encoder signals.

$$s_1(t) = e_1(t) \cos(\omega_1 t + \theta), \qquad 0 \le t \le T$$

$$s_0(t) = e_0(t) \cos(\omega_0 t + \theta), \qquad 0 \le t \le T \qquad (7.4.1)$$

where $e_i(t) \ge 0$ for all t and represents the envelope of the signal $s_i(t)$. The frequencies ω_1 and ω_0 are arbitrary subcarrier baseband frequencies and the phase θ is the corresponding phase angle during a bit period. If $\omega_1, \omega_0 \gg 2\pi/T$, the energy in the signals is again given approximately as

$$E = \tfrac{1}{2} \int_0^T e_i^2(t) \, dt, \qquad i = 0, 1 \qquad (7.4.2)$$

A coherent decoder for this pair of signals would require exact knowledge of the phase θ of the received signals. Consider instead the decoder in Figure 7.14a, which uses only envelope waveforms, rather than the complete signal waveform for binary decisioning. That is, after bandpass filtering at each of the subcarrier frequencies, the decoder recovers the envelope of the received signal and then matched filters each of the envelope functions. It does this in two parallel channels, each matched to one of the signal envelopes. Decoding is achieved by determining which channel produces the largest envelope sample. Thus the system in Figure 7.14a performs a comparison test but uses only the signal envelopes for matching rather than the complete signal. Note that the subcarrier phase is lost in the envelope detection processes and that its absolute value is not needed at the decoder. Such a decoder is said to be *noncoherent.*

First we may note that the decoder can be simplified. The matched filters have particular impulse responses that operate on the envelope of the filtered $x(t)$. From our earlier discussion we know that bandpass filtering of an amplitude modulated subcarrier is equivalent to filtering the envelope with the low frequency version of the bandpass filter. Thus the matched filtering can be incorporated directly into the bandpass filter in Figure 7.14a by proper shaping of the bandpass filter transform. This means the noncoherent decoder can be replaced by the equivalent decoder shown in Figure 7.14b and the sampling can be done directly at the output of the envelope detector. The sample producing the largest output voltage then decodes the bit. We remark that the noncoherent decoder presented here

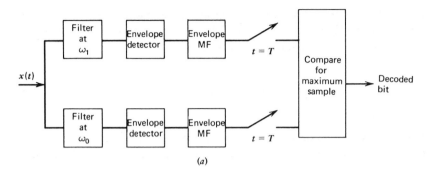

(a)

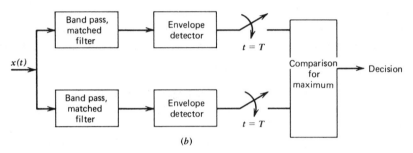

(b)

Figure 7.14. Noncoherent decoders (a) with envelope matched filtering, (b) with bandpass matched filtering.

is in fact the optimal decoder for minimizing error probability when the arbitrary phase in (7.4.1) is assumed a uniform random phase over $(0, 2\pi)$ [3, 4].

The noncoherent bit error probability is more complicated now since the decoder is nonlinear, because of the envelope detection operation. In particular, sampler statistics are no longer Gaussian even when the input noise is Gaussian, as it was for the linear decoder. The bit error probability is then

$$PE = \text{Prob}\left[\left(\begin{array}{c}\text{sample of envelope}\\ \text{of correct channel}\end{array}\right) \leq \left(\begin{array}{c}\text{sample of envelope}\\ \text{of incorrect channel}\end{array}\right)\right]$$

$$= \int_0^\infty \int_{\xi_1}^\infty p(\xi_1, \xi_2)\, d\xi_2\, d\xi_1 \qquad (7.4.3)$$

where $p(\xi_1, \xi_2)$ is the joint probability density of the sampled envelope values ξ_1, ξ_2 in the correct and incorrect channel, respectively. Unfortunately, this joint density is somewhat complicated in general, since ξ_1 and

ξ_2 are not independent and their density depends on θ in (7.4.1). It has been shown [7, 8], however, that PE in (7.4.3) is minimized when the signals in (7.4.1) are orthogonal; that is,

$$\int_0^T s_1(t)s_0(t) \, dt = 0 \qquad (7.4.4)$$

For this case the sampler values ξ_1 and ξ_2 no longer depend on θ and become pairwise statistically independent. Equation (7.4.3) then reduces to

$$PE = \int_0^\infty p(\xi_1) \int_{\xi_1}^\infty p(\xi_2) \, d\xi_2 \, d\xi_1 \qquad (7.4.5)$$

where now ξ_1 and ξ_2 are envelope samples of Gaussian processes. As discussed in (B.6.15) and (B.6.17) of Appendix B, these samples will have the density

$$p(\xi_1) = \frac{\xi_1}{\rho^2} e^{-(\xi_1^2 + \rho^4)/2\rho^2} I_0(\xi_1), \qquad \xi_1 \ge 0 \qquad (7.4.6a)$$

$$p(\xi_2) = \frac{\xi_2}{\rho^2} e^{-\xi_2^2/2\rho^2}, \qquad \xi_2 \ge 0$$

$$\rho^2 = \frac{2E}{N_{b0}} \qquad (7.4.6b)$$

and $I_0(x)$ is the imaginary Bessel function of zero order. Equation (7.4.6a) is the so-called *Rice* density associated with the probability density of an envelope sample of narrowband Gaussian noise with an additive signal term of energy E. Equation (7.4.6b) is the *Rayleigh* density for an envelope sample of narrowband Gaussian noise alone. When (7.4.6) is used in (7.4.5) the result integrates to

$$PE = \tfrac{1}{2} e^{-(E/2N_{bo})} \qquad (7.4.7)$$

where E is the signal energy in (7.4.2). Equation (7.4.7) is plotted in Figure 7.15 as noncoherent orthogonal. The coherent results of Figure 7.7 are also included. Note that noncoherent orthogonal binary systems are about 4 dB worse than coherent antipodal at $PE = 10^{-3}$ and roughly 3 dB worse at $PE = 10^{-5}$. This represents the price being paid for not making use of the phase angle in (7.4.1); that is, using coherent decoding. It is interesting that

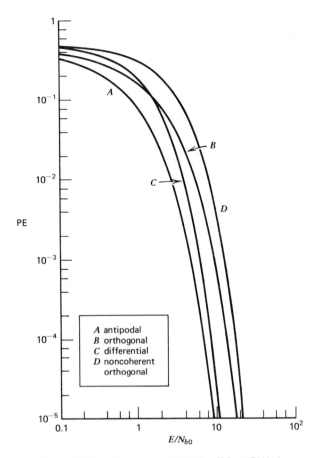

Figure 7.15. Bit error probabilities (PE vs E/N_{b0}).

at low error probabilities noncoherent orthogonal performs about the same as coherent orthogonal.

We can examine the orthogonality assumption in (7.4.4) by substituting from (7.4.1)

$$\int_0^T s_1(t)s_0(t)\, dt = \tfrac{1}{2} \int_0^T e_1(t)e_0(t) \cos (\omega_1 - \omega_0)t\, dt = 0 \qquad (7.4.8)$$

Note that the integration does not depend on θ as long as $\omega_1, \omega_0 \gg 2\pi/T$. Any pair of signals such that (7.4.8) is zero will constitute a pair of orthogonal signals. The following are particular examples of such signals, useful for noncoherent operation.

Pulse-Position-Modulation (PPM) Pulses

$$e_1(t) = \left(\frac{4E}{T}\right)^{1/2}, \quad 0 \le t \le \frac{T}{2}$$

$$= 0, \qquad\qquad \frac{T}{2} \le t \le T$$

$$e_0(t) = 0, \qquad\qquad 0 \le t \le \frac{T}{2}$$ (7.4.9)

$$= \left(\frac{4E}{T}\right)^{1/2}, \quad \frac{T}{2} \le t \le T$$

$$\omega_1 = \omega_0$$

The signals use subcarrier bursts at the same frequency, but each located in one of two possible time intervals, one representing a one, the other a zero, as shown in Figure 7.16a. The signals are orthogonal and have equal energy, but require delay shifting of an oscillator burst. Each requires a transmission bandwidth wide enough to pass a $T/2$ sec burst of oscillator frequency, and therefore a bandwidth of approximately $4/T$ Hz about the subcarrier. The decoder matched filters correspond to tuned bandpass filters, gated during the time interval of the corresponding signal. Comparison of the envelope samples at $t = T/2$ and $t = T$ yields the bit decision,

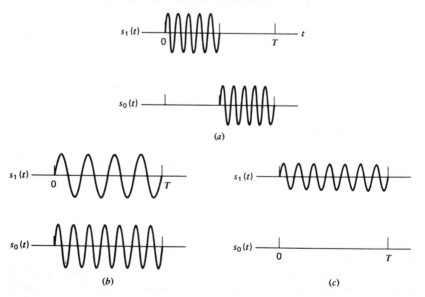

Figure 7.16. Noncoherent signaling formats. (a) PPM, (b) FSK, (c) on-off keying.

with the bit error probability given in (7.4.7). Note that the decoder bases its decision on which of two time intervals has the largest envelope sample.

Frequency-Shift Keying (FSK)

$$e_1(t) = e_0(t) = \left(\frac{2E}{T}\right)^{1/2}, \qquad 0 \le t \le T$$

$$\omega_1 - \omega_0 = 2\pi\left(\frac{k}{T}\right), \qquad k = \text{integer} \qquad (7.4.10)$$

The preceding signal pair (Figure 7.16b) is identical to the FSK signals in (7.2.15), except that we do not assume knowledge of the phase angle θ. By separating the frequencies by an exact multiple of the bit period, we can form signals that are orthogonal, have equal energy, and have the same advantage of ease of encoder generation. Whereas in coherent FSK the phase angle θ had to be exactly phase synchronized at the receiver, the signals here produce orthogonal signals without any knowledge of the value of θ. The bandwidth is the same as that for coherent FSK in Section 7.2. For this signal pair the individual channel matched filters in Figure 7.14 correspond to tuned bandpass filters. Hence the noncoherent decoder for FSK binary transmission corresponds to a receiver that attempts to determine which of two frequency channels produces the larger envelope sample. The frequencies are separated enough so that the transmitted frequency will not be received in the incorrect channel. (Contrast this result with the case of coherent FSK, where we purposely place the signaling frequencies close together to take advantage of slight negative correlation.) The bandpass filters must be matched to the envelopes of the FSK signals, and therefore have bandwidths of approximately $1/T$, centered at ω_1 and ω_0. The ease of encoding again makes FSK extremely popular for noncoherent binary signaling.

A problem with this noncoherent FSK decoder is the necessity of knowing the frequencies ω_1 and ω_0 exactly. Since the bandpass filters are only $1/T$ Hz wide, Doppler effects and associated oscillator instabilities may shift the expected signaling frequencies outside the decoder filter bandwidths. In practice this is compensated for by increasing the bandpass filter bandwidth so as to be sure to encompass the frequency uncertainty. This means the filters are no longer truly matched, and the input noise to the decoder channels is increased. Analytically, the envelope densities in (7.4.6) are no longer valid, and the envelope sample comparison becomes a generalized energy comparison. That is, each channel of the decoder actually measures the received energy of its channel, and bit decisioning is now based on the channel with the most energy. The densities in (7.4.6) are

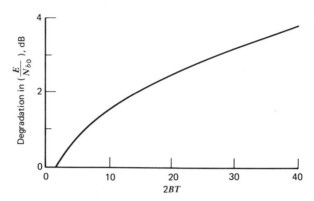

Figure 7.17. Equivalent increase in E/N_{b0} to account for bandpass filter bandwidth extension in Figure 7.14b with noncoherent FSK (B = filter bandwidth, T = bit time).

now replaced by families of chi-squared densities [9], and bit error probability requires the integration over these densities. Although the resulting PE no longer integrates to a result as simple as (7.4.7), some results are known [9], and fairly accurate approximations have been developed involving modified Erfc functions. Equation (7.4.7) can also be used to approximate the resulting PE, provided that a modification to the effective E/N_{b0} is made. Figure 7.17 shows how the true parameter E/N_{b0} must be degraded, as the bandpass filter bandwidth is increased, before using in (7.4.7) to obtain an approximation to the true PE. For $B = 1/T$, the system of course reduces to the noncoherent envelope decoder considered earlier.

The two channel system in Figure 7.14 can be used to decode noncoherent FSK with the minimal PE. However, another technique often used is actually to frequency detect the digital FSK modulation using a standard frequency discriminator. The decoder treats the FSK signal as a subcarrier frequency between ω_1 and ω_0. This subcarrier is then FM demodulated to recover the PCM modulation and bit decisions are based on the sign of the recovered bits. Such a system is shown in Figure 7.18 and is called an *FSK frequency discriminator decoder*. The system is again a form of noncoherent decoder since the subcarrier phase angle θ does not affect the FM detec-

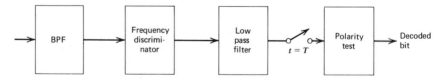

Figure 7.18. FSK discriminator decoder.

tion. The low pass filter can be a bit integrator to provide matched filtering of the demodulated PCM signal, or it can be a simple low pass filter for rejecting demodulator noise. The prediscrimination bandpass filter must be wide enough so as not to bandlimit the FSK signal, while reducing noise for the discriminator threshold. An exact analysis of the overall system is complicated by the fact that the output of the frequency discriminator is not truly a Gaussian process. We can, however, proceed by assuming the discriminator operates above threshold at all times. The FSK signal at the input to the bandpass filter has amplitude C. To model the filter effect we use the quasilinear analysis of Section 3.4 and treat the subcarrier at the discriminator input as having the modified amplitude

$$\tilde{C} = C|\tilde{H}_{bp}(\Delta_\omega)| \qquad (7.4.11)$$

where $\tilde{H}_{bp}(\omega)$ is the low frequency version of the bandpass filter transfer function and Δ_ω is the effective subcarrier deviation

$$\Delta_\omega = \frac{\omega_1 - \omega_0}{2} \qquad (7.4.12)$$

The demodulated output signal is given by the sum of the modulating PCM signal, having amplitude $\pm\Delta_\omega$, and the discriminator frequency noise. If the low pass filter is ideal over bandwidth $1/T$ Hz, the noise at the sampler input has variance

$$\sigma_n^2 = \frac{2N_{b0}(2\pi)^2}{3(\tilde{C})^2 T^3} \qquad (7.4.13)$$

A bit decision based on the sign of the sampler voltage will produce a bit error probability of

$$PE = \tfrac{1}{2} \text{Erfc} \left[3\beta^2 \frac{C^2 T}{2N_{b0}} |\tilde{H}_{bp}(\Delta_\omega)|^2 \right]^{1/2} \qquad (7.4.14)$$

where $\beta = \Delta_\omega T/2\pi$ is the FM modulation index with respect to the bit rate frequency $1/T$. The result illustrates two effects of discriminator decoding. As β increases, the first term implies an increase in the demodulated signal energy and improved detection (i.e., the FM improvement). However, as Δ_ω is further increased, the predetection bandpass filter reduces the FSK subcarrier amplitude according to (7.4.11), which eventually degrades the detection. We see therefore that a value of modulation index (deviation) will exist such that minimal PE will occur for FSK discriminator decoding. This optimal value will depend on the predetection filter function and can be determined by finding the value Δ_ω that maximizes the argument in (7.4.14). This maximization can be easily carried out (Problem 7.19) for several common types of filter functions, and the results are summarized in

Table 7.2. Tabulation of Optimal Modulation Indices and Deviation for FSK Systems with Discriminator Decoding (B_n = predetection filter noise bandwidths, T = bit time)

Filter Function $\|\tilde{H}_{bc}(\omega)\|^2$	Optimal Modulation Index β	Optimal Frequency Separation
Gaussian, $e^{-\omega^2/16\pi B_n^2}$	$0.31 B_n T$	$0.7 B_n$
Butterworth, $\dfrac{1}{1 + \left(\dfrac{\omega}{2\pi B_n}\right)^{10}}$	$0.40\, B_n T$	$0.81 B_n$
Ideal 1, $\|\omega\| \le 2\pi B_n$ 0, elsewhere	$0.5 B_n T$	B_n

Table 7.2. The optimal indices and frequency separations are stated in terms of the noise bandwidth B_n of the predetection filters. This latter bandwidth must be selected so that the baseband CNR (given by $C^2/4N_{b0}B_n$) satisfies the discriminator threshold. We again see that in each case a particular frequency separation is specified for maximizing PE with FSK discriminator decoders. This is in contrast with the noncoherent envelope detection decoders, where any frequency separation producing orthogonality is usable.

The resulting PE for discriminator decoding is plotted in Figure 7.19, using (7.4.14) and the optimal values of β in Table 7.2. In each case, operation was adjusted for a discriminator threshold of 12 dB. Since $CNR = (E/N_{b0})/2B_n T$, this requires adjustment of $B_n T$ at each value E/N_{b0}. The corresponding curves for coherent FSK (orthogonal) and noncoherent envelope detection are also superimposed. We see that discriminator decoding does slightly poorer than noncoherent decoding, although the actual difference will depend on the choice of threshold value. More rigorous analysis of FSK discriminator decoding has taken into account the below threshold operation of the discriminator [10, 11] and the extension to phase lock loop frequency detection [12].

On-Off Keying

$$e_1(t) = \left(\frac{2E}{T}\right)^{1/2}, \qquad 0 \le t \le T$$

$$e_0(t) = 0 \tag{7.4.15}$$

Here a binary zero is transmitted as no signal, whereas a one is transmitted

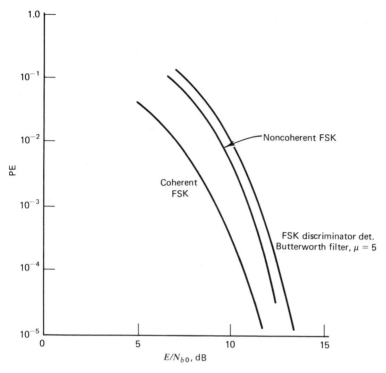

Figure 7.19. PE for FSK decoding.

as a T sec subcarrier burst of frequency ω_1 (Figure 7.16c). The subcarrier frequency and phase can be arbitrary, and the signal pair is always orthogonal. The required bandwidth is only half that of PPM signaling. The signals, however, do not have equal energy, and require continual on-off keying of the encoder subcarrier oscillator (i.e., not continuous operation). Since one signal is zero, only a single channel decoder in Figure 7.14 is needed. Bit decisioning is achieved by comparing the envelope sample to a threshold voltage. Since the signals do not have equal energy, the bit error probability in (7.4.7) must be modified to determine decoder performance. In particular, the probability of an error for a zero bit is different from that of a one bit. When a zero is sent, only noise is received and the error probability is given by the probability that an envelope sample of bandpass filtered Gaussian noise exceeds a threshold value of Y. This becomes

$$\text{PE}|\text{zero bit} = \int_{Y'}^{\infty} p(\xi_2) \, d\xi_2 \qquad (7.4.16)$$

where Y' is the normalized threshold $Y/(2E/N_{0b})^{1/2}$ and $p(\xi_2)$ is the envelope sample density in (7.4.6*b*). Similarly, the probability of an error when a one is sent is given by the probability that an envelope sample of a sine wave in Gaussian noise is less than the threshold. Thus

$$\text{PE} \mid \text{one bit} = \int_0^{\Upsilon'} p(\xi_1) \, d\xi_1 \qquad (7.4.17)$$

where $p(\xi_1)$ is given in (7.4.6*a*). Both of these integrals can be written in terms of the Marcum Q-function, defined as

$$Q(a, b) \triangleq \int_b^\infty \exp\left[-\frac{a^2 + x^2}{2} \right] I_0(ax) x \, dx \qquad (7.4.18)$$

The average bit error probability for noncoherent on-off keying can therefore be written as

$$\text{PE} = \tfrac{1}{2}[\text{PE} \mid \text{zero bit}] + \tfrac{1}{2}[\text{PE} \mid \text{one bit}]$$

$$= \tfrac{1}{2}Q[0, \Upsilon'] + \tfrac{1}{2}\left[1 - Q\left[\left(\frac{2E}{N_{b0}} \right)^{1/2}, \Upsilon' \right] \right] \qquad (7.4.19)$$

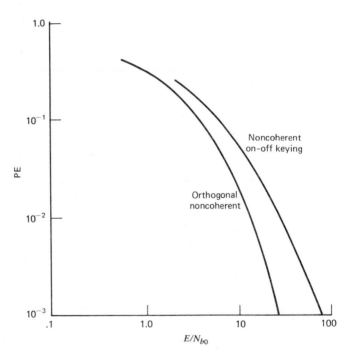

Figure 7.20. Error probability for noncoherent on-off signals [threshold $\Upsilon = 0.5$ $(2E/N_{b0})^{1/2}$].

Equation (7.4.19) is plotted in Figure 7.20 for $Y' = 0.5$. The noncoherent orthogonal result (FSK or PPM) with the same signal energy E is superimposed. At the higher E/N_{b0} values, the on-off system suffers from the lack of usable signal energy during the zero bit transmission. The Q function has been tabulated [13] and recursive computational methods have been developed for its evaluation [14].

7.5 Differential PSK (DPSK) Binary Signaling

The noncoherent encoding scheme avoided the necessity of achieving and maintaining the phase coherence required in the coherent PSK system. Another way to bypass this problem while still performing binary encoding is by the use of *differential PSK* (DPSK). In this scheme, antipodal PSK signaling is again used, but the bits are encoded by relating them to the previous bit. If a binary one is to be sent during a bit interval, it is sent as a PSK signal with the same phase as the previous bit. If a binary zero is sent, it is transmitted with the opposite phase (i.e., the negative version) of the previous bit. Thus ones and zeros are transmitted as a phase change or no phase change from the previous bit. The source sequence is therefore being sent as a sequence of PSK signals, but the phase of each is obtained by the preceding rule. As an example, if we denote the PSK signals as

$$f_1(t) = \left(\frac{2E}{T}\right)^{1/2} \sin(\omega_s t + \theta), \qquad 0 \le t \le T$$

$$f_0(t) = \left(\frac{2E}{T}\right)^{1/2} \sin(\omega_s t + \theta \pm \pi), \qquad 0 \le t \le T \qquad (7.5.1)$$

and if we assume the last bit sent was $f_1(t)$, then the bit sequence 1 0 0 1 will be sent as the waveform sequence $\{f_1, f_0, f_1, f_1\}$. Note that we continue to send one bit every T sec (except for the first bit, which requires an initial reference waveform), and is therefore a form of binary encoding.

An apparent advantage of DPSK can be observed by noting that decoding can be achieved by correlating each T sec waveform with the previous bit, as shown in Figure 7.21a. Without noise the signal correlation produces the value $+E$ if a one is sent and correlates to $-E$ if a binary zero is sent, for any value of the phase angle θ. Since these correlation values are identical to the signal correlation values produced in coherent PSK, it at first appears that DPSK is providing exact coherent operation without phase referencing. However, when noise is added, it combines with both bits, and performance is significantly worse than that of coherent PSK. This can be attributed to the fact that two noisy waveforms are actually being correlated, and the idealized values of $\pm E$ are not really being achieved.

A more rigorous approach to DPSK analysis is to recognize that we are really transmitting each bit with the binary signal pair

$$s_1(t) = \begin{cases} (f_1, f_1), \\ \text{or} \qquad 0 \le t \le 2T \\ (f_0, f_0), \end{cases} \qquad (7.5.2)$$

$$s_0(t) = \begin{cases} (f_1, f_0), \\ \text{or} \qquad 0 \le t \le 2T \\ (f_0, f_1), \end{cases}$$

where f_i is the signal $f_i(t)$ in (7.5.1) and (f_i, f_j) denotes $f_i(t)$ followed by $f_j(t)$. The first T sec of each waveform is actually the last T sec of the previous waveform. Note that $s_1(t)$ and $s_0(t)$ each can have either of two possible forms. We see these signals all have energy $2E$, and the signal pair has correlation

$$\int_0^{2T} s_1(t) s_0(t) \, dt = \int_0^T (\pm f_1(t))^2 \, dt - \int_0^T (\pm f_1(t))^2 \, dt = 0 \qquad (7.5.3)$$

for any combination. The DPSK signal set in (7.5.2) therefore represents a pair of orthogonal signals, $2T$ sec long with arbitrary phase angle θ. Decoding of these signals should correspond to noncoherent envelope detection with four channels matched to each of the possible envelope waveforms, two representing a one and two representing a zero, as in Figure 7.21*b*. However, the two waveforms representing each bit are negatives of each other, and the envelope sample of each will be the same. Hence we need only construct a single noncoherent channel for $s_1(t)$ and a single channel for $s_0(t)$, each matched to either of the two noncoherent waveforms in (7.5.2).† The DPSK decoder is therefore reduced to a standard two channel noncoherent decoder, as in Figure 7.21*c*. For orthogonal signals this operates with the bit error probability in (7.4.7). Since the signals have twice the bit interval energy,

$$\text{PE} = \tfrac{1}{2} e^{-(E/N_{b0})} \qquad (7.5.4)$$

where E is the bit energy in T sec $(E = P_m T)$ and N_{b0} is again the baseband one-sided noise level. This result is superimposed in Figure 7.15. We see that DPSK achieves the same PE as noncoherent orthogonal at half the bit energy E. Thus DPSK is 3 dB better than noncoherent operation, while still avoiding the phase referencing operation. At low $(\text{PE} \le 10^{-5})$ error prob-

---

† The decoder is actually averaging over the two waveforms representing each bit, which is the necessary operation for minimizing PE when θ is purely random. Since the two channels produce the same envelope sample, the averaging is equivalent to observing either one, and only a single decoding channel is necessary for $s_1(t)$ for $s_0(t)$.

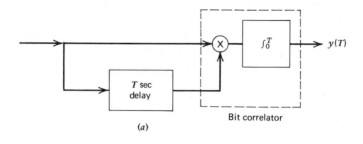

(a)

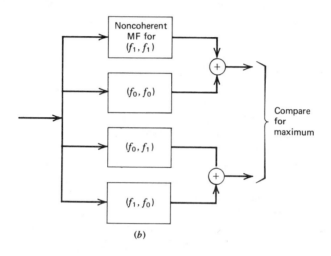

(b)

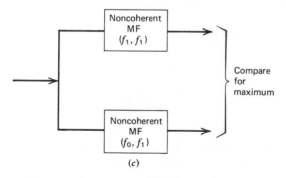

(c)

Figure 7.21. DPSK decoder models. (a) Correlation decoder, (b) four-channel noncoherent decoder, (c) equivalent two-channel decoder.

abilities, DPSK is only about 1 dB poorer than coherent PSK. However, the result must be accepted with care, since PE may not be a good measure of performance with DPSK, since errors tend to accumulate in pairs. That is, a burst of noise during a single T sec interval will likely cause two bits to be decoded in error. In addition, the entire decoder is more complex than the single decoder analyzed. This is due to the fact that each T sec of arriving waveform must be used twice concurrently—as the end of the present bit and as the beginning of the next bit. This requires two separate noncoherent decoders to be operated in parallel with a T sec offset in the channel processing. The alternative is to use T sec waveform storage, but this slows the decoding bit rate, and the system reduces to an effective single decoder noncoherent system with bits $2T$ sec long.

7.6. Decoding with Nonwhite Noise

The results derived so far have been based on the condition that the baseband noise is a white noise process (i.e., has a flat spectrum over all baseband frequencies). As noted in Chapter 4, this is not always the case, as, for example, when the RF system uses FM. We may therefore inquire as to the modification that must be made to our earlier results when the baseband noise cannot be modeled as a white noise process.

Let the Gaussian baseband noise have the general power spectrum $S_{nb}(\omega)$. We again assume a filter-sampler decoder of the type in Figure 7.2. Since the noise is nonwhite, the advantage of the matched filter decoder and antipodal signaling is no longer obvious. If we reexamine (7.1.15) we see that the effect of having nonwhite noise will alter the value of σ_y^2 used in our earlier development. The bit error probability for the decoder is still given by (7.1.22), as

$$\text{PE} = \tfrac{1}{2}\,\text{Erfc}\,(\Lambda) \qquad (7.6.1)$$

where again

$$\Lambda = \frac{|m_1 - m_0|}{2\sqrt{2}\,\sigma_y} \qquad (7.6.2)$$

and σ_y has the modified variance inserted. Here m_1 and m_0 are again defined as in (7.1.14). Our objective is to design again a system that achieves minimal PE. This requires maximization of Λ with the new value of σ_y. If the binary signals are constrained to have fixed energy E as before, we can easily establish that $|m_1 - m_0|$ will be maximized for any decoder filter function if the signals $s_1(t)$ and $s_0(t)$ are antipodal (Problem 7.21). For the antipodal set $s_1(t) = -s_0(t) = s(t)$, $0 \le t \le T$, we have

$$m_1 - m_0 = 2 \int_0^T s(t)h(T-t) \, dt$$

$$= \frac{2}{2\pi} \int_{-\infty}^{\infty} H(\omega)F_s(\omega) e^{j\omega T} \, d\omega \tag{7.6.3}$$

where $H(\omega)$ is the decoder filter transfer function and $F_s(\omega)$ is the transform of $s(t)$. Again only a single channel decoder is needed, since the signaling set is antipodal. Thus (7.6.2) becomes

$$\Lambda = \frac{\left| (1/2\pi) \int_{-\infty}^{\infty} H(\omega)F_s(\omega) e^{j\omega T} \, d\omega \right|}{\left[(2/2\pi) \int_{-\infty}^{\infty} |H(\omega)|^2 S_{nb}(\omega) \, d\omega \right]^{1/2}} \tag{7.6.4}$$

To establish the maximization of Λ, it is convenient to denote

$$u_1(\omega) \triangleq H(\omega)(S_{nb}(\omega))^{1/2}$$

$$u_2(\omega) \triangleq \frac{F_s(\omega)}{\sqrt{S_{nb}(\omega)}} e^{j\omega T} \tag{7.6.5}$$

where $\sqrt{S_{nb}(\omega)}$ is the function obtained by taking the square root of $S_{nb}(\omega)$ at every ω. We then see that the numerator in (7.6.4) is simply

$$\left| \frac{1}{2\pi} \int_{-\infty}^{\infty} u_1(\omega)u_2(\omega) \, d\omega \right| \tag{7.6.6}$$

Applying the Schwarz inequality to this numerator establishes that

$$\Lambda = \frac{\left| (1/2\pi) \int_{-\infty}^{\infty} u_1(\omega)u_2(\omega) \, d\omega \right|}{\sqrt{2}\sigma_y}$$

$$\le \frac{(1/2\pi) \left[\int_{-\infty}^{\infty} |u_1(\omega)|^2 \, d\omega \int_{-\infty}^{\infty} |u_2(\omega)|^2 \, d\omega \right]^{1/2}}{\left[(2/2\pi) \int_{-\infty}^{\infty} |u_1(\omega)|^2 \, d\omega \right]^{1/2}}$$

$$= \left[\frac{1}{4\pi} \int_{-\infty}^{\infty} \frac{|F_s(\omega)|^2}{S_{nb}(\omega)} \, d\omega \right]^{1/2} \tag{7.6.7}$$

Furthermore, this upper bound on Λ occurs only if $\mu_1(\omega) = \mu_2^*(\omega)$, or if

$$H(\omega)\sqrt{S_{nb}(\omega)} = \frac{F_s^*(\omega)}{\sqrt{S_{nb}(\omega)}} e^{-j\omega T} \tag{7.6.8}$$

where * denotes complex conjugate. Solving for the decoder filter function then yields

$$H(\omega) = \frac{F_s^*(\omega)}{S_{nb}(\omega)} e^{-j\omega T} \tag{7.6.9}$$

This is the required filter transfer function that minimizes PE for antipodal signaling in nonwhite noise. Note that the desired filter function now depends on the noise spectrum, as well as that of the signal $s(t)$. This means that, in the case of nonwhite noise, the desired filter is related to both the signal and the interfering noise. To obtain a physical interpretation of (7.6.9), it is convenient to rewrite as

$$H(\omega) = \left[\frac{1}{\sqrt{S_{nb}(\omega)}}\right]\left[\frac{F_s^*(\omega)}{\sqrt{S_{nb}(\omega)}} e^{-j\omega T}\right] \qquad (7.6.10)$$

The optimal decoder filter now appears as the cascade of two separate filters, each having transfer function given by the preceding bracketed terms. Without regard to physical realizability, we can intepret the first filter as one that filters the total baseband power spectrum with the function

$$\left|\frac{1}{\sqrt{S_{nb}(\omega)}}\right|^2 = \frac{1}{S_{nb}(\omega)} \qquad (7.6.11)$$

When applied to the noise portion of the baseband waveform, we see that this first filter generates an output spectrum that is of unit level for all ω. Thus the first filter effectively "whitens" the baseband noise and is referred to as a *whitening filter*. In the process of whitening the noise, however, the baseband signal $\pm s(t)$ is modified to a new signal $\pm \hat{s}(t)$, where $\hat{s}(t)$ has the transform

$$F_{\hat{s}}(\omega) = \left[\frac{1}{\sqrt{S_{nb}(\omega)}}\right]F_s(\omega) = \frac{F_s(\omega)}{\sqrt{S_{nb}(\omega)}} \qquad (7.6.12)$$

We now see that the second filter function [second bracket in (7.6.10)] is related to this spectrum. Recall that if a real function $f(t)$ has transform $F(\omega)$, then the function $f(-t)$ has transform $F(-\omega) = F^*(\omega)$, and $f(T-t)$ has transform $F^*(\omega) e^{-j\omega T}$. From this we see that the second filter function has an impulse response matched to the signal $s(t)$ having the transform in (7.6.12). That is, the second filter is a matched filter for the modified signal at the whitening filter output. The optimal decoder filter for nonwhite noise can therefore be considered as the cascade of a filter that first whitens the baseband noise followed by a filter that is matched to the modified signal produced by the whitening (Figure 7.22). Physically, the whitening filter may not be realizable as a separate filter (e.g., if $S_{nb}(\omega) = 0$ beyond some upper cutoff frequency, the whitening filter must have infinite gain beyond this frequency). However, when the required filter in (7.6.9) is constructed as a single filter function, the resulting filter is usually realizable, or can at least be adequately approximated.

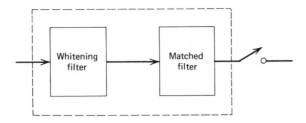

Figure 7.22. Matched filter decoder for nonwhite noise and antipodal signals.

If the required filter function is implemented, then the resulting value of Λ given in (7.6.7) is

$$\Lambda_{\max} = \left[\frac{1}{4\pi} \int_{-\infty}^{\infty} \frac{|F_s(\omega)|^2}{S_{nb}(\omega)} \, d\omega \right]^{1/2} \tag{7.6.13}$$

Note that the maximum value no longer depends only on the signal energy E, as for the case of white noise, but instead is integrably related to the spectrum of both signal and noise. This means the actual antipodal signal spectrum will determine the resulting value of $\Lambda_{\max}$ and therefore the decoder performance in (7.6.1). (Compare this with the result for white noise, where we concluded that any antipodal signal set was as good as any other with the same energy.) We now see that one antipodal signal pair is preferable over another if it produces a larger value of $\Lambda_{\max}$. This immediately suggests that we attempt to determine the signal $s(t)$ such that $\Lambda_{\max}$ is maximized, under the condition that $s(t)$ have energy E. That is, we want to maximize over $F_s(\omega)$ the integral

$$\int_{-\infty}^{\infty} \frac{|F_s(\omega)|^2}{S_{nb}(\omega)} \, d\omega \tag{7.6.14a}$$

subject to the energy constraint

$$\frac{1}{2\pi} \int_{-\infty}^{\infty} |F_s(\omega)|^2 \, d\omega = E \tag{7.6.14b}$$

We could proceed by formal application of calculus of variations. However, the solution follows simply by noting

$$\frac{1}{4\pi} \int_{-\infty}^{\infty} \frac{|F_s(\omega)|^2}{S_{nb}(\omega)} \, d\omega \le \left(\frac{1}{2S_{\min}} \right) \int_{-\infty}^{\infty} |F_s(\omega)|^2 \, d\omega = \frac{E}{(2S_{\min})} \tag{7.6.15}$$

where $S_{\min}$ is the minimum value of $S_{nb}(\omega)$ over all ω. The upper bound occurs only if

$$|F_s(\omega)| = 0, \qquad \omega \notin I_{\min}$$

$$\neq 0, \qquad \omega \in I_{\min} \tag{7.6.16}$$

where I_{min} is the specific frequencies where S_{min} occurs. I_{min} may be a point, a set of points, an interval, or a collection of intervals on the ω axis. The condition in (7.6.16) means that the spectrum of the signal should be placed, in any manner, over the frequency set where the noise spectrum is minimum. If I_{min} is the entire ω axis [i.e., $S_{nb}(\omega)$ is flat], then any signal spectrum will satisfy (7.6.15) and any signal with energy E achieves the bound, a result we have already derived. In summary, then, performance is optimized with matched filter decoders and nonwhite noise if the antipodal signal set is selected so that its interfering noise is least. However, there may be other systems constraints that may prevent complete freedom in spectral selection and thereby prevent attaining the bound in (7.6.15).

Notice in (7.6.15) that if $S_{min} = 0$ (the noise spectrum is zero somewhere on the ω axis), then the upper bound on Λ becomes infinite. This in turn means the bit error probability in (7.6.1) is zero. Theoretically, perfect bit transmission can be achieved under this condition by proper design. Obviously, if we utilize a portion of the frequency axis where no noise is present, the decoder will always correctly decide if $s(t)$ or $-s(t)$ is being received for any signal of energy E. Operation under such a condition is called *singular detection*, and has been generalized to other domains of signal representation [15]. That is, if any coordinate system can be determined for signal representation such that at least one coordinate contains no noise, singular detection can always be achieved by binary signaling over that particular coordinate.

Our most important application for the nonwhite noise result is for the case of FM usage of the RF link. Here the baseband noise following RF demodulation has the quadratic spectrum of (4.9.1). Since the spectrum is zero only at $\omega = 0$, singular detection can only be achieved by using a signal concentrated at $\omega = 0$. This would require a constant (dc) signal (delta function at $\omega = 0$), violating our condition that the signaling waveform be only T sec long. However, the result does allow us to conclude that the signaling scheme should concentrate its spectrum around zero frequency, as in the case of straight PCM signals. Hence binary transmission over an RF link using FM demodulation favors standard PCM|FM modulation formats.

7.7. Bandlimiting Effects

We have been assuming that the bit signals at the decoder input are identical to the signals formed at the transmitter encoder. We know from practice that this again is an idealized condition and that in fact decoder signals will often be distorted during transmission. This most predominant cause is the filtering applied to the signal during modulation, transmission, and demodulation. The most severe filtering generally comes from base- band filtering, following the demodulator and just prior to the decoder. For PCM pulsed signals this usually corresponds to a low pass filter designed to pass only a bandwidth approximately equal to the bit rate in hertz. For PSK signals, which basically represent an oscillator amplitude modulated by a PCM pulsed sequence, the baseband filter is usually a bandpass filter tuned to the subcarrier frequency, with a width of approximately twice this bit rate. By our AM-filter conversion, we can again consider this to be an effective low pass filtering, with a bandwidth equal to half the bandpass bandwidth, on the PCM sequence itself.

Bandlimiting and filtering causes two primary effects—waveform distortion and bit spreading. The first causes removal or reduction of portions of the bit waveform spectrum, thereby reducing its energy content for decoding. The second produces waveform spreading in time (recall the basic inverse relation between signal frequency and time extent, from Problem 1.6), causing the signal in one bit interval to interfere with the bits in the subsequent intervals. This latter effect is referred to as *intersymbol interference*. To examine both these effects consider again the baseband subsystem model in Figure 7.23*a*, similar to that used in Chapter 5. We

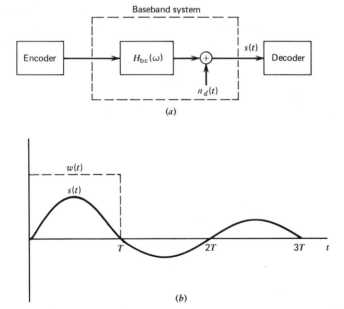

Figure 7.23. Filtered baseband subsystem. (*a*) Block diagram, (*b*) filtered waveforms.

assume $H_{bc}(\omega)$ represents the equivalent baseband filtering applied to the transmitted bit waveforms up to the decoder input, and $n_d(t)$ represents the baseband Gaussian noise at the decoder input. The spectra characteristics of this noise depend on the specific model intended. If the filter represents filtering applied prior to demodulation, then $n_d(t)$ is the same as the demodulator noise $n_b(t)$, and its spectrum is given by $S_{nb}(\omega)$, as before. If the filter represents postdemodulation baseband filtering (which is generally the case of most interest), it will then filter the demodulator noise as well, and the spectrum of $n_d(t)$ is then $S_{nb}(\omega)|H_{bc}(\omega)|^2$. We assume the encoder signals consist of antipodal PCM waveforms, which we write as $\pm\sqrt{(E/T)}w(t)$, where

$$w(t) = 1, \qquad 0 \le t \le T$$

$$= 0, \qquad \text{elsewhere} \qquad (7.7.1)$$

The decoder is taken as the integrate-and-dump matched filter associated with the encoder signals. We can write the filtered decoder input during any specific bit interval $(0, T)$ as

$$x(t) = \left(\frac{E}{T}\right)^{1/2} \sum_{j=0}^{\infty} d_j s(t+jT) + n_d(t) \qquad (7.7.2)$$

where $d_j = \pm 1$, d_0 is the bit during $(0, T)$, and d_j is the bit j intervals in the past. The waveform $s(t)$ represents the filtered version of the pulse $w(t)$ in (7.7.1). Since $s(t)$ is no longer of infinite bandwidth, it is no longer limited to T sec and is instead spread in time (Figure 7.23b). Note this will have the transform

$$F_s(\omega) = W(\omega)H_{bc}(\omega) \qquad (7.7.3)$$

where $W(\omega)$ is the transform of $w(t)$. The integrate and dump decoder produces the sample value

$$y(T) = \int_0^T x(t)\,dt$$

$$= \sqrt{E}m_0 + \sqrt{E}m_I + \int_0^T n_d(t)\,dt \qquad (7.7.4)$$

where

$$m_0 = \frac{1}{\sqrt{T}} \int_0^T d_0 s(t)\,dt \qquad (7.7.5a)$$

$$m_I = \frac{1}{\sqrt{T}} \sum_{j=1}^{\infty} d_j \int_0^T s(t+jT)\,dt \qquad (7.7.5b)$$

The first term in (7.7.4) is the effect of the present bit, whereas m_I accounts for the combined intersymbol interference from past bits. The last term is the integrated decoder noise and represents a zero mean Gaussian variable with variance

$$\sigma_n^2 = \frac{1}{2\pi} \int_{-\infty}^{\infty} S_d(\omega) |W(\omega)|^2 \, d\omega \tag{7.7.6}$$

Note that $y(T)$ has a mean value $\sqrt{E}(m_0 + m_I)$, which depends on the past bits, and a variance σ_n^2. For a particular bit history $\mathbf{d} = (d_0, d_1, d_2, \ldots)$, the resulting decoder error probability PE($\mathbf{d}$) follows from (7.2.3) as

$$\text{PE}(\mathbf{d}) = \tfrac{1}{2} \text{Erfc}\left[\Lambda(\mathbf{d})\right] \tag{7.7.7}$$

where now

$$\Lambda(\mathbf{d}) = \left[\frac{E}{N_{b0}} \left(\frac{m_0 + m_I}{2\sigma_n/N_{b0}}\right)^2\right]^{1/2} \tag{7.7.8}$$

We see that m_0 accounts for the waveform distortion, whereas m_I is the intersymbol effect of previous bits, adding a constant term to the present mean m_0. This constant, however, depends on the channel filter pulse response $s(t)$ and on the particular sequence of past bits. This term may have the same sign as the m_0 term, thereby aiding detection, or it may be opposite in sign, causing further decoding degradation. To determine the exact filter effect on the bit error probability of a given bit, it is necessary to average (7.7.7) over all past bit sequences $\mathbf{d}$. A simpler indication of the effect is to examine (7.7.7) for the worst case bit sequence, that is, the past bit sequence for which m_I causes the largest degradation to m_0. The worst case sequence can be easily determined in (7.7.5b) (Problem 7.23), and the resulting value of m_0 and m_I can be evaluated by inverse transforming (7.7.3) and integrating in (7.7.5). Hartmann [16] has computed this worst case degradation to Λ for a bandlimiting channel filter,

$$H_{\text{bc}}(\omega) = 1, \qquad |\omega| \leq 2\pi B_m$$
$$= 0, \qquad \text{elsewhere} \tag{7.7.9}$$

and his result is shown in Figure 7.24 as a function of B_m. Also included is the worst case result for a single pole RC filter with 3 dB bandwidth B_m, making use of the results of Problem 1.23. Both results clearly exhibit the worst case degradation to Λ as B_m is narrowed. The corresponding error probability in (7.7.7) yields a worst case result, and although somewhat pessimistic, it does indicate an upper bound to the true performance.

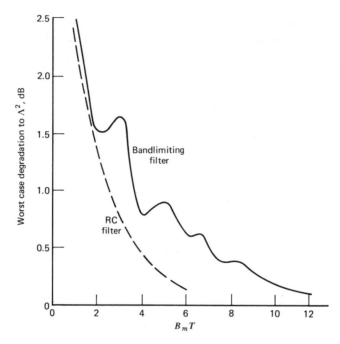

Figure 7.24. Bandlimiting and filtering effects on binary PE (B_m = bandwidth of bandlimiting filter and 3 dB bandwidth of first order RC filter).

7.8. Designing for the Bandlimited Channel

The degradation due to the channel filter is caused by the mismatch between the decoder and bandlimited signals. We may therefore inquire if, given a known bandlimiting channel filter $H_{bc}(\omega)$ (either as part of the transmission channel or imposed by system constraints), there is a formal design procedure that best compensates for the filtering. To examine this, consider again the bandlimiting filter of (7.7.9). Let us allow an arbitrary encoder signal $\pm z(t)$ to be transmitted for antipodal signaling and again assume the decoder has the general form in Figure 7.2 composed of a filter $H(\omega)$ followed by a sampler. We would like to determine an optimal decoder filter and encoder signal $z(t)$ for combating the channel bandlimiting and yielding minimal error probability. Let $Z(\omega)$ be the transform of $z(t)$ and assume it to be confined to energy E. The output of the decoding filter can again be written as in (7.7.2)

$$y(t) = \left(\frac{E}{T}\right)^{1/2} \sum_{j=0}^{\infty} d_j s(t+jT) + n_y(t) \tag{7.8.1}$$

where now $s(t)$ is the filtered version of $z(t)$ and therefore has the transform

$$F_s(\omega) = Z(\omega)H_{bc}(\omega)H(\omega)$$

$$= Z(\omega)H(\omega), \qquad |\omega| \le 2\pi B_m \qquad (7.8.2)$$

Note that $F_s(\omega)$ is also bandlimited to B_m Hz because of the channel filtering. The noise $n_y(t)$ has spectrum

$$S_{ny}(\omega) = S_d(\omega)|H(\omega)|^2 \qquad (7.8.3)$$

At the sampling instant, $t = T$, $y(T)$ is

$$y(T) = \sqrt{(E/T)}\left[d_0 s(T) + \sum_{j=1}^{\infty} d_j s[(j+1)T]\right] + n_y(T) \qquad (7.8.4)$$

Again we have the effect of the present bit (first term), the intersymbol effect of past bits (summation), and the baseband noise. However, we immediately see that the entire intersymbol effect depends only on $s(t)$ at $t = jT$, $j \ge 2$. This means that any bandlimited waveform $s(t)$ such that

$$s(jT) = 0, \qquad j \ge 2$$

$$\ne 0, \qquad j = 1 \qquad (7.8.5)$$

automatically removes intersymbol interferences. If the waveform is not bandlimited, then of course any waveform confined to $(0, T)$ would suffice, such as those considered throughout this chapter. Since the sinc function

$$\mathrm{sinc}\left(\frac{\pi(t-T)}{T}\right) \triangleq \frac{\sin\left(\pi(t-T)/T\right]}{\pi(t-T)/T} \qquad (7.8.6)$$

satisfies (7.8.5), a general solution occurs if $s(t)$ has the form

$$s(t) = s_a(t)\,\mathrm{sinc}\left[\frac{\pi(t-T)}{T}\right] \qquad (7.8.7)$$

where $s_a(t)$ is an arbitrary time function. The transform of (7.8.7) is the convolution of the transform of the arbitrary function $s_a(t)$ and the transform of (7.8.6). That is,

$$F_s(\omega) = F_{s_a}(\omega) \oplus F_T(\omega) \qquad (7.8.8)$$

where $F_T(\omega) = 1$, $|\omega| \le 2\pi/2T$, and zero elsewhere. We see that $s(t)$ has a bandwidth of B_m Hz, as required by (7.8.2), only if $B_m \ge 1/2T$, and only if $s_a(t)$ has a bandwidth of $B_m - (1/2T)$ Hz. That is, the convolution of a bandlimited spectrum of width $1/2T$ with one bandlimited to $B_m - (1/2T)$ Hz would produce an overall spectrum bandlimited to B_m Hz. Thus if we are transmitting T sec bits over a channel bandlimited to B_m Hz, with

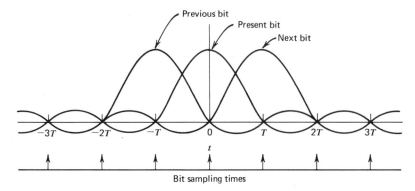

Figure 7.25. Signaling at the Nyquist rate without intersymbol interference.

$B_m \geq 1/2T$, then intersymbol interference can be completely removed by any decoder filter output waveform of the form in (7.8.7), provided $s_a(t)$ is a time function bandlimited to $B_m - (1/2T)$ Hz. This means that for any $B_m \geq 1/2T$ there are an infinite number of solutions that can be used for intersymbol elimination, one for each possible waveshape $s_a(t)$ whose bandwidth is $B_m - (1/2T)$ Hz.

Of particular importance is the fact that a channel bandwidth B_m no larger than $1/2T$ is needed. If $B_m = 1/2T$, then $s_a(t)$ has zero bandwidth (delta function at $\omega = 0$) and corresponds to a constant. The decoder filter output corresponds to signals of the form of (7.8.6), in which each bit has a single peak voltage value when all other bit waveforms pass through zero (Figure 7.25). A bit transmission rate of $1/T$ bits/sec is therefore maintained with no intersymbol effect. Conversely, we can state that a channel bandlimited to B_m can theoretically transmit at a rate of $2B_m$ bits/sec without intersymbol effects. This is referred to as the *Nyquist rate* of data transmission. Of course, the intersymbol interference is removed only if we sample exactly at each $t = iT$. A slight offset in the decoder timing generates intersymbol values that can accumulate rather quickly. Thus transmission at the Nyquist rate places severe requirements on decoder timing. For this reason it may be advantageous to use a larger B_m for a given time T (or conversely, a lower rate for a given bandwidth B_m) in order to utilize $s_a(t)$ in (7.8.7) to cause a faster decay in the interference values caused by timing offsets. [Roughly speaking, $s_a(t)$ adds a decaying factor to $s(t)$ that decreases as the reciprocal of its bandwidth.] Thus by proper selection of $s_a(t)$, timing requirements can be eased at the expense of information rate. The amount by which the bandwidth B_m exceeds the minimal (Nyquist) bandwidth of $1/2T$ is called the *excess* bandwidth.

It should also be pointed out that the frequency function corresponding to the signal in (7.8.6), used when communicating at the Nyquist rate, is that of an ideal filter. This means the combined effect of all the channel filtering on the transmitted pulse waveform must produce a waveform with an ideal frequency function at the decoder input, a result that can only be approached asymptotically. Thus binary transmission with no excess bandwidth presupposes ideal frequency functions. By allowing some excess, the required function can depart from the idealized result and can therefore be more realistically constructed. A convenient frequency function of this type that produces a waveform as in (7.8.7) is the *raised cosine* function given by

$$F_s(\omega) = 1, \qquad\qquad 0 \le \omega \le \frac{\pi}{T}(1-\alpha)$$

$$= \frac{1}{2}\left(1 - \sin\frac{T}{2\alpha}\left(\omega - \frac{\pi}{T}\right)\right), \qquad \frac{\pi}{T}(1-\alpha) \le \omega \le \frac{\pi}{T}(1+\alpha)$$

$$(7.8.9)$$

with $0 \le \alpha \le 1$. This frequency function is sketched in Figure 7.26*a*, and its corresponding time function is shown in Figure 7.26*b*. We note this pulse function passes through zero at all multiples of T for any α, as required for eliminating the intersymbol interference. The parameter α indicates the amount of bandwidth excess used, with $\alpha = 0$ corresponding to the ideal filter function, and $\alpha = 1$ producing a slower roll-off and a doubling of the Nyquist bandwidth. However, as $\alpha \to 0$ the pulse function is more oscillatory and spread, making exact timing mandatory. As $\alpha \to 1$, the response decays quickly and therefore produces signals that are easier to handle. We emphasize that (7.8.9) is just one of many possible frequency functions whose transform eliminates intersymbol interference.

Assume now that the decoder signal frequency function has been selected for a bandwidth B_m and bit time T. We now must find combinations of $Z(\omega)$ and $H(\omega)$ to produce this $F_s(\omega)$. With intersymbol removed, Λ in (7.7.8) becomes

$$\Lambda = \left[\frac{Es_a(T)/T}{2\sigma_{ny}^2}\right]^{1/2} \qquad (7.8.10)$$

where, from (7.7.6),

$$\sigma_{ny}^2 = \frac{1}{2\pi}\int_{-\infty}^{\infty} S_d(\omega)|H(\omega)|^2 \, d\omega \qquad (7.8.11)$$

This immediately suggests a design procedure whereby Λ is maximized with respect to $|H(\omega)|^2$, subject to a constraint on the energy in $z(t)$, that is,

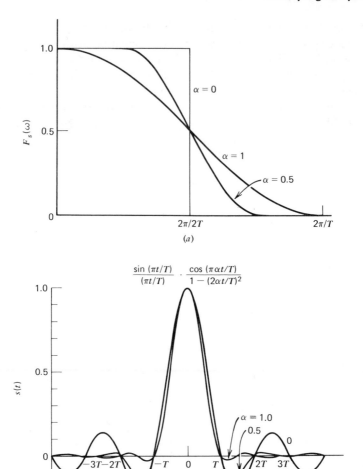

Figure 7.26. The raised cosine pulse. (*a*) Frequency function, (*b*) time function.

on the condition

$$\int_{-\infty}^{\infty} |Z(\omega)|^2 \, d\omega = \int_{-\infty}^{\infty} \left|\frac{F_s(\omega)}{H(\omega)}\right|^2 d\omega = \text{constant} \qquad (7.8.12)$$

The problem is now identical to the minimization considered in Section 5.2. The solution for the decoder filter follows as

$$|H(\omega)|^2 = \frac{|F_s(\omega)|^2}{S_d(\omega)}, \qquad |\omega| \le 2\pi B_m \qquad (7.8.13)$$

The required $Z(\omega)$ needed to produce the desired $F_s(\omega)$ is then

$$Z(\omega) = \frac{F_s(\omega)}{H(\omega)}, \qquad |\omega| \leq 2\pi B_m \qquad (7.8.14)$$

We therefore achieve a maximization of Λ (minimum error probability) without intersymbol interference. Note $H(\omega)$ in (7.8.13) is basically achieving noise reduction, whereas the transmitted encoder signal $z(t)$, whose transform is given in (7.8.14), is selected to generate the desired $s(t)$ in (7.8.5). Our results, of course, reduce to all our earlier conclusions if we have no bandwidth restrictions. In this case $B_m \to \infty$ and $s_a(t)$ can be selected to be any waveform time limited to T sec. We therefore always satisfy (7.8.5), and the optimal decoding filter reduces to our nonwhite noise matched filter derived earlier in (7.6.9). Further discussions of bandlimiting on digital data transmissions can be found in References 17–20.

So far we have treated intersymbol interference as an effect that must be eliminated. This requires the design of ideal filter functions and perfectly timed systems if transmission is to be with Nyquist bandwidths, or requires excess bandwidth if more realistic filters are used. An alternate approach is to use realizeable roll-off filters with the Nyquist bandwidth and, rather than eliminate intersymbol effects, produce a controlled amount of interference. The decoder can then be designed around this effect. Suppose we consider a decoder signal waveform $s(t)$ which, at the bit sampling instants, has the values

$$s(jT) = s_0, \qquad j = 0, 1$$
$$= 0, \qquad j \geq 2 \qquad (7.8.15)$$

Note that since two samples of $s(t)$ are nonzero we are now allowing some intersymbol interference to be generated from the previous bit (but no other bit). This means that in sampling a given bit for decoding, the sample value of the previous bit is added (Figure 7.27). Thus, in the absence of noise, the present sample value will be $\pm 2s_0$ if the previous bit matched the present bit, and will be zero if not. Having already decided the previous bit, a decision can be made on the present bit by determining whether the sample value is closer to zero or to $\pm 2s_0$.† Thus the intersymbol interference has not prevented bit detection. The decoder however must now make a

---

† If the two nonzero sample values in (7.8.15) were not equal, but had values s_0 and s_1, respectively, then the decision requires distinguishing among the decoding sample values $\pm(s_0 + s_1)$ and $\pm(s_0 - s_1)$. This decision is easier to make, in the presence of noise, when $s_1 = s_0$. The latter condition is referred to as "opening the eye" of the pattern made by the waveform and its intersymbol interference.

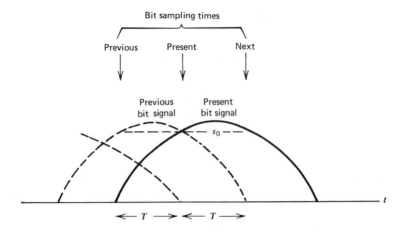

Figure 7.27. Partial response signalling. Previous bit intersymbol interference.

three-level decision (whether the sample is between $\pm s_0$) rather than the usual binary decision (whether the sample is positive or negative). The former is a more difficult decision, and a penalty incurs in error probability for using a nonideal Nyquist bandwidth. This encoding-decoding procedure is called *partial response* signalling since the ideal Nyquist filter response is not being entirely utilized. It is also called *duobinary* signalling [20]. A convenient waveform that satisfies the condition in (7.8.15) while using a nonideal Nyquist bandwidth is the pulse signal whose transform is given by the cosine function

$$F_s(\omega) = 2\,T\,\cos\left(\frac{\omega T}{2}\right), \qquad |\omega| \le \frac{\pi}{T}$$
$$= 0, \qquad\qquad\qquad \text{elsewhere} \qquad (7.8.16)$$

This transform and its corresponding time function are shown in Figure 7.28. Note the frequency function is nonideal, and occupies bandwidth $1/2T$ with no excess, while time samples of its time response at multiples of T sec, starting with the point $t_d - (1/2T)$, satisfy (7.8.15). Thus partial response binary signalling can be produced if the binary waveform at the decoder input is a waveform whose spectrum has been shaped to that given in (7.8.16). This again means the effect of all link filtering, from transmitter to decoder, must combine to yield a waveform with exactly $F_s(\omega)$. This often becomes a nontrivial design task since the transmitter filters are usually preconstrained to perform acceptable out of band rejection, while receiver filters must also perform satisfactory noise removal. Note that in using (7.8.16), the sample values used in decoding are less by a factor of

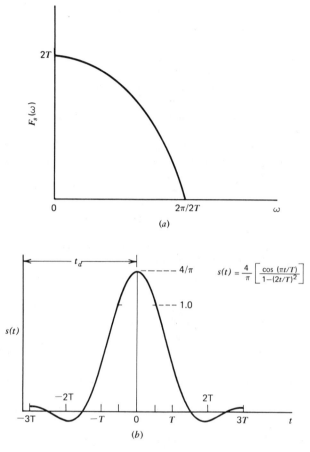

Figure 7.28. The cosine partial response pulse. (*a*) Frequency function, (*b*) time function. Here t_d is the time delay from the time of pulse insertion.

$4/\pi$ than the peak pulse value that could have been used if the intersymbol interference was removed. Hence an effective 2 dB loss in detection energy occurs when this form of partial response signalling is employed instead of idealized binary transmission without intersymbol interference.

References

1. Van Trees, H. *Detection, Estimation, and Modulation Theory*, Part 1, Wiley, New York, 1968, Chap. 2.
2. Helstrom, C. *Statistical Theory of Signal Detection*, Pergamon, Elmsford, N.Y., 1960.

3. Whalen, A. *Detection of Signals in Noise*, Academic, New York, 1971, Chap. 6.

4. Hancock, J. and Wintz, P. *Signal Detection Theory*, McGraw-Hill, New York, 1966, Chap. 3.

5. Gronmeyer, S. and McBride, A. "MSK and Offset QPSK Modulation," *IEEE Trans. Comm.*, vol. COM-24, August 1976, pp. 809–819.

6. DeBoda, R. "Coherent Demodulation of FSK with Low Deviation Ratio," *IEEE Trans. Comm.*, vol. COM-20, June 1972, pp. 429–435.

7. Weber, C. *Elements of Detection and Signal Design*, McGraw-Hill, New York, 1968, Chap. 19.

8. Viterbi, A. *Principles of Coherent Communications*, McGraw-Hill, New York, 1966.

9. Urkowitz, H. "Energy Detection of Unknown Deterministic Signals," *Proc. IEEE*, vol. 17, April 1967, pp. 523–531.

10. Schilling, D., Hoffman, L., and Nelson, E. "Error Rates for Digital Signals Demodulated by an FM Discriminator," *IEEE Trans. Comm. Tech.*, vol. CT-14, August 1967.

11. Taub, H. and Schilling, D. *Principles of Communication System*, McGraw-Hill, New York, 1971, Chap. 9.

12. Lindsey, W. *Synchronization Theory in Control and Communications*, Prentice-Hall, Englewood Cliffs, N.J., 1972, Chap. 15.

13. Marcum, C. "Tables of the Q Function," *Rand Corp. Research Memo RN* 399, January 1950.

14. Brennan, L. and Reed, I. "A Recursive Method for Computing the Q Function," *IEEE Trans. Info. Theory*, vol. IT-11, April 1965, p. 312.

15. Selin, I. *Detection Theory*, Princeton University Press, Princeton, N.J., 1965.

16. Hartmann, H. "Degradation of SNR Due to Filtering," *IEEE Trans. Aero. Electron. Syst.*, vol. AES, January 1969, pp. 22–32.

17. Carlson, B. *Communication Systems*, 2nd Edition, McGraw-Hill, New York, 1975, Chap. 10.

18. Ziemer, R. and Tranter, W. *Principles of Communications*, Houghton Mifflin, Boston, 1976.

19. *Transmission Systems for Communications*, Bell Telephone Laboratories, Winston-Salem, N.C., 1970.

20. Lucky, R., Salz, J., and Weldon, E. *Principles of Data Transmission*, McGraw-Hill, New York, 1968.

Problems

1. (7.1) Prove that if $r(\tau) = \int s(t)s(t+\tau)\,dt$, then its Fourier transform is given by $F_r(\omega) = |F_s(\omega)|^2$, where $F_s(\omega)$ is the Fourier transform of $s(t)$. [*Hint:* Carry out the steps similar to Equations (1.6.12)–(1.6.14) in Chapter 1.]

2. (7.1) Prove that the maximum likelihood binary decoding test will yield the minimal value of PE over all possible decoding tests we can perform on the voltage sample $y(T)$. [*Hint:* Write $\text{PE}|1 =$

$\int p_1(y_T)P(0|y_T)\,dy_T$ and show that $PE = \frac{1}{2}(PE|1) + \frac{1}{2}(PE|0)$ is minimized if $P(i|y_T)$ is selected according to the maximum likelihood rule.]

3. (7.1) Determine how the maximum likelihood decoding rule of (7.1.17) must be modified when the transmitted bits are not equally probable. Show this is equivalent to an adjustment of the selected threshold Y.

4. (7.1) Integrate by parts to establish the useful bounds

$$\text{Erfc}\,(x) \le \frac{1}{\sqrt{2\pi x}}\,e^{-x^2/2} \le e^{-x^2/2}, \qquad x > 0$$

$$\text{Erfc}\,(x) \ge \frac{1}{\sqrt{2\pi x}}\,e^{-x^2/2}\left[1 - \frac{1}{2x^2}\right], \qquad x > 0$$

5. (7.1) Derive an RC low pass network that will approximate a matched filter for a pulse. Indicate the condition on the elements and time constants to improve the approximation.

6. (7.1) Let $s(t)$ be a signal, $0 \le t \le T$, with transform $F_s(\omega)$. Show that the matched filter for $s(t)$ has the frequency function $H(\omega) = F_s^*(\omega)e^{-j\omega T}$, where $*$ denotes conjugate.

7. (7.1) Let $s(t)$ be written as an orthonormal expansion

$$s(t) = \sum_{i=1}^{k} a_i \Phi_i(t), \qquad 0 \le t \le T$$

Show that a MF for $s(t)$ takes the form of a bank of weighted correlators.

8. (7.1) Let $s(t)$ be a general signal bandlimited to B Hz and time limited to T sec. Show that a matched filter for any $s(t)$ of this class can always be designed as a tapped delay line with adjustable gains in conjunction with an ideal low pass filter. Show the final system. [*Hint:* Use the sampling theorem and the associated signal expansion for $s(t)$.]

9. (7.1) In using a pair of matched correlators in (7.1.33) for binary decoding, show that if the binary signals $s_1(t)$ and $s_0(t)$ have identical waveshape over a subinterval τ in $(0, T)$, then this τ sec of integration need not be included in the correlation operation.

10. (7.1) Show that the time waveform of the output of a filter matched to a signal $s(t)$ traces out the correlation function of $s(t)$ when the latter is applied at the input.

11. (7.2) A digital system samples at the rate of 1000 samples per second and quantizes to 16 possible levels. If the quantization words are sent by PCM encoding, estimate the baseband bandwidth needed to send the encoded waveform.

12. (7.2) Compute and sketch the spectral density in (7.1.9) for the Manchester coded antipodal PCM waveform.

13. (7.2) A NRZ code with T sec bits and amplitude A is FM onto a carrier and ideally frequency detected along with additive white Gaussian noise. The FM detector operates above threshold and is followed by an ideal low pass filter over bandwidth B Hz and a sampler that samples at $t = T$. (a) Assume $B \gg 1/T$ and write the expression for PE. (b) Will this be the minimal possible PE? Explain.

14. (7.2) A PM transmitter uses subcarrier PSK for binary digital transmission over a white Gaussian noise channel. The receiver employs product demodulation with carrier tracking followed by matched filtering detection for the bits. A prescribed value of PE and rms loop tracking error is given for the system. By showing appropriate design equations explain how the required transmitter power should be determined. Assume linear tracking loop, perfect bit timing and phase coherence, and known transmission losses and gains.

15. (7.2) A PCM encoded waveform with bit energy E is received with bandlimited white Gaussian noise of level N_{b0} and bandwidth B_m. Write the expression for the bit error probability if the decoder filter is removed $(H(\omega) = 1)$.

16. (7.2) Rewrite (7.2.15) as a single carrier at frequency $\omega_0 + (\omega_1 - \omega_0)/2$ with binary modulation as a frequency shift between $\pm\Delta$, where $\Delta = (\omega_1 - \omega_0)/2$. Use this to show that MSK signals can be written as a modulated quadrature expansion. Define the modulation on each component.

17. (7.3) A binary system is to maintain a specified PE. It uses a timing subsystem that produces a timing error of 10% of the bit period. Determine how much the transmitted bit energy E must be increased to overcome the timing error in (a) a PCM system, (b) a PSK system where $\omega_s T = 1$, (c) a PSK system with $\omega_s T = 5$.

18. (7.4) Derive (7.4.7) by formally integrating (7.4.5), using (7.4.6). Make use of the fact that the integral of (7.4.6a) over $(0, \infty)$ is one.

19. (7.5) Determine the value of Δ that maximizes the function $\Delta^2 |H(\Delta)|^2$, when $H(\omega) = 1/1 + \omega^2$.

20. (7.5) Determine the bit error probability expression for the DPSK decoder in 7.21a. Assume an energy per bit of E, zero mean noise and the noise in two consecutive bit intervals are uncorrelated, each white with level N_{b0}.

21. (7.6) Given two signals $s_1(t)$ and $s_0(t)$, and m_1 and m_0 defined by (7.1.14). Show that for any arbitrary filter function $|m_1 - m_0|^2 \le |m_1|^2 + |m_0|^2$, and determine the condition on the signals when the equality holds.

22. (7.6) Let a time signal $s(t)$ be represented by its time samples $(s(t_1), s(t_2), \ldots, s(t_N))$. Assume noise is added to $s(t)$ at the decoder input in such a way that one particular time point, say t_j, has no noise at this instant. Explain why this will correspond to a case of singular detection. How should the encoder and decoder operate for this case?

23. (7.7) Show that the worst case degradation occurs in (7.7.7) when the sequence of past bits $(d_1, d_2, \ldots)$ are defined by

$$d_i = \mp d_0 \text{ if } \left[\int_0^T s(t + iT)\, dt \right] \left[\int_0^T s(t)\, dt \right] \ge 0$$

Show that for this condition, the worst case is

$$\sqrt{T}(m_0 + m_I) = \int_0^T s(t)\, dt - \int_T^\infty |s(t)|\, dt$$

CHAPTER 8

BLOCK ENCODED DIGITAL SYSTEMS

In digital systems using block encoding, source bits are transmitted in blocks of bits in sequence to the receiver. These blocks of bits are often referred to as binary words. Each word of k bits is encoded into one of $M = 2^k$ distinct baseband waveforms, and receiver decoding of the entire word is accomplished by attempting to recognize which of the M waveforms is being decoded during each block interval. In this chapter we study the basic characteristics, design procedures, and performance trade-offs in practical block encoded digital systems. We also extend the standard block encoding format to include the more generalized convolutional encoding. One of the most basic questions facing a system designer is whether block or convolutional encoding should be used at all. This decision requires that he be aware of the prime advantages of these techniques and the cost or penalty, if any, required to obtain these advantages. We shall emphasize these points throughout our investigation.

8.1. The Filter-Sampler Block Decoder

We formulate the basic block decoding problem exactly as in the binary case. Blocks of bits are encoded into baseband waveforms for carrier transmission (Figure 8.1). If k bits are used in a word, one of $M = 2^k$ different possible waveforms will occur, similar to the encoding format (6.7.2). Each waveform is T sec long. The generated sequence of waveforms is sent over the carrier link and demodulated to recover the base-

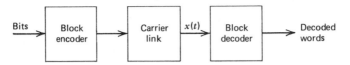

Figure 8.1. Block encoding system diagram.

band signal. The bandwidth required in the baseband subsystem to transmit the encoded signal can be computed just as in (7.1.1)–(7.1.9). If $s_i(t)$, $i = 1, 2, \ldots, M$ are the M encoder baseband signals to be used, and if $F_{s_i}(\omega)$ is their transform and P_i their probability of being used, then the baseband spectrum is given by Problem 8.1.

$$S_m(\omega) = \frac{1}{T} \sum_{i=1}^{M} P_i |F_{si}(\omega)|^2 - \frac{1}{T} \left[\sum_{i=1}^{M} P_i F_{si}(\omega) \right] \left[\sum_{i=1}^{M} P_i F_{si}^*(\omega) \right]$$

$$+ \frac{2\pi}{T^2} \sum_{j=-\infty}^{\infty} \left| \sum_{i=1}^{M} P_i F_{si}\left(\frac{2\pi j}{T}\right) \right|^2 \delta\left(\omega - \frac{2\pi j}{T}\right) \tag{8.1.1}$$

The spectrum is therefore directly related to the individual signal transforms and contains the harmonic lines, just as in the binary case. The receiver decoder processes the received baseband waveform during each T sec interval and decodes a particular block of bits, repeating for each T sec interval. The resulting decoded word sequence is then the recovered source sequence. Decoding timing for the block intervals is provided by the auxiliary timing subsystem. We again assume (1) each bit combination during a block is equally likely, (2) the baseband noise is additive, Gaussian, and white with one-sided level N_{b0}; and (3) the baseband signals satisfy the energy constraint

$$\int_0^T s_i^2(t)\, dt = E = P_m T, \qquad i = 1, 2, \ldots, M \tag{8.1.2}$$

By extending the previous results for a binary system, we consider a receiver decoder given by the M-ary extension of the white noise filter-sampler binary decoder in Figure 7.2. Such a decoder has the block diagram in Figure 8.2, corresponding to M parallel channels, each containing a matched filter for one of the signals. The filter outputs are simultaneously sampled at the end of each block interval, with the largest sample used for word decisioning. The matched filters can be constructed from passive filters or as correlator-integrators, as discussed in Section 7.1. If the baseband noise is not white, the channels would generalize to the whitening filter channels of Section 7.6, although we shall not pursue this

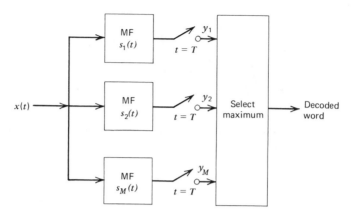

Figure 8.2. Block decoder.

case. The parallel channels compute the parameters

$$y_i = \int_0^T x(t)s_i(t)\,dt, \qquad i = 1, 2, \ldots, M \tag{8.1.3}$$

where $x(t)$ is the noisy decoder input and a comparison is made among the $\{y_i\}$ for decoding. The processing during any block interval is repeated for all the word intervals. The decoder is again a coherent decoder, since it assumes exact knowledge of the waveshape $s_i(t)$ of each signal during each block interval. Physically, the decoder again operates by having each channel of the decoder look for a particular signal of the encoder set, and the largest sample value is selected as the received signal. This again corresponds to a maximum likelihood decoder (Problem 8.2). Note that the maximum y_i in (8.1.3) corresponds to the minimal value of the quantity

$$\int_0^T [x(t) - s_i(t)]^2\,dt = \int_0^T x^2(t)\,dt + E - 2y_i \tag{8.1.4}$$

over all i. The integral on the left is the integrated squared difference between the noisy input $x(t)$ and the signal $s_i(t)$. Hence maximum likelihood decoding (selecting the largest y_i) is equivalent to determining to which $s_i(t)$ the input is "closest."

Performance in block coded systems is usually assessed in terms of the average probability of making a word error during decoding, PWE. This is defined as

$$\text{PWE} = \frac{1}{M} \sum_{j=1}^{M} \text{PWE}|j \tag{8.1.5}$$

where PWE$|j$ is the probability of decoding the incorrect word when the jth word (i.e., the waveform $s_j(t)$] is sent. In many cases, we find exact calculation for PWE difficult to determine and we must often settle for approximation, simulation, or bounding methods. A simple bound often applied in block encoding analysis is the "union bound." This involves use of the fact that if $x_1, \ldots, x_M$ are M random numbers, the probability that x_1 is less than the $M-1$ remaining numbers is bounded from above by the union (sum) of probabilities that x_1 is less than each x_j individually (Problem 8.3). If the numbers $\{x_i\}$ correspond to the $\{y_i\}$ when a particular word is sent, this union rule implies

$$\text{PWE}|j \le \sum_{\substack{q=1 \\ q \ne j}}^{M} \text{Prob}\left[(y_j \le y_q)|j\right] \qquad (8.1.6)$$

where the probability is conditioned on the jth word being sent. The average word error probability then satisfies

$$\text{PWE} \le \frac{1}{M} \sum_{j=1}^{M} \sum_{\substack{q=1 \\ q \ne j}}^{M} \text{Prob}\left[(y_j \le y_q)|j\right] \qquad (8.1.7)$$

By properly pairing terms, we can then write (8.1.7) as

$$\text{PWE} \le \frac{1}{M} \sum_{j=1}^{M} \sum_{\substack{q=1 \\ q \ne j}}^{M} \left\{\tfrac{1}{2}\text{Prob}\left[(y_j \le y_q)|j\right] + \tfrac{1}{2}\text{Prob}\left[(y_q \le y_j)|q\right]\right\} \quad (8.1.8)$$

The term in the braces is the probability of erring in attempting to decide between a pair of the y_i and is therefore the error probability associated with a binary test. Since the latter can usually be evaluated, (8.1.8) is a useful upper bound to PWE. Its value is in the fact that we are guaranteed that performance must be at least this good. Since (8.1.8) involves a binary error probability, many of our results in the preceding chapter are immediately applicable for establishing this bound.

To explore further the PWE in (8.1.5), we examine the behavior of each decoder channel during the decoding operation. The decoder input is written specifically as

$$x(t) = m(t) + n_b(t) \qquad (8.1.9)$$

as in (7.1.11). The ith channel computes y_i in (8.1.3). When $m(t) = s_j(t)$ (i.e., the jth signal is sent), the sample output y_i is

$$y_i|s_j = E\gamma_{ij} + \int_0^T n_b(t)\, s_i(t)\, dt \qquad (8.1.10)$$

where

$$\gamma_{ij} \overset{\Delta}{=} \frac{1}{E} \int_0^T s_i(t)s_j(t)\, dt, \qquad i \neq j$$

$$\gamma_{ii} = 1 \tag{8.1.11}$$

The parameter γ_{ij}, $i \neq j$, is the cross-correlation coefficient of the signals $s_i(t)$ and $s_j(t)$. We see that $y_i | s_j$ is a Gaussian random variable with mean

$$\mathscr{E}[y_i | s_j] = E\gamma_{ij} \tag{8.1.12}$$

and a pair of them have covariance

$$\mathscr{E}\{(y_i | s_j - \mathscr{E}[y_i | s_j])(y_k | s_j - \mathscr{E}[y_k | s_j])\} = \int_0^T \int_0^T \mathscr{E}[n_b(t)n_b(\rho)]s_i(t)s_k(\rho)\, dt\, d\rho$$

$$= \int_0^T \int_0^T \frac{N_{bo}}{2}\delta(t-\rho)s_i(t)s_k(\rho)\, dt\, d\rho$$

$$= \left(\frac{N_{b0}}{2}\right)E\gamma_{ik} \tag{8.1.13}$$

The random variables $\{y_i | s_j\}$ are therefore not statistically independent. The probability of an error in decoding is the probability that an incorrect channel sample value exceeds the correct one. Thus

$$\text{PWE}|j = 1 - \text{Prob}\,[y_j | s_j \geq y_i | s_j], \qquad i = 1, 2, \dots, M, \quad i \neq j \tag{8.1.14}$$

The probability on the right is simply the probability that one Gaussian variable exceeds $M-1$ other Gaussian variables but requires an integration over an M dimensional Gaussian density. That is,

$$\text{PWE}|j = 1 - \int_{-\infty}^\infty \int_{-\infty}^{y_j} \cdots \int_{-\infty}^{y_j} p(y_1, y_2, \dots, y_M | s_j)\, dy_1\, dy_2, \dots, dy_j \tag{8.1.15}$$

when $p(y_1, y_2, \dots, y_M | s_j)$ is the conditional joint Gaussian density of $y_1, \dots, y_M$, given signal $s_j(t)$ is received. (Note the order in the integration, where the variable y_j is integrated last.) For our case the joint Gaussian density of (B.3.3) becomes

$$p(y_1, y_2, \dots, y_M | j) = \frac{1}{(2\pi)^{M/2}|\det Q|} \exp\left[-\tfrac{1}{2}Y_j^{\text{tr}}Q^{-1}Y_j\right] \tag{8.1.16}$$

Here Q is the M by M covariance matrix [matrix of covariance values in (8.1.13)],

$$Q \overset{\Delta}{=} \frac{EN_{b0}}{2}[\gamma_{ij}] \tag{8.1.17}$$

det Q is its determinant, Q^{-1} its inverse, tr denotes *transpose*, and Y_j is the column vector

$$Y_j = \begin{bmatrix} y_1 - \mathscr{E}(y_1|s_j) \\ y_2 - \mathscr{E}(y_2|s_j) \\ \vdots \end{bmatrix} \qquad (8.1.18)$$

Integration of (8.1.15) for each s_j and insertion in the sum in (8.1.5) yields the desired average word error probability. Unfortunately, a general integration of the form (8.1.15) is difficult because of the dependence among the $\{y_i\}$ and the necessity to invert the matrix Q. Nevertheless, we do see that the resulting value of PWE depends only on the parameters E and N_{b0}, and the matrix of signal cross-correlation values γ_{ij}. Thus the selection of the signal set for channel encoding influences the resulting system performance only through its cross-correlation values. In the next section we present several types of common signal sets and the matrices they generate in (8.1.17).

8.2. Block Encoding Signaling Sets

When using a coherent decoder with white noise interference, we have found that the resulting performance in terms of PWE will depend only on the signal correlation values γ_{ij}. We denote the matrix of the correlation values by Γ,

$$\Gamma \triangleq [\gamma_{ij}] \qquad (8.2.1)$$

This signal correlation matrix, along with the values of E and N_{b0}, generates the required sample covariance matrix Q in (8.1.17) by

$$Q = \frac{EN_{b0}}{2} \Gamma \qquad (8.2.2)$$

Since only the correlation values determine performance, all signal set generating the same matrix Γ perform the same. We therefore label signal sets by their corresponding matrix Γ. In the following we list some examples of signal sets commonly used in block encoded systems.

Orthogonal Signals.   The set of M signals $s_i(t)$, $i = 1, 2, \ldots, M$ for which

$$\begin{aligned} \gamma_{ij} &= 0, & i \neq j \\ \gamma_{ii} &= 1, & i = j \end{aligned} \qquad i, j = 1, 2, \ldots, M \qquad (8.2.3)$$

constitute an orthogonal signal set. The matrix Γ then becomes an Mth order identity matrix

$$\Gamma = \begin{bmatrix} 1 & 0 \\ 0 & 1 \end{bmatrix} \triangleq I_M \tag{8.2.4}$$

Orthogonal signal sets have the property that any pair of signals from the set are pairwise orthogonal signals, as defined in (7.2.4).

Biorthogonal Signals. The set of M signals containing $M/2$ pairs of antipodal signals of which each pair is orthogonal is called *biorthogonal* signal set. Thus each signal of the set is the negative of one other signal and orthogonal to all others. By proper ordering of the signals the correlation matrix can always be partitioned into the form

$$\Gamma = \begin{bmatrix} I_{M/2} & -I_{M/2} \\ -I_{M/2} & I_{M/2} \end{bmatrix} \tag{8.2.5}$$

A set of M biorthogonal signals can always be formed by starting with $M/2$ distinct orthogonal signals and adding the negative of each to form the total set.

Equally Correlated Signals. The set of M signals whose cross-correlation among all pairs is identical constitutes a set of equally correlated signals. The correlation matrix is then

$$\Gamma = \begin{bmatrix} 1 & \gamma \\ \gamma & 1 \end{bmatrix} \tag{8.2.6}$$

where

$$\gamma = \frac{1}{E} \int_0^T s_i(t)s_j(t)\, dt, \qquad i \neq j \tag{8.2.7}$$

is the common cross-correlation value. Orthogonal signals are a special case where γ is zero. From our binary study we would expect best performance to occur when γ has its most negative value. However, there is a limit to how negative we can make γ. This can be observed by noting that the signal formed by summing all the signals of the set must have nonnegative energy. This means that

$$\int_0^T \left(\sum_{i=1}^{M} s_i(t) \right)^2 dt = \int_0^T \left[\sum_{i=1}^{M} s_i^2(t) + \sum_{j=1}^{M} \sum_{i=1}^{M} s_i(t)s_j(t) \right] dt$$

$$= ME + (M^2 - M)\gamma E$$

$$= [M + M(M-1)\gamma]E$$

$$\geq 0 \tag{8.2.8}$$

Solving for γ shows it is necessary that

$$\gamma \geq -\frac{1}{M-1} \qquad (8.2.9)$$

Thus M equally correlated signals can have a cross-correlation value no more negative than that given in (8.2.9). Signal sets having this minimal cross-correlation value for γ are called *simplex signal sets*, or *transorthogonal* signal sets. Note that for $M = 2$ (binary encoding case) simplex signals reduce to antipodal signals. It can be shown [1, 2] that for equally correlated signals, and any M,

$$\frac{\partial \text{PWE}}{\partial \gamma} \geq 0 \qquad (8.2.10)$$

This means that indeed γ should be as negative as possible for minimizing word error probability. Hence simplex signal sets achieve the minimal value of PWE over all equally correlated sets. For this reason simplex signals are always of interest and general methods for generating them are desired. This will be considered subsequently. We point out that for $M \gg 1$, $\gamma \approx 0$, and simplex signals are approximately orthogonal signals when the set is large.

8.3. Signal Waveform Generation

We have discussed block signaling waveforms only in terms of their correlation matrix. An engineer involved in actual system implementation is concerned with specific time waveforms used for the block signaling. In this section we present general methods for deriving signal waveforms with the desired correlation matrix. This generalization is achieved by again resorting to signal representations using orthonormal function expansions as discussed earlier in Section 6.7. By this approach the problem of generating signals of any particular type can be unified into a simpler procedure of selecting orthonormal functions.

Let $\{\Phi_q(t)\}$, $q = 1, 2, \nu$, again be an arbitrary set of orthonormal functions over $(0, T)$, as defined in (6.7.3), and write each of the signals of a block encoding set as

$$s_i(t) = \sum_{q=1}^{\nu} s_{iq} \Phi_q(t), \qquad i = 1, 2, \ldots, M \qquad (8.3.1)$$

The constants s_{iq} are the signal coordinates associated with each component function and form the signal vectors of (6.7.5),

$$\mathbf{s}_i = (s_{i1}, s_{i2}, \ldots, s_{i\nu}) \qquad (8.3.2)$$

It then follows from the orthonormality property of the $\{\Phi_q(t)\}$ that

$$E = \int_0^T s_i^2(t)\,dt = \sum_{q=1}^{\nu} s_{iq}^2 \qquad (8.3.3a)$$

$$\gamma_{ij} = \frac{1}{E}\int_0^T s_i(t)s_j(t)\,dt = \frac{1}{E}\sum_{q=1}^{\nu} s_{iq}s_{jq} \qquad (8.3.3b)$$

The value of the preceding integrals can therefore be obtained directly from the vectors s_i of signal components. We see that the correlation γ_{ij} in (8.3.3b) is simply the *inner product* of such vectors, whereas E is the square magnitude of each vector. Thus signal representation in terms of orthonormal expansions allows us to deal with energy and cross-correlation values exclusively in terms of the vectors formed from their components. Furthermore, these vectors can be plotted as points in a ν dimensional vector space to give geometrical location to each signal (Figure 8.3). In this sense, $\sqrt{E}$ is recognized as the distance to the vector point, whereas γ_{ij} is related to the angle between vectors drawn to such points. Thus all signals of equal energy have their vector points on a common hypersphere, and the distance between two such vector points is

$$\left[\sum_{q=1}^{\nu}(s_{iq}-s_{jq})^2\right]^{1/2} = [2E(1-\gamma_{ij})]^{1/2} \qquad (8.3.4)$$

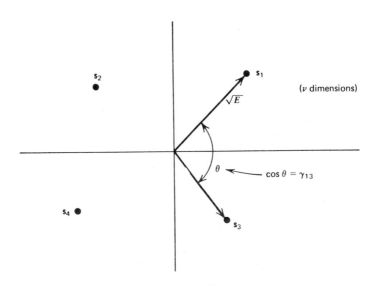

Figure 8.3. Signal vectors in ν dimensional signal space.

Note that signals that have a high cross-relation ($\gamma_{ij} \approx 1$) are closely spaced in vector space, whereas signals with a high negative cross-correlation are widely separated. Finding signals of equal energy and given cross-correlation values is therefore reduced to finding vector points on a hypersphere that are properly separated. Once the vector points are specified, the signal time waveforms in (8.3.1) can be generated from any orthonormal set. Since the orthonormal time function set is arbitrary, many different waveforms can be derived from a specific vector set. Since the selection of signal vectors was previously described in Section 6.7 as a problem of channel encoding, we see here the equivalence between separating signal vectors so as to achieve desirable waveform correlation and channel encoding to improve digital transmission.

Orthogonal function representation allows us to embed the basic channel encoding problem into a geometrical problem in vector space. It is from this equivalence that many interesting and far-reaching notions in encoding theory have been developed. In particular, it allows precise derivation of the relation between ν and M of the signal set. For example, a set of M signals that are orthogonal must be described by orthogonal vector points. However, a vector space of at least M dimensions is needed to have M distinct orthogonal vector points on a hypersphere of fixed radius. Thus $\nu \geq M$, and we must therefore have at least M orthogonal components (expansion functions) in order to describe an M-ary orthogonal signal set. Just as important, however, is the fact that we do not need any more than M components to form this set.

For simplex signals, vectors are needed whose inner products are all given by $\gamma = -1/(M-1)$. This means that the signal vectors must have the property of vector points that are equally spaced on the hypersphere with a separation in (8.3.4) given by

$$[2E(1-\gamma)]^{1/2} = \left[2E\left(1+\frac{1}{M-1}\right)\right]^{1/2} = \left[\frac{2EM}{M-1}\right]^{1/2} \qquad (8.3.5)$$

The last term on the right also happens to be the distance between vertices of a geometrical regular simplex[†] of M vertices inscribed on a hypersphere of radius $\sqrt{E}$. Hence design of simplex signal sets can be equated to determining regular simplex surfaces (from which the signal set gets its name) on hyperspheres and associating a signal vector with each of its vertices. From geometry [3] it is known that regular simplex surfaces of M vertices can only be constructed in spaces of dimension $\nu \geq M - 1$. This means the dimension of the signal need be no larger than $M - 1$. Thus

---

[†] A simplex is a closed surface formed from the intersection of hyperplanes. A regular simplex has equal surface areas for all its outer faces.

simplex signal sets require expansions having one less component than the number of signals. Further geometrical properties of simplex signals can be found in Reference 4.

Biorthogonal signal sets can be simply described by the expansion form of (8.3.1). Since such sets are composed of $M/2$ antipodal, mutually orthogonal signals, they must be of the form

$$s_1(t) = \sqrt{E}\,\Phi_1(t)$$
$$s_2(t) = -\sqrt{E}\,\Phi_1(t)$$
$$s_3(t) = \sqrt{E}\,\Phi_2(t) \qquad\qquad (8.3.6)$$
$$s_4(t) = -\sqrt{E}\,\Phi_2(t)$$
$$\vdots \qquad\qquad \vdots$$

In this case the signaling waveforms are actually the orthonormal functions and the signal vectors take the specific form

$$\mathbf{s}_i = (0, 0, \ldots \pm\sqrt{E}, 0\ 0 \ldots) \qquad\qquad (8.3.7)$$

The signal vectors are therefore all zero, except for a binary component in one position.

The use of binary vectors as the signal vectors is analytically convenient in correlation analysis. Let $\mathbf{s}_i$ have only binary coefficients, $s_{ij} = \pm\sqrt{(E/\nu)}$. Then the correlation between any two such vectors can always be expressed as

$$\gamma_{ij} = \frac{1}{E}\sum_{q=1}^{\nu} s_{iq}s_{jq} = \frac{1}{\nu}\left[\sum^{\mathcal{N}}(1) - \sum^{\mathcal{D}}(1)\right] = \frac{\mathcal{N}-\mathcal{D}}{\nu} \qquad (8.3.8)$$

where $\mathcal{N}, \mathcal{D}$ are the number of components where $\mathbf{s}_i$ and $\mathbf{s}_j$ have like and opposite symbols, and $\mathcal{N}+\mathcal{D} = \nu$. Thus the correlation coefficients depend only on the number of positions where the vectors are the same and opposite, when placed side by side. This reduces the selection of a signal set to simply finding binary sequences with the proper number of like and opposite bits. Note from (8.3.4) that the distance between any two binary vectors is

$$2E(1 - \gamma_{ij}) = 2E\left[1 - \frac{\mathcal{N}-\mathcal{D}}{\mathcal{N}+\mathcal{D}}\right]$$

$$= \frac{4E\mathcal{D}}{\nu} \qquad\qquad (8.3.9)$$

and therefore can be rewritten directly in terms of the number of positions where bits differ. This latter parameter was defined in Section 6.7 as the

Hamming distance between the vectors. We see that finding binary vectors with suitable Hamming distance is equivalent to finding vectors properly separated in the signal vector space. We also note that signal sets with binary vectors can always be written as a $\nu \times M$ matrix of ± 1 values whose rows represent each signal vector. For an orthogonal signal set, $\nu = M$, and we require an $M \times M$ binary square matrix such that any pair of rows differ in exactly $M/2$ columns. This requires M to be even. A general method for deriving such orthogonal binary matrices is as follows. Denote the 2×2 matrix H_1 as

$$H_1 \triangleq \begin{bmatrix} 1 & -1 \\ 1 & 1 \end{bmatrix} \qquad (8.3.10)$$

The preceding is a trivial example of two orthogonal vectors. Now define the nth order extension of H_1 as

$$H_n = \begin{bmatrix} H_{n-1} & \hat{H}_{n-1} \\ H_{n-1} & H_{n-1} \end{bmatrix} \qquad (8.3.11)$$

where $\hat{H}_{n-1}$ represents the complement of H_{n-1}, obtained by replacing ± 1 by ∓ 1. It is easy to see H_n will always have orthogonal rows and can therefore serve as the binary vector set of orthogonal signals. The matrix is called a *Hadamard* matrix, [5] and signal sets generated from these matrices are referred to as *Reed–Muller* signals [6].

Our discussion so far in this section has concentrated on only the vector encoding aspect of waveform selection. That is, finding the signal vectors that are properly separated in vector space. The selection of thé actual waveforms that these vectors will represent depends on the orthonormal set $\{\Phi_q(t)\}$ used in (8.3.1). We pointed out that those orthonormal functions are arbitrary in deriving the signal waveforms. However, we would expect some functions to be more suitable than others in a particular communication system, perhaps because of ease of generation, bandwidth, and so on. The following discussion gives some common examples of expansion functions and the typical waveforms they generate.

Frequency Functions

$$\Phi_i(t) = \left(\frac{2}{T}\right)^{1/2} \sin\left(\frac{2\pi i}{T}\right), \qquad 0 \le t \le T \qquad (8.3.12)$$

These orthonormal expansion functions are harmonic frequencies of fundamental frequency $1/T$ Hz. The waveforms formed from the signal vectors become combinations of such harmonics. Signals can be physically generated from a bank of phase locked oscillators (note that the functions all have fixed phase with respect to the fundamental period). By combining the oscillator outputs with amplitudes defined by the desired vector set, the signaling set can be formed. For orthogonal signals, the signal vector set must itself be orthogonal. Consider the particular orthogonal vector set denoted by

$$\mathbf{s}_1 = (\sqrt{E}, 0, 0, \ldots)$$

$$\mathbf{s}_2 = (0, \sqrt{E}, 0, \ldots)$$

$$\mathbf{s}_M = (0, 0, \ldots, \sqrt{E})$$

(8.3.13)

That is, $\mathbf{s}_i$ is a vector with a nonzero entry in the ith position. The corresponding signals generated from frequency functions then become

$$s_i(t) = \left(\frac{2E}{T}\right)^{1/2} \sin\left(\frac{2\pi i}{T}\right), \quad \begin{matrix} i = 1, \ldots, M \\ 0 \le t \le T \end{matrix}$$

(8.3.14)

representing a set of multiple frequency shift keyed (MFSK) signals. Each signal of the set is a separate harmonic, and encoding is achieved by transmitting a different frequency burst according to the block word. Thus MFSK signals are a special class of signals formed from orthonormal frequency functions.

The basebandwidth that must be allowed for transmitting frequency functions is that necessary to pass the spectrum of all harmonics. Each harmonic burst occupies a spectrum identical to that of a PSK signal in Figure 7.9, and a bandwidth of approximately $2/T$ Hz about its frequency. Hence a total signaling bandwidth of approximately

$$B_m = (\nu + 1)\frac{1}{T} \text{ Hz}$$

(8.3.15)

must be provided. Note that the required bandwidth increases directly with the number of functions generating the signal set.

Pulsed Functions

$$\Phi_i(t) = \left(\frac{\nu}{T}\right)^{1/2}, \quad (i-1)\frac{T}{\nu} \le t \le i\frac{T}{\nu}, \quad i = 1, 2, \ldots \nu$$

(8.3.16)

The orthonormal functions are square pulses, placed in ν adjacent time slots spanning $(0, T)$. The signals are formed from the vectors as a sequence of amplitude modulated pulses. If an orthogonal signal set is formed from the orthogonal vector in (8.3.13), then $\nu = M$ and the signals correspond to narrow pulses placed in one of M time slots in $(0, T)$. This is referred to as a *pulse position modulation* (PPM) set.

An interesting case occurs when the vector components are binary; that is, $s_{ij} = \pm\sqrt{E}$. Then each vector $\mathbf{s}_i$ is itself a sequence of ν binary symbols. The waveforms formed from the pulsed functions then represent a binary pulsed waveform. Such waveforms have the advantage of beng easily generated by flip-flop circuits, and thus conveniently processed by common digital hardware. If the binary vectors are selected to be precisely the digital words from the source encoder, then the pulse waveform becomes the voltage equivalent of the source sequence. When we speak of processing the digital symbols of a source sequence, we often really mean processing the equivalent pulsed waveform generated in this fashion.

The required basebandwidth is that necessary to transmit these narrow pulses. Since a τ sec pulse has a bandwidth of approximately $1/\tau$ Hz (see Figure 1.10), the orthonormal functions in (8.3.16) require a bandwidth of approximately

$$B_m = \frac{\nu}{T}\,\text{Hz} \tag{8.3.17}$$

Again we see that the required bandwidth increases with the number of pulse positions used.

Quadrature Functions

$$\Phi_1(t) = \left(\frac{2}{T}\right)^{1/2} \cos \omega_s t,$$
$$0 \le t \le T \tag{8.3.18}$$
$$\Phi_2(t) = \left(\frac{2}{T}\right)^{1/2} \sin \omega_s t$$

Here only two ($\nu = 2$) orthonormal functions are to be used for the signal representation. The signals of the set are all of the form

$$s_i(t) = s_{i1}\left(\frac{2}{T}\right)^{1/2} \cos \omega_s t - s_{i2}\left(\frac{2}{T}\right)^{1/2} \sin \omega_s t \tag{8.3.19}$$

where $E = s_{i1}^2 + s_{i2}^2$. If we convert the coordinates (s_{i1}, s_{i2}) to the new coordinate θ_i by the transformation

$$\theta_i = \tan^{-1}\left(\frac{s_{i2}}{s_{i1}}\right), \qquad \begin{matrix} s_{i1} = \sqrt{E}\cos\theta_i \\ s_{i2} = \sqrt{E}\sin\theta_i \end{matrix} \tag{8.3.20}$$

then (8.3.19) can be rewritten as

$$s_i(t) = \left(\frac{2E}{T}\right)^{1/2} [\cos\theta_i \cos\omega_s t - \sin\theta_i \sin\omega_s t]$$
$$= \left(\frac{2E}{T}\right)^{1/2} \cos[\omega_s t + \theta_i] \tag{8.3.21}$$

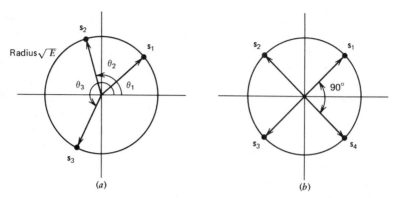

Figure 8.4. Signal vectors as two-dimensional phasors. (*a*) MPSK signal set, (*b*) QPSK, or quadraphase, signal set.

The signals are therefore phase shifted sine waves, and the signal vectors become phasors in polar coordinates (Figure 8.4*a*). Each signal becomes a point on a circle of radius $\sqrt{E}$ at angle θ_i. A set of M signals appears as M such points, and is called a *polyphase*, or *multiple phase shift keyed* (MPSK), signal set. MPSK signals have cross-correlation

$$\gamma_{iq} = \cos{(\theta_i - \theta_q)} \tag{8.3.22}$$

and therefore depend only on the angular difference between the signals. If the phase angles are equally spaced over $(0, 2\pi)$, then $\theta_i = 2\pi i/M$, and

$$\gamma_{iq} = \cos{\left[\frac{2\pi}{M}(i - q) \right]} \tag{8.3.23}$$

Note that polyphase signal sets are not, in general, equally correlated. For $M = 2$, equally spaced polyphase signals reduce to the antipodal PSK signals of Section 7.3. For $M = 3$ they would constitute a simplex signal set. For $M = 4$ they correspond to four phasors, spaced every 90° in $(0, 2\pi)$ (Figure 8.4*b*). Such a signal set is referred to as a *quadraphase* or *QPSK* set. Note that QPSK signals can also be viewed as two pairs of antipodal phasors placed orthogonally to each other and therefore also represent a biorthogonal signal set. QPSK signaling, which has become quite popular in modern digital systems, is considered in more detail in Section 8.5.

Since the phasor signals represent a T sec burst of sinusoidal signal, the bandwidth required is identical to that of a PSK signal. Thus quadrature signals have a frequency spectrum

$$S_m(\omega) = 2E \left[\frac{\sin{(\omega - \omega_s)T/2}}{(\omega - \omega_s)T/2} \right] \tag{8.3.24}$$

This occupies a bandwidth of approximately $2/T$ Hz, around the subcarrier frequency ω_s. Note the basebandwidth does not depend on M. This is a basic advantage of polyphase signaling, since the required bandwidth does not increase as the size of the signal set increases.

Representing signal waveforms by orthonormal expansions also gives us more insight into decoder simplification. Formally the coherent decoder computes the parallel correlation of the input $x(t)$ with a coherent version of each of the M signal waveforms, as defined in (8.1.3). However, use of the expansion in (8.3.1) allows us to write the ith channel correlation instead as

$$
\begin{aligned}
y_i &= \int_0^T x(t)s_i(t)\, dt \\
&= \sum_{q=1}^{\nu} s_{iq} \int_0^T x(t)\Phi_q(t)\, dt
\end{aligned}
\tag{8.3.25}
$$

The integrals are common to all i. Therefore the only difference between y_i and y_q is in the manner in which these integrals are weighted prior to summing. Thus the M channel coherent decoder can be reduced to an equivalent operation in which the ν integral values in (8.3.25) are computed simultaneously, and the resulting outputs weighted and summed to form each y_i (Figure 8.5a). Only ν correlators are now needed, followed by M parallel store-and-sum circuits. The decoder is therefore greatly simplified if $\nu \ll M$, but the arithmetic operations of storing and summing generally require some type of decoder quantization. This reduces the accuracy of the $\{y_i\}$ computation. (Recall that the $\{y_i\}$ will be compared for the maximum.) In addition, the summing must be done after the T sec integrations and within the next word time. This introduces decoding delay and may limit the word transmission rate if the speed of computation is constrained. However, the overall complexity of the decoder is critically related to the degree of quantization used, especially if M is quite large. (Recall that M increases exponentially with block size k.)

An important and useful special case occurs if the orthonormal functions are the pulsed sequences in (8.3.16). The integrals in (8.3.25) reduce to ν disjoint integrations, each over successive T/ν sec intervals. The integrations can therefore be performed by a single integrator operated and read out successively (Figure 8.5b). Each output of the integrator is weighted in the M store-and-sum circuits and added to the previous sum. The $\{y_i\}$ are therefore computed in real time, and no bit delay occurs. Typically, decoder quantization is used to perform the summations, and the degree of quantization is important here. In two interval quantization only the sign of the integrator output is stored. Since the integration is over exactly one

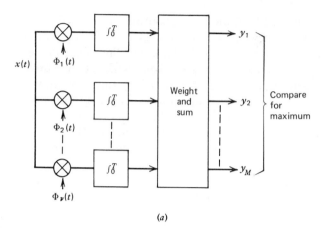

(a)

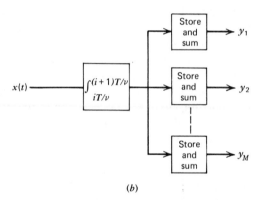

(b)

Figure 8.5. Alternative block decoder implementation (a) in terms of general orthonormal expansion functions, (b) for pulsed orthonormal functions.

pulse of the binary waveform, this is equivalent to making a binary decision concerning the polarity of that pulse, that is, detecting its binary symbol. Thus two level decoder quantization is equivalent to simply storing the bit decisions of the binary waveform prior to making a word decision. This type of operation is called *hard decisioning*. In higher level quantization, the integrator output is quantized more accurately in computing the necessary y_i. This is referred to as *soft decisioning*. Decoder simplification of the form of Figure 8.5b makes the handling of word lengths on the order of $k = 10$ ($M = 1024$) quite practical, and such systems have been implemented.

8.4. Performance of Block Coded Systems

Evaluation of word error probability for particular signal sets formally requires computation of the multidimensional Gaussian integral in (8.1.15). For general signal sets this is hampered by the necessity to invert Q in (8.1.17). For the more common signal sets of Section 8.2 these probabilities have been extensively investigated in the literature, and various simplifying techniques have been inserted to aid evaluation. We review these results here.

Orthogonal Signals. When the signal set is orthogonal the straightforward evaluation of PWE becomes somewhat tractable. This is due to the fact that for the orthogonal case of (8.2.4), the joint density in (8.1.16) factors into a product of individual Gaussian densities. Stated alternatively, for the orthogonal case the Gaussian random variables $y_i | s_j$ become uncorrelated (and therefore independent), leading to the density simplification. Furthermore, each such independent Gaussian variable has variance

$$\sigma^2 = \frac{N_{b0}E}{2} \tag{8.4.1}$$

and mean values in (8.1.12) of

$$\mathscr{E}\{y_j | s_j\} = E$$
$$\mathscr{E}\{y_i | s_j\} = 0, \qquad i \neq j \tag{8.4.2}$$

Thus (8.1.15) can be simplified to

$$\text{PWE}|j = 1 - \int_{-\infty}^{\infty} G\left(y_j, E, \frac{N_{b0}E}{2}\right)\left[\prod_{\substack{i=1 \\ i \neq j}}^{M} \int_{-\infty}^{y_j} G\left(y_i, 0, \frac{N_{b0}E}{2}\right)dy_i\right] dy_j \tag{8.4.3}$$

where

$$G(y_j, m, \sigma^2) \triangleq \frac{1}{\sqrt{2\pi}\sigma} \exp\left[-\frac{(y_j - m)^2}{2\sigma^2}\right] \tag{8.4.4}$$

The integral over each y_i can be written in terms of the Erfc function, and (8.4.3) simplifies to

$$\text{PWE}|j = 1 - \int_{-\infty}^{\infty} G(y, 0, 1)\left[1 - \tfrac{1}{2}\text{Erfc}\left(y + \sqrt{\frac{E}{N_{b0}}}\right)\right]^{M-1} dy \tag{8.4.5}$$

where we have substituted $y = (y_j - E)/(N_{b0}E/2)^{1/2}$ as the integration variable. Since the right side of (8.4.5) does not depend on j, (8.4.5) is also the average word error probability PWE, obtained from substituting into (8.1.5). Thus we have established

$$\text{PWE} = 1 - \int_{-\infty}^{\infty} \left[1 - \tfrac{1}{2} \text{Erfc} \left(y + \sqrt{\frac{E}{N_{b0}}} \right) \right]^{M-1} \frac{e^{-y^2/2}}{\sqrt{2\pi}} \, dy \qquad (8.4.6)$$

We emphasize that (8.4.6) assumes orthogonal, equiprobable, equal energy signals and white noise. Note that PWE depends only on the two parameters M and E/N_{b0}. It is immediately evident that since the bracket in (8.4.6) is less than one, PWE increases as M gets larger for a fixed E/N_{b0}. A plot of (8.4.6) is shown in Figure 8.6, exhibiting quantitatively this degradation in PWE with increasing M. It should be emphasized that PWE in this figure corresponds to the block error probability (i.e., the probability that a block will be decoded incorrectly). We see from the preceding discussion that system performance apparently degrades as we increase block size. This can be attributed to the fact that as we increase M for a given E/N_{b0}, we are increasing the number of possible signals occurring in a fixed T sec block time. The decoder therefore has a more difficult decision and tends to make more errors. However, it may not be meaningful to compare block systems of different values M at the same value of E/N_{b0} because the higher M systems, while possibly making more word errors, are sending more bits in the same T sec interval. This suggests that the bit transmission rate be normalized before comparisons are made. That is, we compare PWE when all systems transmit source bits at the same rate, while operating at the same baseband power P_m. The transmission bit rate of an M-ary system is

$$\mathcal{R} = \frac{k}{T} = \frac{\log_2 M}{T} \qquad \text{bits/sec} \qquad (8.4.7)$$

The right-hand side increases with M for fixed T. To have all M-ary systems operate at the same rate $\mathcal{R}_0$, it is necessary to increase T as M increases. That is, a particular M-ary system should use a block time, T_M, given by

$$T_M \triangleq \frac{\log_2 M}{\mathcal{R}_0} \qquad (8.4.8)$$

This means that different M systems must operate with different block time intervals if they are to have the same rate $\mathcal{R}_0$. However, this also means that for a given M,

$$\frac{E}{N_{b0}} = \frac{P_m T_M}{N_{b0}} = \left[\frac{P_m}{N_{b0} \mathcal{R}_0} \right] \log_2 M \qquad (8.4.9)$$

Since the bracket involves all fixed parameters, E/N_{b0} increases as $\log_2 M$.

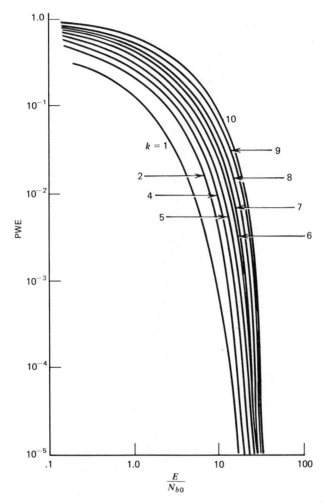

Figure 8.6. Word error probabilities for orthogonal block signaling ($M = 2^k$, $k =$ block size).

Thus M-ary systems operating at fixed bit rate $\mathcal{R}_0$ and transmitting power P_m should be compared at values of the parameter E/N_{b0}, as given earlier. When this adjustment is taken into account, the curves of Figure 8.7 are derived, showing PWE as a function of $P_m/N_{b0}\mathcal{R}_0$, where $P_m/\mathcal{R}_0$ is the effective baseband energy per bit. The results now show an actual improvement in performance with increasing block size. Thus increasing block size yields lower word error probabilities when transmitted at the same rate with the same power.

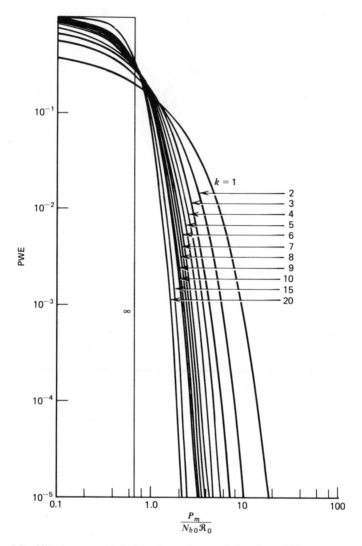

Figure 8.7. Word error probabilities for orthogonal signals and bit rate normalized to $\mathcal{R}_0$ bits/sec (k = block size).

A curve has been included for the case $M = \infty$, obtained by examining the limiting behavior of (8.4.6), when (8.4.9) is substituted. By properly taking the limits as $M \to \infty$ (Problem 8.11), we obtain

$$\text{PWE} \xrightarrow[M \to \infty]{} \begin{cases} 1, & \text{if } \dfrac{P_m}{N_{b0}\mathcal{R}_0} < \log_e 2 \\[4mm] 0, & \text{if } \dfrac{P_m}{N_{b0}\mathcal{R}_0} \geq \log_e 2 \end{cases} \tag{8.4.10}$$

Thus continual increase of the block size produces uniform PWE improvement, approaching the limit in (8.4.10). We see that as long as the energy per bit is large enough to satisfy (8.4.10), continual increase of M will eventually allow digital transmission with zero error probability. This means that we approach *perfect* transmission, in spite of the decoder noise, by using infinitely long block sizes, under the preceding conditions.

The result has another interesting interpretation. The condition required in (8.4.10) can also be stated

$$\mathcal{R}_0 \leq \frac{P_m}{N_{b0} \log_e 2} = \frac{P_m}{N_{b0}(0.69)} \tag{8.4.11}$$

This now appears as a condition on the rate $\mathcal{R}_0$ at which we transmit. The term on the right is the infinite bandwidth channel capacity (theoretical maximum allowable bit rate) for an additive white Gaussian noise channel having signal power P_m, given earlier in (6.8.2). Hence (8.4.10) and (8.4.11) state that we can transmit with zero error probability over a noisy channel as long as the transmitted bit rate is less than the channel capacity. This is, in effect, a restatement of the basic Shannon theorem of information theory, which we stated in Section 6.8. We immediately see the physical drawbacks to this theorem (which are often obscured in theoretical derivations). To achieve the perfect transmission predicted by Shannon's theorem we must transmit with infinite block lengths. This means the decoder becomes infinitely complex and, since word interval $T_M \to \infty$, an infinite delay occurs in the decoding of any particular bit. We point out that, on the other hand, if the energy per bit is not large enough, as required by (8.4.10), we are certain to make word errors if we continually increase block size.

Although the preceding argument indicates the theoretical optimality of block encoding with infinite word lengths, the advantage of block encoding vis-à-vis binary encoding with finite word lengths is not obvious. To investigate this trade-off, consider a specific binary word of k bits. This can be sent as k binary antipodal transmissions, or by a single k bit block transmission. In the binary case the probability of not receiving every bit in the word correctly is

$$\text{PWE} = 1 - \left[1 - \tfrac{1}{2}\,\text{Erfc}\left(\frac{E_b}{N_{bo}}\right)^{1/2} \right]^k \qquad (8.4.12)$$

where E_b is the energy per bit. Since the Erfc value can be read from Figure 7.6, the binary PWE can be easily evaluated for any k. An orthogonal block coded system, using words of length k and transmitting at the same bit rate, has its PWE given in Figure 8.7 at the same value of $E_b = P_m/\mathcal{R}_0$. When these results are compared, the block coded system produces the smaller PWE. This effect can best be exhibited by determining how much less bit energy is needed in block coding to achieve the same PWE as a binary system sending the same number of bits. The result is shown in Figure 8.8 for several values of PWE, as a function of word size k. Note that a power savings of about 5 dB is obtained using only $k = 5$ bit blocks. The improvement increases more slowly for higher k values, being only about 6 dB for $k = 10$. This comparison illustrates a basic fact of bit detection. A binary system sending a k bit interval can be envisioned as having a decoding scheme that makes k separate bit decisions to decode the word. A block coded system, in the other hand, integrates the total word energy and makes a single decision at the word end. We see clearly that the single decision, using all the available energy, is superior to a sequence of subword decisions. This is even more apparent as we let $k \to \infty$. The block coded system has the asymptotic behavior of (8.4.10). However,

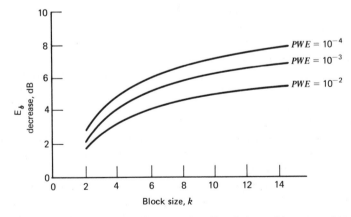

Figure 8.8. Decrease in required block encoding E_b relative to binary to achieve same PWE ($k =$ block size).

(8.4.12) indicates that $\mathrm{PWE} \to 1$ for a binary system for any value of E_b. This means we are making so many bit decisions to decode a long word that we are in fact bound eventually to make an error.

It is often desirable to convert word error probability in block coding to an equivalent bit error probability. This can be accomplished by determining the probability that a given bit of the word will be incorrect after incorrect decoding. With orthogonal signals the incorrectly decoded word is equally likely to be any of the remaining $M-1$ words, and a given bit will be decoded as any of the bits in the same position of each other word. In M equally likely patterns a given bit will be a one or a zero $M/2$ times. The chance of it being the incorrect bit in the $(M-1)$ incorrect patterns is then $(M/2)/(M-1)$. This means that the probability of a given bit being in error is then this probability times the probability that the word was in error. Hence the equivalent bit error probability PE is related to the word error probability PWE by

$$\mathrm{PE} = \frac{1}{2}\left(\frac{M}{M-1}\right)\mathrm{PWE} \tag{8.4.13}$$

when using orthogonal signaling.

The PWE for the orthogonal signal set can be approximated for large M by use of the union bound of (8.1.8). Since the binary test between any two decoding channels is equivalent to an orthogonal coherent test, we have

$$\mathrm{PWE} \le \frac{M^2 - M}{M}\left[\tfrac{1}{2}\,\mathrm{Erfc}\left(\frac{E}{N_{b0}}\right)^{1/2}\right] \tag{8.4.14}$$

This serves as a simple upper bound, which becomes increasingly more accurate as M and E/N_{b0} is increased. Note that by use of Problem 7.4, (8.4.9), and (8.4.13) we can establish that for large M, the block coded bit error probability is bounded by

$$\mathrm{PE} \le \frac{M}{4}\,e^{-(E/N_{b0})}$$

$$= (\tfrac{1}{4})\exp-\left[\left(\frac{P_m}{\mathscr{R}_0 N_{b0}}\right) - \log_e 2\right]\log_2 M \tag{8.4.15}$$

We again see that as $M \to \infty$, the bit error probability PE, with orthogonal block encoding, must necessarily go to zero as long as (8.4.11) is true. This again verifies perfect transmission under these conditions.

Biorthogonal Signals. Evaluation of PWE using (8.1.15) for the biorthogonal signal set of (8.2.5) is hindered by the singularity of the matrix Γ, preventing matrix inversion. This can be avoided by recalling that

M biorthogonal signals are actually $M/2$ pairs of antipodal signals. The M channel decoder can be reduced to $M/2$ channels, since a channel matched to any particular signal is likewise negatively matched to its negative version. Word decisions can be made by first determining which channel sample y_i has the largest magnitude, then determining the polarity of this sample. For a decoder implemented in this way only $M/2$ outputs y_i are needed, and PWE can be obtained directly from this reduced system, thus avoiding the singularity problem. It can be shown [7] that the PWE for this system is given by

$$\text{PWE} = 1 - \int_{-(E/N_{b0})^{1/2}}^{\infty} \frac{e^{-x^2}}{\sqrt{\pi}} \left[1 - \text{Erfc}\left(x + \left(\frac{E}{N_{b0}}\right)^{1/2}\right) \right]^{(M/2)-1} dx \qquad (8.4.16)$$

Equation (8.4.16) is plotted in Figure 8.9 for the case $M = 4$. Performance is comparable to that of orthogonal signals but we retain the advantage of a reduced decoder. An accurate approximation in biorthogonal signaling can be obtained from the union bound in (8.1.8). In this case we have $(M/2) - 1$ orthogonal binary tests, and one antipodal test, for determining the bound. Hence

$$\text{PWE} \lesssim \frac{M-2}{2} \text{Erfc}\left[\left(\frac{E}{2N_{b0}}\right)^{1/2}\right] + \tfrac{1}{2}\text{Erfc}\left[\left(\frac{E}{N_{b0}}\right)^{1/2}\right] \qquad (8.4.17)$$

The bound is quite accurate at $\text{PWE} \leq 10^{-2}$ and can be easily evaluated to estimate performance.

For biorthogonal signals the bit error probability is no longer related to the word error probability by (8.4.13) because when a word error is made it is no longer equally likely to be any of the remaining words, as in the orthogonal case. Instead the bit error distribution depends on the manner in which the word error occurs. If the decoder decides the correct maximum magnitude bit but errs in the polarity decision, every bit of the word is incorrect. If the decoder decides the wrong magnitude, bit errors are equally distributed. The probability of a given bit being in error must then be obtained by averaging over the probability of each of these types of errors. This is given approximately by

$$\text{PE} \cong \tfrac{1}{2}\text{Erfc}\left[\left(\frac{E}{N_{b0}}\right)^{1/2}\right] + \frac{1}{2}\left(\frac{M-2}{2}\right)\text{Erfc}\left[\left(\frac{E}{2N_{b0}}\right)^{1/2}\right] \qquad (8.4.18)$$

which becomes increasingly accurate with large M and E/N_{b0}. Biorthogonal signals have the same limiting performance as $M \to \infty$ as orthogonal signals.

Equally Correlated Signals. The word error probability for signal sets that have equal cross-correlation can be obtained from the known results of the orthogonal case. This follows since equally correlated signals with energy E and cross-correlation γ generate the same PWE as an orthogonal set with energy $E(1-\gamma)$ (Problem 8.12). The PWE for the latter is obtained from (8.4.6) and Figure 8.7. Hence we need only adjust the effective energy in determining the corresponding PWE for a given M. For simplex signals, $\gamma = -1/(M-1)$, and the effective energy is increased to $E(1-\gamma) = EM/M - 1$. Performance of simplex signals is therefore better than for orthogonal signals because of the slight increase in effective energy but this advantage becomes negligible at large M. A plot of the simplex signal set performance for $M = 4$ is superimposed in Figure 8.9.

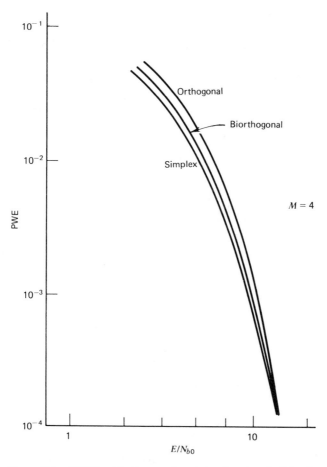

Figure 8.9. PWE for biorthogonal and simplex signals ($M = 4$).

Other types of M-ary signaling, such as M level differential encoding, have also been proposed for digital systems. We shall not pursue their study here. The interested reader is referred to the discussion of such systems in Lindsey and Simon [7].

8.5. Polyphase Signaling

Because polyphase signals can be easily generated and require a bandwidth essentially independent of the number of signals of the set, they are a popular signaling method in modern systems. The signals can be generated by merely phase keying an oscillator into the proper phase state during each block interval. Each signal of the set can be written as

$$s_i(t) = \left(\frac{2E}{T}\right)^{1/2} [\cos \theta_i \cos \omega_s t + \sin \theta_i \sin \omega_s t], \qquad 0 \le t \le T \quad (8.5.1)$$

where ω_s is the oscillator frequency and θ_i is its phase shift. Decoding is achieved by coherent correlation of the receiver baseband signal with each member of the set, selecting the largest correlator output as the signaling phase state. Each correlator computes

$$y_i = \int_0^T x(t) s_i(t)\, dt$$

$$= \left(\frac{E}{T}\right)^{1/2} [X \cos \theta_i - Y \sin \theta_i] \qquad (8.5.2)$$

where

$$X \triangleq \int_0^T x(t) \cos \omega_s t\, dt$$

$$Y \triangleq \int_0^T x(t) \sin \omega_s t\, dt \qquad (8.5.3)$$

We see that (8.5.2) is identical to writing

$$y_i = C \cos (\theta_i - \psi) \qquad (8.5.4)$$

where $C^2 = 2E(X^2 + Y^2)/T$ and

$$\psi \triangleq \tan^{-1} \frac{Y}{X} \qquad (8.5.5)$$

The decoding decision as to which y_i is largest is therefore equivalent to determining which $(\theta_i - \psi)$ is smallest, over all i. That is, determining which θ_i the ψ is closest to. The decoder can therefore be reduced to a pair of

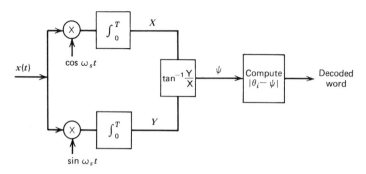

Figure 8.10. Decoder for polyphase signals.

channels in which X and Y in (8.5.3) are separately determined, and the decoder decisioning is based only on a computation of ψ in (8.5.5) and a comparison to $\{\theta_i\}$ (Figure 8.10). In essence, the decoder detects the phase of the input signal [as defined in (8.5.5)] for decisioning. Note that the decoder uses quadrature mixing followed by an integrate-and-dump detector. The quadrature signals, $\cos \omega_s t$ and $\sin \omega_s t$, could be generated from a common free-running oscillator in which the quadrature component is achieved by an additional 90° phase shift. However, this oscillator must be perfectly phase coherent with the received oscillator frequency to compute X and Y in (8.5.3).

The PWE can be evaluated by formally computing the signal correlation matrix from (8.3.22) and again evaluating (8.1.15). However, a simpler procedure is to determine PWE by computing the probability that the angle ψ is not closer to the correct θ_i. For the case of equally spaced values for θ_i over $(0, 2\pi)$ this has been determined [7] to be

$$\text{PWE} = \frac{M-1}{M} - \tfrac{1}{2} \text{Erf}\left[\left(\frac{E}{N_{b0}}\right)^{1/2} \sin\left(\frac{\pi}{M}\right)\right]$$
$$- \frac{1}{\sqrt{\pi}} \int_0^{(E/N_{b0})^{1/2} \sin(\pi/M)} \exp\left[-y^2\right] \text{Erf}\left[y \cot\left(\frac{\pi}{M}\right)\right] dy \qquad (8.5.6)$$

For $M = 2$ (binary signaling) the signal set is antipodal and (8.5.6) reduces to our earlier result. When $M \gg 1$, PWE is adequately approximated by neglecting the last term, which simplifies PWE to

$$\text{PWE} \cong \text{Erfc}\left[\left(\frac{E}{N_{b0}}\right)^{1/2} \sin\left(\frac{\pi}{M}\right)\right] \qquad (8.5.7)$$

We immediately see a basic disadvantage to polyphase signaling with large M. The signal vector points (phasors on the circle of $\sqrt{E}$) become crowded together and accurate detection becomes difficult. This effect is most obvious by noting that for large M, $\sin(\pi/M) \approx \pi/M$ so that the required signal energy E must be increased as M^2 to maintain a given PWE. For this reason, polyphase signals are generally restricted to low values of M ($M = 4, 6, 8$). Although these values of M do not correspond to long block sizes and therefore may not produce significant improvement over binary transmission, polyphase signals do allow increased bit rate without increase in the required transmission bandwidth. This is because the required signal bandwidth is essentially not dependent on M and is approximately equal to the reciprocal of the block interval for any M.

For $M = 4$ (QPSK), PWE in (8.5.6) reduces to

$$\text{PWE} \doteq \text{Erfc}\left[\left(\frac{E}{2N_{b0}}\right)^{1/2}\right] - \tfrac{1}{4}\text{Erfc}^2\left[\left(\frac{E}{2N_{b0}}\right)^{1/2}\right] \qquad (8.5.8)$$

This could have also been derived by recalling that QPSK is actually a biorthogonal signal set, corresponding to two pairs of antipodal signals orthogonal to each other. If we choose the four angles θ_i as ($45°$, $135°$, $225°$, $315°$), then deciding which θ_i is closest to ψ is identical to determining which quadrant ψ lies in. However, this quadrant will be determined by simply the signs of X and Y. If $X, Y \geq 0$, $0 \leq \psi \leq 90$, if $X < 0$, $Y > 0$, $180 \leq \psi \leq 90$, and so on. This sign detection is simply a binary test performed separately on X and Y, and QPSK decoding is achieved by simply performing separate zero threshold tests on each (Figure 8.11). In this sense, (8.5.8) is simply the two bit word error probability associated with independent bit decisions, each with energy $E/2$. If we envision QPSK encoding as the simultaneous encoding of each bit of a two bit word onto separate orthogonal signals, the equivalent QPSK bit error probability is identical to that of an antipodal binary system sending two successive bits each with the same bit energy but with each bit $T/2$ sec long. Thus QPSK achieves the same PE as a binary system with the same bit rate and same

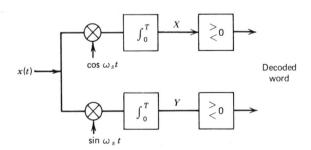

Figure 8.11. Decoder for QPSK ($M = 4$) polyphase signals.

energy per bit, but utilizes only one-half the required bandwidth. (Conversely, it transmits at twice the bit rate with the same bandwidth and PE.)

A difficulty with QPSK is the necessity of making large (180°) phase changes almost instantaneously in changing between certain phase states. Physical requirements on phase modulators prevent these instantaneous changes, so that the bit rate is effectively slowed down, reducing an inherent advantage of QPSK. This problem can be somewhat combated by using so-called *offset QPSK* encoding. If we again envision QPSK as a biorthogonal encoding scheme, we can consider the first bit of every two bit block of source bits to be a binary sequence modulated on one quadrature signal in (8.5.1), and the second bit of each block as another sequence for the other quadrature component. We list these sequences as

$$\mathbf{a} = \{a_1, a_2, a_3, \dots\}$$
$$\mathbf{b} = \{b_1, b_2, b_3, \dots\} \tag{8.5.9}$$

where a_i, b_i, are ± 1. Each pair of bits (a_j, b_j) therefore determine the phase angle θ_i in QPSK via

$$\theta_i = \tan^{-1} \left(\frac{b_j}{a_j} \right) \tag{8.5.10}$$

The sequences **a** and **b** must therefore be in bit alignment. To obtain offset QPSK, we continue to use the same quadrature encoding, except that we offset one sequence in (8.5.9), by one-half a bit period from the other (Figure 8.12). The transmitted phase is still given by (8.5.10) but halfway during a bit interval b_j changes to b_{j+1} while a_j remains the same (or vice versa). Either the next bit b_{j+1} has the same polarity as b_j, in which case θ_i does not change, or it has opposite polarity, in which case θ_i jumps to the adjacent quadrant. Hence a change of at most $\pm 90°$ can occur every half bit period by staggering, or offsetting, the bit sequence in (8.5.9). This achieves control of the maximum phase change, limiting it to a 90° phase

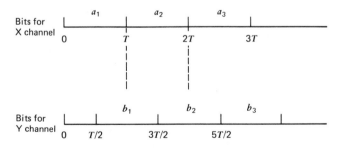

Figure 8.12. Bit alignment for offset QPSK.

shift. The quadrature decoders in Figure 8.11 must also be offset by $T/2$ for individual bit decisioning. Since the channels are orthogonal, the offset quadrature signal will not affect individual bit detections of the in phase channel.

Just as in PSK, QPSK decoders must be phase referenced. To determine the effect of imperfect referencing, consider (8.5.3) when the decoder quadrature signals have an added phase shift θ_e relative to the proper phase (assumed zero). Then

$$X(\theta_e) = \int_0^T x(t) \cos [\omega_s t + \theta_e] \, dt$$

$$= \int_0^T x(t)[\cos \theta_e \cos \omega_s t - \sin \theta_e \sin \omega_s t] \, dt$$

$$= X \cos \theta_e - Y \sin \theta_e \qquad (8.5.11)$$

and, similarly,

$$Y(\theta_e) = Y \cos \theta_e + X \sin \theta_e \qquad (8.5.12)$$

where X, Y are the values generated from perfect referencing. Thus the effect of a phase error θ_e is to convert the desired values of X, Y to $X(\theta_e)$, $Y(\theta_e)$. In addition to the typical cosinusoidal degradation characterizing phase referencing errors in PSK, we now see that θ_e also causes a cross coupling into the orthogonal channel, proportional to $\sin \theta_e$. QPSK is therefore degraded more by imperfect phase referencing than is binary PSK. Note that the cross coupling is destructive for one channel but is constructive for the other. To determine its effect on bit error probability PE, we average over both channel bit errors. Since X and Y have mean values of $\pm (ET/2)^{1/2}$, this becomes

$$PE(\theta_e) = \tfrac{1}{4} \text{Erfc} \left[\left(\frac{E}{2N_{b0}} \right)^{1/2} (\cos \theta_e + \sin \theta_e) \right]$$

$$+ \tfrac{1}{4} \text{Erfc} \left[\left(\frac{E}{2N_{b0}} \right)^{1/2} (\cos \theta_e - \sin \theta_e) \right] \qquad (8.5.13)$$

A plot of (8.5.13) for several values of θ_e is shown in Figure 8.13.

In offset QPSK, bit transitions in the other channel may occur at the middle of the bit interval of each channel. If no transition occurs, performance during that bit is identical to QPSK in (8.5.13). When a transition occurs, the cross coupling term (involving $\sin \theta_e$) changes polarity at midpoint, and the cross coupling during the first half bit is canceled by that during the second half of the interval. Hence performance is identical to that with no interference at all, and therefore is the same as for indepen-

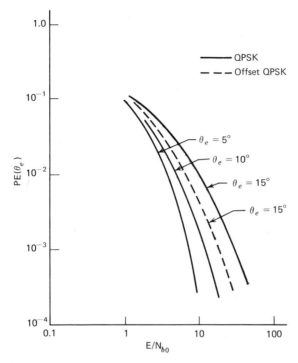

Figure 8.13. Effect of phase error on QPSK performance.

dent binary PSK channels with phase reference error θ_e. Since a bit transition occurs with a probability of $1/2$, the offset QPSK bit error probability with phase error θ_e is

$$\text{PE}_{\text{OQPSK}}(\theta_e) = \tfrac{1}{2}\text{PE}_{\text{QPSK}}(\theta_e) + \tfrac{1}{2}\text{PE}_{\text{PSK}}(\theta_e) \qquad (8.5.14)$$

where $\text{PE}_{\text{QPSK}}(\theta_e)$ is given in (8.5.13) and $\text{PE}_{\text{PSK}}(\theta_e)$ is given in (7.2.14). This equation is also shown in Figure 8.13. The result shows slightly better performance for the offset system with a given phase referencing error.

Although offset QPSK achieves an advantage in phase shift reduction, its frequency spectrum will still be fairly wide because of effective pulse modulation in each quadrature channel by the **a** and **b** sequences in (8.5.9). This means that any transmission bandlimiting effects filtering off the tails of the spectrum must be accounted for. (Remember the prime advantage of QPSK is that it allows doubling of the bit rate with the same spectrum as PSK, so its prime application is in bandlimited situations.) This suggests that an advantage can be attained by using modulation formats in each quadrature channel that are not perfectly pulsed but that avoid significant

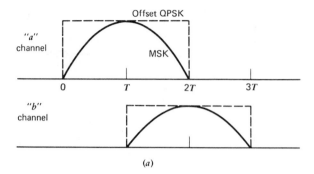

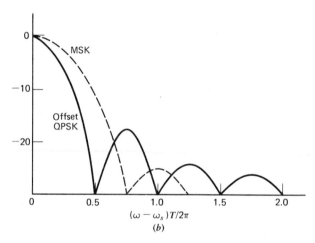

Figure 8.14. Comparison of MSK and offset QPSK. (*a*) Waveforms, (*b*) spectral densities (ω_s = carrier frequency).

spectrum tails and energy loss caused by bandlimiting. Consider using rounded half-sine wave pulses as the modulating signals in each quadrature channel, as shown in Figure 8.14*a*. We let T be the bit time so that each QPSK word is $2T$ sec long, and the **a** and **b** sequences are offset by T sec. The offset QPSK waveform is then

$$m(t) = a_i \cos\left[\frac{\pi}{2T}(t-(2i+1)T)\right] \cos \omega_s t$$

$$+ b_{i-1} \cos\left[\frac{\pi}{2T}(t-2iT)\right] \sin \omega_s t \qquad (8.5.15)$$

for $2iT \le t \le (2i+1)T$, and

$$m(t) = a_i \cos\left[\frac{\pi}{2T}(t-(2i+1)T)\right] \cos \omega_s t$$

$$+ b_i \cos\left[\frac{\pi}{2T}(t-(2i+2)T)\right] \sin \omega_s t \qquad (8.5.16)$$

for $(2i+1)T \le t \le (2i+2)T$. However, by applying trigonometric identities, this can be written equivalently as

$$m(t) = \cos\left(\omega_s t + a_i \frac{\pi}{2T} t + \psi_{i1}\right), \qquad 2iT \le t \le (2i+1)T$$

$$= \cos\left(\omega_s t + b_i \frac{\pi}{2T} t + \psi_{i2}\right), \qquad (2i+1)T \le t \le (2i+2)T$$

$$(8.5.17)$$

where ψ_{i1} and ψ_{i2} are constant phase angles, depending on the data bits. We see that (8.5.17) is equivalent to a carrier at frequency ω_s, deviated between $\pm 1/4T$ Hz according to the bit stream $(a_1, b_1, a_2, b_2, ..)$. In Section 7.2 we defined this as a minimum shift keyed (MSK) signaling format. Thus offset QPSK signals with half-cosine modulating bit pulses in each quadrature channel are identical to binary MSK signaling with the same bit stream and a controlled phase variable. The spectrum of MSK signals was shown in Figure 7.11, and in Figure 8.14b it is compared to the corresponding spectrum of pulsed offset QPSK. Note that MSK signals have a slightly wider main hump but fewer out-of-band tails than for offset QPSK. Thus offset QPSK with cosine pulses is advantageous in bandlimited channels, provided that the slightly larger spectral hump can be accommodated. The offset QPSK coherent decoders now correspond to matched filters in each quadrature channel, each properly matched to the offset half-cosine sub-carrier burst representing each bit. However, we should also point out that since cosinusoidal QPSK correspond to MSK signaling, and since the latter is a form of FSK, the cosine QPSK waveforms can also be decoded by noncoherent frequency discriminator detection if the phase synchronization is lost. Further discussions of offset QPSK signals, and related bandwidth consideration, can be found in References 8–11.

8.6. Noncoherent Signaling (MFSK)

Our discussion of block coded systems has considered only coherent decoding. In each case it was tacitly assumed that each member of the block signaling set can be generated coherently at the receiver. This means,

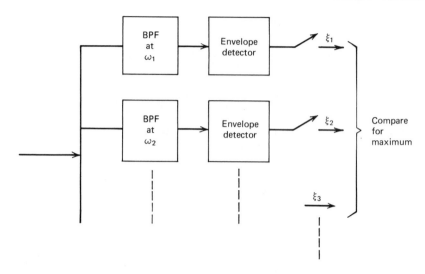

Figure 8.15. Noncoherent block decoder for MFSK.

for example, that in using the frequency shift set (MFSK) in (8.3.14), the phase angle of each signaling frequency was phase synchronized at the decoder. Although MFSK signaling is advantageous because of its relative ease of generation, the phase referencing requirement does complicate the overall receiver design. For this reason there is interest in using MFSK signals with noncoherent decoders, that is, decoders that perform by ignoring the phase information. Such a noncoherent M-ary decoder can be constructed as an extension of the binary, noncoherent envelope detectors of Section 7.6. Such a system is shown in Figure 8.15.

The signal set is given by

$$s_i(t) = \left(\frac{2E}{T}\right)^{1/2} \sin\left[\omega_s t + \frac{2\pi i}{T}t + \theta\right], \qquad i = 1, 2, \ldots, M$$

$$0 \le t \le T \qquad (8.6.1)$$

corresponding to a set of M frequencies $1/T$ Hz apart, so as to be orthogonal over $(0, T)$ for any arbitrary phase angle θ. The selection of ω_s allows us to locate the frequency set anywhere on the frequency axis. Note that the total signal set occupies a basebandwidth of M/T Hz, which must be provided in the transmitter and receiver subsystems. The decoder utilizes the noncoherent decoding bank in Figure 8.15. The bandpass filters are tuned matched filters for each frequency and the envelope samples $\{\xi_i\}$ are used for comparison decisioning as in the binary case. In essence, the decoder is simply a frequency detector composed of a filtering bank, and word decisioning is based on which frequency is being received after T sec. By using envelope samples, the phase angle θ in (8.6.1) does not affect decisioning.

Performance of the noncoherent MFSK decoder is obtained by evaluating the probability that the correct envelope sample is not the largest. Since the frequencies are separated so as to produce orthogonal signals, the deisred probability is the probability that the envelope sample of a filtered

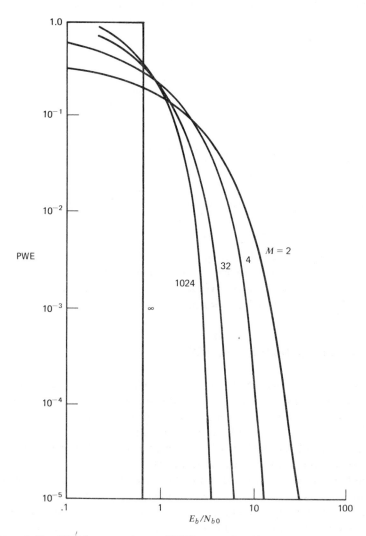

Figure 8.16. PWE for noncoherent MFSK, normalized bit rate $\mathcal{R}_0$. ($E_b = P_m/\mathcal{R}_0$.)

sine wave plus Gaussian noise does not exceed $M-1$ envelope samples of Gaussian noise alone. The probability density of these samples was given earlier in (7.4.6) and it follows that

$$\text{PWE} = 1 - \int_0^\infty u e^{-(u^2+\rho^2/2)} I_0(u\rho) \left[\int_0^u \eta e^{-\eta^2/2} d\eta \right]^{M-1} du \quad (8.6.2)$$

where $\rho^2 \triangleq 2E/N_{b0}$ and E is defined in (8.6.1). Equation (8.6.2) is the noncoherent MFSK equivalent of the coherent orthogonal result in (8.4.6). The integration can be carried out by noting that

$$\left[\int_0^u \eta e^{-\eta^2/2} d\eta \right]^{M-1} = [1 - e^{-u^2/2}]^{M-1}$$

$$= \sum_{q=0}^{M-1} (-1)^q \binom{M-1}{q} e^{-qu^2/2} \quad (8.6.3)$$

Substitution then yields

$$\text{PWE} = 1 - e^{-\rho^2/2} \sum_{q=0}^{M-1} (-1)^q \binom{M-1}{q} \int_0^\infty u e^{-u^2(q+1)/2} I_0(\rho\mu) du$$

$$= 1 - e^{-\rho^2/2} \sum_{q=0}^{M-1} (-1)^q \binom{M-1}{q} \frac{\exp[(\rho^2/2)/(1+q)]}{1+q} \quad (8.6.4)$$

When the parameter ρ^2 is normalized to the same bit rate $\mathcal{R}_0$, using (8.4.9), the curves in Figure 8.16 are generated. The results exhibit the same uniform improvement with M as in the coherent case, except that the noncoherent performance is several decibels poorer at each M. Note we again obtain the limiting behavior of (8.4.10) as $M \to \infty$.

8.7. Amplitude Shift Keying (ASK)

Another popular block encoding signaling scheme uses pulse transmission with *amplitude shift keying* (ASK). In this format, a pulse is sent with one of M distinct amplitudes to represent each of the M block words of the set. Block encoding is achieved by encoding the words into specific pulse amplitudes. ASK is therefore a form of baseband PCM transmission where more than two amplitudes are used. For this reason ASK signaling is also referred to as *pulse amplitude* modulation (PAM) signaling. The prime advantage of ASK is that the basebandwidth needed to send any level signal need be no larger than that needed to send a single pulse. Hence any size signal set M can be accommodated within a single pulse bandwidth, and ASK has bandwidth advantages similar to polyphase signaling.

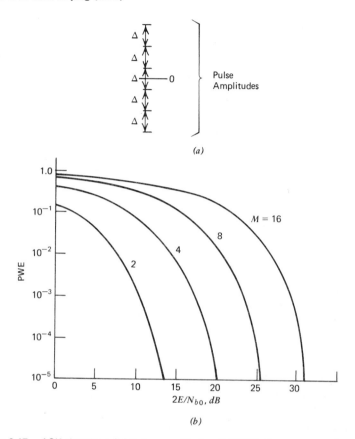

Figure 8.17. ASK signaling. (*a*) Pulse amplitudes, (*b*) PWE. E = average pulse energy.

ASK differs from the previous signaling schemes of Section 8.2 since the signals of the set no longer have equal energy. The ASK signal set can be denoted as

$$s_i(t) = s_i, \qquad 0 \le t \le T, i = 1, 2, \ldots, M \qquad (8.7.1)$$

where $\{s_i\}$ is a set of distinct pulse amplitudes, assumed to be equally spaced with width Δ, and symmetric about zero, as shown in Figure 8.17*a*. The *i*th signal has energy $s_i^2 T$, and the entire signal set has average energy

$$E = \frac{1}{M} \sum_{i=1}^{M} s_i^2 T = \frac{2T}{M} \sum_{i=1}^{M/2} \left[\frac{(2i-1)\Delta}{2} \right]^2$$

$$= \frac{\Delta^2 T (M^2 - 1)}{12} \qquad (8.7.2)$$

ASK word decoding is accomplished by integrating over the received baseband pulse, and determining which value of $s_i T$ is closest to the integrator value (Problem 8.15). A word decoding error will occur if the decoder noise causes the integrator value to lie closer to an amplitude value different from that transmitted. For the symmetric ASK signal set, a decoding error will occur if the integrated decoder noise has a value exceeding $\pm \Delta T/2$ for the inner amplitudes, exceeds $\Delta T/2$ for the most negative amplitude, and is less than $-\Delta T/2$ for the highest amplitude. For additive Gaussian white noise with level N_{b0} the integrated noise has variance $N_{b0}T$. The probability of a word decision, assuming all levels equally likely, is then

$$\text{PWE} = \frac{1}{M}\left[(M-2)\,\text{Erfc}\left(\frac{\Delta T/2}{\sqrt{N_{b0}T}}\right) + 2(\tfrac{1}{2})\,\text{Erfc}\left(\frac{\Delta T/2}{\sqrt{N_{b0}T}}\right)\right]$$

$$= \left(\frac{M-1}{M}\right)\text{Erfc}\left(\frac{\Delta T/2}{\sqrt{N_{b0}T}}\right) \tag{8.7.3}$$

We can rewrite this in terms of the average pulse energy by substituting from (8.7.2)

$$\text{PWE} = \left(\frac{M-1}{M}\right)\text{Erfc}\left[\left(\frac{3E}{(M^2-1)N_{b0}}\right)^{1/2}\right] \tag{8.7.4}$$

Equation (8.7.4) is plotted in Figure 8.15*b* as a function of the average pulse energy to noise level E/N_{b0} for several values of M. We note the obvious degradation as the number of pulse levels M is increased while maintaining fixed bit energy. Conversely, if PWE is to remain constant, it is necessary to increase average energy as the number of pulse amplitudes is increased. We emphasize that this effect is uniform in M, and we always degrade performance by increasing M with ASK signaling.

ASK signals occupy a frequency spectrum corresponding to a single pulse, and therefore extend over approximately $1/T$ Hz. Such a spectrum is low pass in form, and therefore susceptible to low frequency noise interference. This latter problem can be avoided by using ASK subcarrier pulses, in which T sec bursts of a subcarrier are used with multiple amplitudes for the block signals. This results in a basebandwidth that is $2/T$ Hz wide, but spectrally shifted to the subcarrier frequency. A further modification is to use quadrature signaling, as in (8.3.18), in which separate ASK is used on each quadrature component. This permits twice the number of bits to be sent simultaneously over the same bandwidth as a single ASK subcarrier link. Phase coherent decoding of the quadrature signals will then avoid the interference from the orthogonal channel. Such a system is called *quadrature ASK* (QASK), and is the amplitude modulated version of QPSK. ASK and QASK have the disadvantage that their operation requires variable peak power levels. In addition, they require phase coherency, accurate timing, and suffer from power degradation and intersymbol interference if the transmitted signals are distorted and bandlimited. Further modifications in the form of partial response shaping of the QASK pulses can be employed to combat these latter effects.

8.8. Convolutional Encoding

In block encoding the source sequence is partitioned into disjoint blocks for encoding. Each block represents a fixed group of bits, and the blocks are each encoded distinctly and transmitted sequentially. However, handling long block words in this fashion at high data rates tends to be somewhat cumbersome and computationally costly, and often places limitations on the ultimate speed of bit transmission. With the appearance of new requirements for high rate data transmission in modern and future systems, there has been interest in devising new methods of source sequence encoding that still maintain the inherent error probability improvement of block encoding. One technique that has developed from these investigations, and has been successfully applied, is the method of *convolutional encoding*. Like block encoding, convolutional encoding involves the encoding of sequences of source bits into a sequence of binary transmission symbols, but in a somewhat continuous manner rather than in distinct blocks of a given size. This continuity is established by effectively sliding the location of what we previously considered a block down the source sequence, rather than jumping to disjoint bit groupings. This results in a continuous interleaving of bit blocks, and long source streams can be encoded without the blocking operation. The formation of the output sequence is accomplished with sufficient inherent structure so that decoding can be achieved with much less computational burden than lengthy block codes. Studies of such systems have exhibited error probability improvement comparable to or exceeding that observed with standard block encoding.

Convolutional encoding is generated by high speed shift registers having the typical form shown in Figure 8.18. The source bits to be encoded are shifted into the register in a clocked sequence and dumped out at the end. The register stages are binary summed in V separate modulo 2 adders, each connected to certain register taps. At preselected clock times, a commutating switch reads out the output of the V adders in sequence, forming a V bit word at this clock time. As the source sequence shifts in, the clocking of the commutator is repeated and another V bit output

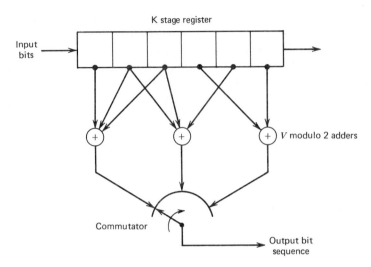

Figure 8.18. Shift register convolutional encoder. The modulo-2 additions performed once following each bit shift.

occurs, with the succession of V bit words forming the encoded output. In general, the commutator clocking is performed after, say, k bits have shifted into the register (and k bits have been dumped out the end). For each k source bits there are therefore V output bits produced, and the encoder operates at an output rate of V/k times the input rate. This is called a *rate k/V* encoder. Typically, $k = 1$ (a commutation performed after each input source bit) and $V = 2$ or 3. When encoding is accomplished in this manner, the output sequence can be written as a convolution of the input sequence and an encoder weighting function; hence the name convolutional coding.

As a given source bit shifts through a K stage register we see that it will affect $(K/k)V$ successive output bits via the register summation logic. This parameter, KV/k, is called the *constraint length*, or *memory time*, of the encoder. (Recall that in block encoding each source bit influenced only one specific block of channel bits.) The manner in which the V adders are connected determines the structure of the output sequence. If the summations are such that the k source bits shifted during each clock time of a rate k/V encoder appear as the first k bits of the V bit output, the convolutional code is said to be *systematic*. Systematic codes have the interpretation of being formed by appending extra encoder bits to each k bit source group, similar to adding parity bits as in block encoding, except the added bits are related to past source bits through the constraint length. Various studies have been made to evaluate the separation, or *minimum free distance*, of codes generated by specific register tap connections [12–15]. In general, it has been found that, for a given register size, nonsystematic encoders tend to have larger minimum free distances than systematic ones, but the latter are more immune to decoding error accumulation.

Graphical portrayal of the convolutional encoding operation is accomplished by the use of a *trellis* diagram (Figure 8.19a). Each column of nodes of the trellis represents the state of the register prior to any input. If a register has K stages, then the first $K - k$ bits in the register determine its state. Only $K - k$ bits are needed, since the end k bits are dumped out as the next k input bits occur, and therefore do not affect the encoder output at the subsequent commutation. Thus the trellis has 2^{K-k} nodes in a column. Successive columns refer to successive commutation times. Branches connecting nodes indicate the change of register state as a particular input of k bits is shifted in and a commutation is performed. A branch must exist at each node for each possible k bit input. Branches are labeled by the k bit input to which they correspond and by the V bit commutator output sequence produced by the $K - k$ bits of the previous state and the input k bits as they fill the register. Thus for a rate k/V encoder there will be 2^k such emanating branches from each possible register state. If we start with the register in some initial state, not all nodes are possible until $K - k$ bits have been shifted in and the register is free of its initial condition. After Lk input bits we will have progressed a distance of L columns into the trellis, producing LV output symbols. After the initial state has been dumped out of the register, the trellis diagram is identical at each column. Figure 8.19b shows a specific trellis diagram for the encoder shown, for which $k = 1$, $V = 2$, $K = 3$, and the register is started in the all zero state. Note that a register can be returned to any desired state (node of the trellis) by simply affixing the $K - k$ bits representing that state to the input sequence. This is commonly called *clearing* the register.

From the trellis diagram of any convolutional encoder we see that, starting with any state of the register, a given sequence of input bits traces out a path through the trellis. The output sequence that will be produced are the branch symbols along this path. Thus convolutional encoding of a source sequence can be equated with following paths through the trellis associated with its encoder implementation. In this sense, decoding can likewise be interpreted as attempting to find the input sequence along the encoding path from the noisy observation of the output sequence. This graphical analogy is helpful in understanding the various decoding procedures that have been proposed with convolutional coding.

The output of the convolutional encoder is transmitted by channel encoding each V bit commutator sequence into a waveform. We represent these waveforms by the set $s_i(t)$, $i = 1, 2, \ldots, 2^V$, $0 \le t \le T_V$. Thus channel encoding of the convolutional encoder binary output corresponds to mapping each V bit binary sequence into a unique waveform from the set $\{s_i(t)\}$. This is identical to simply relabeling each branch of the encoder trellis diagram by its correspnding $s_i(t)$ waveform instead of its V bit binary sequence. Now paths through the trellis correspond to sequences of waveforms. As we proceed to input source bits, the convolutional encoder generates a function of time that is transmitted as the baseband signal. The required encoder waveform bandwidth follows from the $s_i(t)$ set, as in (8.1.1).

Let us now confine attention to the case $k = 1$. A long L bit source input sequence is convolutionally channel encoded into a sequence of L waveforms. An encoder baseband waveform sequence exists for each possible input bit sequence. We denote the jth such sequence as

$$m_j(t) = (s_{j1}(t), s_{j2}(t), \ldots, s_{jL}(t)), \qquad j = 1, 2, \ldots, 2^L \qquad (8.8.1)$$

where each $s_{jq}(t)$ corresponds to one of the $\{s_i(t)\}$. At the receiver decoder, maximum likelihood decoding of each possible waveform sequence from the noisy baseband waveform $x(t)$ requires a computation of the cor-

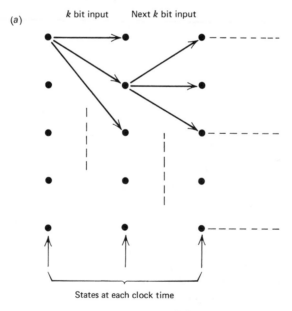

States at each clock time

Figure 8.19. Encoder trellis diagram. (*a*) (*above*) General structure, (*b*) (*on page 377*) an example.

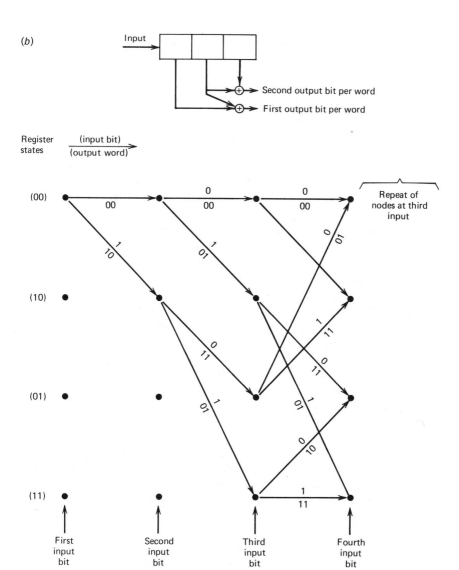

Figure 8.19. (b)

responding y_j, as in Figure 8.2. Thus

$$y_j = \int_0^{LT_V} x(t)m_j(t)\,dt$$

$$= \sum_{q=1}^{L} \int_{(q-1)T_V}^{qT_V} x(t)s_{jq}(t)\,dt \qquad (8.8.2)$$

Formally, we see that each y_j requires a sum of input integrations where each integration is with respect to a particular waveform $s_i(t)$, mainly that occurring in the qth position of $m_j(t)$ in (8.8.1). However, this is also the qth branch waveform in the trellis corresponding to the jth path. Hence we view (8.8.2) as the cumulative sum of branch integrals associated with a particular trellis path. The branch integral is referred to as the *metric* of the branch. The required y_j in (8.8.2) is, equivalently, the sum of branch metrics along a particular input path of the trellis. Note that the metric is with respect to a given decoder input $x(t)$. Thus formal decoding of a convolutionally encoded source sequence corresponds to effectively associating a metric with each branch of the trellis and selecting the input sequence associated with the path having the largest cumulative metric. This type of decoding is also referred to as *tree searching*. Note that the branch metrics between two columns can be computed sequentially every T sec, as we move through the trellis during the tree search.

If the register size K and input sequence length L is large, there will be many branches and paths in the trellis to be considered, and maximum likelihood tree search decoding places a burden on decoder computation and storage requirements. Various decoder algorithms have been suggested to reduce the required computation and storage. The *Viterbi decoding* algorithm [16, 17, 18] performs the maximum likelihood test with (8.7.2), but with considerable reduction in computation. This is achieved by taking advantage of the fact that at any given node of the trellis only the path with the largest metric into that node need be preserved. Any other path into that node will necessarily have a smaller total metric no matter what the ensuring path beyond the node is. Hence only 2^{K-1} candidate paths need be retained in the decoder computation at each node column as we proceed into the trellis. For each candidate path the corresponding input and metric to that point must be stored. During the next T_V sec, a branch metric is computed for each of the branches emanating from each node and added to the candidate path metric up to that node, but again only the resulting paths with the highest metric leading into each node of the next column need be retained. Thus at each step of the decoding, only the most likely inputs are carried through the tree search. It is precisely these intermediate decisions during the entire decoding operation that give the convolutional encoding its computational advantage over long length block encoding while achieving comparable performance.

If the source sequence was terminated so as to force the trellis to a single node, only the one most likely path will eventually be decided. The input sequence stored for that path will then be read out as the decoded source sequence.

In spite of the computational saving of the Viterbi decoding algorithm, there remain some basic disadvantages. The amount of computation depends directly on the number of nodes in a column, which varies exponentially with the register size K. This becomes critical if we recall that a given source bit is effectively spread over the encoder constraint length (the latter also depending on K), and bit error probability reduction will require increased values of K. In addition, no decoding is achieved until the final node is reached and the most likely path selected, no matter how long the sequence. Hence there is an inherent decoding delay that may become significant for long sequences. For these reasons, nonmaximum likelihood convolutional decoding methods have been investigated to overcome these problems [19–22]. However, these methods tend to operate satisfactorily in adequate SNR but degrade rapidly in noisy environments.

Computing exact expressions for performance in terms of error probability with convolutional encoding is extremely difficult because of the inherent complexity of the signal structures. Bounds on performance in terms of the minimum free distance of the encoder have been derived [23], and extensive simulation studies have been reported on operating systems. For a given decoder SNR, bit error probability depends directly on the amount of storage and computation capability that is implemented into the decoder. As this decoding capability is increased, longer and more precise tree searching can be accomplished, improving the decoder performance. Two of the most important parameters in this regard are the size of the encoder register K and the degree of storage quantization used in the tree search. The former determines the encoder constraint length, and the latter determines the accuracy in distinguishing path metrics. Figure 8.20 shows the reported performance [23, 24] of bit error probability PE for various register sizes, as a function of energy per bit $P_m T_V / V N_{b0} = E_b / N_{b0}$, assuming single bit quantization. Also shown is the dependence on the degree of quantization for a $K = 5$ register size. Note that single bit quantization is about 2 dB worse than three bit quantization, but the latter is not much worse than infinite quantization [24]. We see also that increasing register length K by one gains about 0.5 dB at PE $\leq 10^{-4}$. The curve shows that the performance of convolutional encoding, in terms of PE, is significantly superior to that of binary transmission in Figure 7.7. However, care must be used in accepting these values, since hidden within results of this type

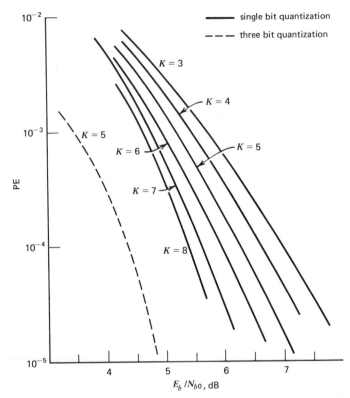

Figure 8.20. Convolutional encoding performance using Viterbi decoding (from [23, 24]). Error probability versus E_b/N_{bo} for hard decisioning and rate $\frac{1}{2}$ encoding ($K =$ register size).

are important parameters such as buffer overflow and limitations to source bit rates. In addition, we must also be aware of the possibility of a catastrophic accumulation of bit errors when a single bit error occurs. Such a condition implies the selection of an incorrect path in tree search decoding, for which there exists the possibility of never returning to the correct path, thereby decoding all the remaining bits in error. Necessary and sufficient conditions are known for avoiding this phenomena [25] but require limitations on the encoder selection, thereby again decreasing some freedom of the system designer. For example, it has been shown that systematic encoding can never produce infinite error accumulations, but nonsystematic codes may produce better overall performance. We also point out that since convolutional decoding uses word correlations in (8.8.2), it is susceptible to the same degradations caused by imperfect phase referencing and word timing as in block signaling.

8.9. A Review of Coding Alternatives

Perhaps the most basic decisions facing a digital system designer is the selection of the particular encoding and decoding scheme to implement. Such a selection, of course, depends on many influencing factors (cost, bandwidth, available hardware, application, etc.), so that no general rule of thumb can be stated for making this decision. We have presented in Chapters 7 and 8 various alternatives that may be considered and have been successfully applied in modern communication systems. We have shown that performance in terms of bit error probability (which we had shown in Chapter 6 to be a contributing factor in A-D and D-A reconstructed SDR) can only be improved on at the expense of additional complexity in some other system aspect. Which of these is the most important and which of these is adaptable and in the control of the designer ultimately decide design format.

Let us review the digital transmission alternatives. Assume a source produces a bit sequence that is to be transmitted with rate $\mathcal{R}_0$ bits/sec over a baseband channel having flat baseband noise. The decoder has available a constrained average baseband signal power of P_m W. One encoding technique is simply to transmit the source bits in sequence at rate $\mathcal{R}_0$, using any of the binary methods of Chapter 7. We know we achieve the minimal bit error probability PE by using coherent antipodal signaling. The performance of such a system is sketched again in Figure 8.21 and shows the idealized attainable performance as a function of the bit energy ratio $E_b/N_{b0} = P_m/\mathcal{R}_0 N_{b0}$. Other binary transmission methods produce degradations of this performance, as we noted earlier. The curve also shows the ultimate PE performance of any system, as predicted by the Shannon theorem, which we know can only be approached with infinite bandwidth and infinite block length encoding. If the PSK performance is unsatisfactory for our application (i.e., we want smaller PE at a fixed P_m or we want a specified PE with less P_m), we must operate in the shaded region of Figure 8.21. Since no other binary scheme can improve on the coherent antipodal signaling, operation in the shaded region can only be achieved by block encoding and improved decoding. The question now is one of determining the degree and method of encoding, recognizing the price to be paid for such improvement.

A first alternative is to retain the decoder implementation, but to channel encode the source bits into a bit sequence with bit error protection, such as the (n, k) error correction encoding discussed in Section 6.7. We again transmit the encoded symbols by antipodal signaling, using bit-by-bit decoding, but taking advantage of the error correction and/or detection capability of the encoding. Note that we are effectively using binary

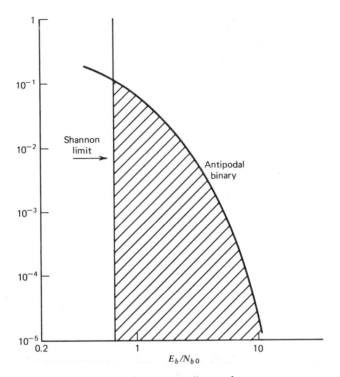

Figure 8.21. Binary encoding performance.

($M = 2$) signal encoding in which the signal vectors correspond to the actual binary symbols being transmitted. However, since n binary symbols must be sent in the time it takes to send k source symbols, each binary symbol waveform can only be kT/n sec long. Hence binary symbol detection is made with energy $P_m(kT/n) = (P_m/\mathcal{R}_0)(k/n)$ per bit. An original source block of k bits will be decoded correctly only if the encoded n bits are all binary decoded correctly or with correctable errors. Encoded words for which an error has been detected but not corrected will be discarded, or repeated, but will not be used in error. Thus we must distinguish between undetected and uncorrected source word errors, and it becomes important in design to decide if detected (and discarded) word errors are sufficient for practical operation. This would tend to be the case when the source rate is low and source words can be repeated, as in command-verify systems. An (n, k) code set having a given code set distance d will have the capability of correcting $(d-1)/2$ bit errors and detecting $(d-1)$ bit errors in a word of n bits. The uncorrelated and undetected n bit word error probabilities are then

$$\text{uncorrected PWE} = \sum_{j=(d-1)/2}^{n} (\text{PE})^j (1-\text{PE})^{n-j} \binom{n}{j} \tag{8.9.1}$$

$$\text{undetected PWE} = \sum_{j=d-1}^{n} (\text{PE})^j (1-\text{PE})^{n-j} \binom{n}{j} \tag{8.9.2}$$

where $\binom{n}{j}$ is the binomial coefficient, and

$$\text{PE} = \tfrac{1}{2} \text{Erfc} \left[\left(\frac{P_m(k/n)}{\mathcal{R}_0 N_{b0}} \right)^{1/2} \right] \tag{8.9.3}$$

Since a source word decoded in error will be any other source word with equal probability, a given bit will be in error with a probability of $1/2$ when a word error is made. Hence the encoded source bit error probability will be one-half the PWE in (8.9.1) and (8.9.2). These results are shown in Figure 8.22 for the case of a (15, 10) linear code, which is known to have single bit error correction and double bit error detection capability. The curves are superimposed on the previous antipodal results. Note that we have effectively moved into the desired encoding region of Figure 8.21. However, the transmission of n binary symbols in the time of k source bits

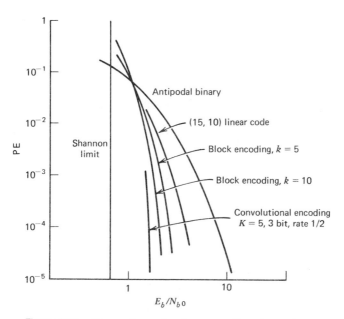

Figure 8.22. Comparison of various encoding alternatives.

Table 8.1. Summary of Encoding Schemes

Type of System	Encoder Output	Decoding	Properties	Required Bandwidth
Binary	k bits in T sec	Bit by bit	Standard systems, PSK, FSK, PCM, easily implemented	$B_u = \frac{k}{T}$
Error correction encoding	n bits representing k source bits	Bit by bit	Error correction and detection capability	$\left(\frac{n}{k}\right) B_u$
Block encoding	One of 2^k waveforms	Bank of 2^k correlators	Complex decoder, improved error probability	$\left(\frac{2^k}{k}\right) B_u$
Convolutional encoding (rate k/V)	Vk bit words representing k source bits	Vk correlators plus logic and search circuitry	Improved error rates, random decoding time computational problems	$\frac{Vk}{k} B_u = V B_u$

means the transmission bandwidth must be increased by a factor of n/k. These facts are summarized in Table 8.1.

If the preceding error correction is not satisfactory, either we must utilize longer codes, with higher levels of bit correction but increased bandwidth occupancy as well, or we must resort to block encoding. In the latter case the decoder must be restructured to perform block decoding with a channel matched to each possible encoded word, as discussed in Section 8.1. We now achieve the performance curves shown in Figure 8.22 for several values of block size k, assuming orthogonal signaling and converting PWE to PE by use of (8.4.13). Recall, however, that a decoder of 2^k channels is required to achieve this performance and the k bits are being encoded into signal vectors of 2^k components. Hence the system bandwidth is now increased to $(2^k/k)$ times that of antipodal binary signaling (Table 8.1). However, we see the improved performance obtained by this latter scheme. Note that we are now within several decibels of the ultimate bound in achieving a bit error probability in this range.

Still greater improvement is possible only by extending to larger block or word sizes, with the added complexity, or by resorting to convolutional encoding. Figure 8.22 shows several points extracted from our earlier results in Section 8.7. For a rate k/V encoder the transmission rate and bandwidth must be increased by the factor V/k over that of the binary case. Although a definite superiority of convolutional encoding is shown in terms of the PE parameter, it must be remembered that a particular decoding algorithm is needed to achieve this performance. In addition, one must be aware of catastrophic occurrences, such as buffer overflow and long accumulations of decoding errors. Nevertheless, the continual advancement into the encoding region of Figure 8.21 in modern systems is quite apparent, and the techniques and procedures for further improvement are now well understood. It is to be expected that with the advancement in the technology of high speed computation and digital processing, and its subsequent adaptation to communication links, we shall continue to make further progress toward the utopic Shannon limit.

8.10. Bandwidth Efficiency—Bits Per Cycle

The previous section compared signaling formats at fixed information rates and their ability to trade off performance (PE) and bit energy, at the expense of system bandwidth. By expending bandwidth, a given level of performance was achievable at the minimum possible energy level while always communicating at a fixed bit rate. Such trade-offs are important when power is at a premium while system bandwidth is readily available, as for example in a deep space communication link. There is also, however, a significant interest in links in which channel bandwidth is constrained, and power may be expendable, as in Earth-based links. In this case, the objective is to design systems that transmit the most information over the bandlimited channel, at the expense of transmitter power, while maintaining a fixed level of performance. A comparison measure for such bandlimited operation is the *bandwidth efficiency*, defined as the ratio of the bit rate transmitted to the bandwidth used. This is equivalent to the number of transmitted bits per cycle of channel bandwidth. Thus, if a system transmits bits at rate $\mathcal{R}$ bits/sec over a bandwidth of B Hz, its efficiency if $\mathcal{R}/B$ bits/cycle. In any particular system, improvement in bandwidth efficiency invariably requires an increase of transmitter power levels if the receiver performance is to be maintained. Hence, the ultimate interest here is in finding signaling schemes that achieve the most profitable trade-off of power for bandwidth efficiency.

Here again, the information theoretical work of Shannon in Section 6.8 is immediately helpful. Since channel capacity $\mathcal{C}$ is the maximum bit rate that can be sent over a channel with bandwidth B and signal power P, then the ratio of $\mathcal{C}/B$ is in fact the largest bandwidth efficiency factor that can be achieved. Thus, from (6.8.1), for any communication link,

$$\text{Bandwidth efficiency} \leq \frac{\mathcal{C}}{B} = \log_2\left(1 + \frac{P}{N_0 B}\right) \qquad (8.10.1)$$

Equation (8.10.1) is plotted in Figure 8.23 as a function of $\text{SNR} = P/N_0 B$, and indicates the maximum possible efficiency for a white Gaussian noise channel of noise level N_0. However, as we have already seen in this chapter, the theoretical performance predicted by Shannon is invariably attainable only with rather complex design. We may therefore inquire how the more practical systems of this chapter compare to each other in terms of bandwidth efficiency, and to the bound in (8.10.1).

Any digital signaling scheme that transmits $\log_2 M$ bits in T sec, using a bandwidth of B Hz always operates at a bandwidth efficiency of

$$\frac{\mathcal{R}}{B} = \frac{\log_2 M}{BT} \text{ bits per sec/Hz} \qquad (8.10.2)$$

For a fixed M, we see that signals with the smaller BT products are the most bandwidth efficient. In Section 8.3 we showed that signals with dimension ν utilize bandwidths given approximately by $B \approx \nu/T$. Hence signals with high dimension are generally not bandwidth efficient. Signals with the smallest BT product are the class of ASK signals, with $BT = 1$, and subcarrier quadrature signal sets, with $BT = 2$. The latter requires twice the bandwidth but allows separate transmissions over each quadra-

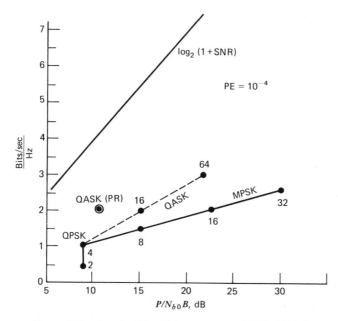

Figure 8.23. Bandwidth efficiency versus SNR ($=P/N_0B$).

ture component, as in QASK and polyphase signaling, thereby doubling the number of transmitted bits. Both QASK and MPSK, however, must increase their power level to maintain a desired PE, as is evident from (8.5.7) and (8.6.4). Points for QASK and MPSK at several values of M are superimposed in Figure 8.23 for $PE = 10^{-4}$. Note that both systems are well below the Shannon limit, while QASK benefits from improved signal detectability. Both these systems assume subcarrier bandwidths of $2/T$ Hz. By employing partial response signaling, this bandwidth can be reduced to twice the Nyquist bandwidth, thereby doubling the efficiency for QASK signals with $M = 4$, as shown. This is achieved at a slight increase in power level, and further improvements are possible by resorting to higher level partial response signals [26]. This demonstrates that any attempt to approach the Shannon limit will invariably require an increase in decoder complexity. Bandwidth efficiency curves, as in Figure 8.23, are also convenient for signal selection. For example, if we are designing a system that is to send 90 Mbits through a 45 MHz subcarrier channel, an efficiency factor of two is required. We see that 16-level MPSK, 4-level (per quadrature component) QASK, and 2-level QASK with partial response are candidates. The trade-off in required power and complexity must then be made.

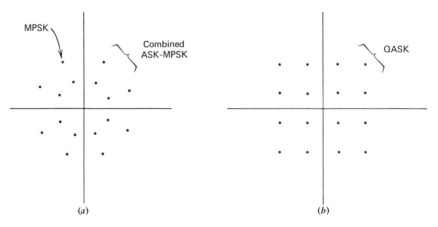

Figure 8.24. Signal space for bandwidth efficient encoding. (*a*) Combined MPSK signal sets, (*b*) QASK signal sets.

The inherent signal design problem for improved bandwidth efficiencies can also be described from our earlier signal space concepts, where the signals again correspond to vector points. The minimal number of dimensions we wish to consider is two, and the objective is to use as large a number of signal vector points M as possible, Figure 8.24. The average energy of the signals is the average distance to all vector points, while performance (PE) is related to the ability to detect the signals, which in turn is directly dependent on their relative distance apart. As we place more signals in our space, the points will necessarily be more clustered (harder to distinguish), unless we increase the energy to allow more separation. Polyphase signal points lie on circles of fixed radius (Figure 8.24*a*), while QASK signals are located at the coordinates of the quadrature amplitudes, thereby filling in squares of signal points. Combined ASK and MPSK signaling schemes can be defined from such diagrams [27–29]. We emphasize the direct contrast here with the encoding objective of Section 8.9. There the objective was to use as high a dimension (bandwidth) as possible to allow large signal separations with a given power level. In bandwidth efficiency design, the objective is to use as few dimensions as possible, while achieving sufficient signal spacing through increased power levels.

References

1. Weber, C. *Elements of Detection and Signal Design*, McGraw-Hill, New York, 1968, Chap. 14.

2. Viterbi, A. *Principles of Coherent Communications*, McGraw-Hill, New York, 1966, Sec. 8.8.

3. Balakrishnan, A. "A Contribution to the Sphere Packing Problem of Communication Theory," *Math. Anal. Appl.*, vol. 3, December 1961, pp. 485–506.

4. Weber, loc. cit., Chaps. 14–17.

5. Golomb, S., et al. *Digital Communications with Space Applications*, Prentice-Hall, Englewood Cliffs, N.J., 1964.

6. Reed, I. "A Class of Multiple Error-Correcting Codes," *IEEE Trans. Info. Theory*, vol. PGIT-4, September 1954, pp. 38–49.

7. Lindsey, W. and Simon, M. *Telecommunications System Design*, Prentice-Hall, Englewood Cliffs, N.J., 1973.

8. Rhodes, S. A. "Effect of Hardlimiting on Bandlimited Transmission with Offset QPSK Modulation," *Proc. Nat. Telecommunication Conf.*, 1972, pp. 20F/1–7.

9. Rhodes, S. A. "Effect of Noisy Phase Reference on Coherent Detection of Offset QPSK," *IEEE Trans. Comm.*, vol. COM-22, August 1974, pp. 1046–1055.

10. Gronmeyer, S. and McBride, A. "MSK and Offset QPSK Modulation," *IEEE Trans. Comm.*, vol. COM-24, August 1976, pp. 809–819.

11. Simon, M. and Smith, J. "Offset Quadrature Communications with Decision Feedback Synchronization," *IEEE Trans. Comm.*, vol. COM-22, October 1974, pp. 1576–1584.

12. Bussgang, J. "Properties of Binary Convolutional Codes Generators," *IEEE Trans. Info. Theory*, vol. IT-11, January 1965, pp. 90–100.

13. Heller, J. A. "Short Constraint Length Convolutional Codes," Jet Propulsion Lab., Pasadena, Calif., SPS 37-54 vol. III, December 1968, pp. 171–177.

14. Odenwalder, J. "Optimal Decoding of Convolutional Codes," PhD. Dissertation, Univ. of Calif. at Los Angeles, Los Angeles, Calif., 1970.

15. Layland, J. and McEliece, R. "An Upper Bound on Free Distance of a Code," Jet Propulsion Lab., Pasadena, Calif., SPS 37-62, vol. III, April 1970, pp. 63–65.

16. Viterbi, A. "Error Bounds for Convolutional Codes and Optimum Decoding Algorithm," *IEEE Trans. Info. Theory*, vol. IT-13, April 1967, pp. 260–269.

17. Viterbi, A. and Odenwalder, J. "Further Results on Optimal Decoding of Convolutional Codes," *IEEE Trans. Info. Theory*, vol. IT-15, November 1969, pp. 732–734.

18. Omura, J. "On the Viterbi Decoding Algorithm," *IEEE Trans. Info. Theory*, vol. IT-15, January 1969, pp. 177–179.

19. Wozencraft, J. and Reiffen, B. *Sequential Decoding*, Wiley, New York, 1961.

20. Reiffen, B. "Sequential Decoding for Discrete Channels," *IEEE Trans. Info. Theory*; vol. IT-8, April 1962, pp. 208–220.

21. Jacobs, I. "Sequential Decoding for Efficient Communication from Deep Space," *IEEE Trans. Comm. Tech.*, vol. COM-15, August 1967, pp. 492–501.

22. Massey, J. *Threshold Decoding*, MIT Press, Cambridge, Mass., 1963.

23. Heller, J. and Bucker, E. "Error Probability Bounds for Systematic Convolutional Codes," *IEEE Trans. Info. Theory*, vol. IT-16, March 1970, pp. 219–224.

24. Viterbi, A. "Convolutional Codes and Their Performance in Communication Systems," *IEEE Trans. Comm.*, vol. COM-19, October 1971, pp. 751–772.

25. Massey, J. "Catastrophic Error Propagation in Convolutional Codes," *Proc. Eleventh Midwest Symp. Circuit Theory*, Notre Dame, Ind., May 1968.

26. Lucky, R., Salz, J., and Weldon, E. *Principles of Data Transmission*, McGraw-Hill, New York, 1968.

27. Thomas, C. M., Weldner, M., and Durrani, S. "Digital Amplitude Phase Keying with M-ary Alphabets," *IEEE Trans. Comm.*, vol. COM-22, February 1974, pp. 168–180.

28. Salz, J., Sheehan, J., and Paris, D. "Data Transmission By Combined AM-PM," *Bell Syst. Tech. J.*, September 1971, pp. 174–193.

29. Miyauchi, K., Seki, S., and Ishio, H. "New Techniques For Generating and Detecting Multilevel Signals," *IEEE Trans. Comm.*, vol. COM-24, February 1976, pp. 263–267.

Problems

1. (8.1) Compute the baseband encoded waveform spectrum in (8.1.1) by formulating the problem as in (7.1.1), determining the equivalent of the correlation $R_m(\tau)$ in (7.1.3) and (7.1.1), and then Fourier transforming.

2. (8.1) Extend the maximum likelihood decoding rule to the block encoded case. Prove that selecting the largest y_i decodes the most probable block and also minimizes the average PWE. Follow the same argument as in Section 7.1 and Problem 7.2.

3. (8.1) Prove formally the union bound inequality; that is, prove that if $x_1, x_2, \ldots, x_m$ are random numbers, then

$$\text{Prob}\,[x_1 \leq x_2, x_3, \ldots, x_m] \leq \sum_{i=2}^{m} \text{Prob}\,[x_1 \leq x_i]$$

4. (8.1) Show how the PWE in (8.1.15) must be altered if the y_i are quantized before comparison for block decoding.

5. (8.1) Develop the general block decoder for the nonwhite noise case. Assume the design objective is to maximize the Λ of the correct decoding channel.

6. (8.2) The bandwidth of a signal set is sometimes defined as

$$B \triangleq \frac{\text{rank of } \Gamma}{2T}$$

where Γ is the signal set correlation matrix in (8.2.1). From this definition, show that for (a) an orthogonal set, $B = M/2T$; (b) a biorthogonal set, $B = M/4T$; (c) a simplex set, $B = (M-1)/2T$; (d) a polyphase set, $B = 1/2T, M = 2$, and $B = 1/T, M > 2$.

7. (8.2) Consider a block encoded system operating with a given PWE. Show that $PE = PWE/\log_2 M$ is a lower bound to the corresponding bit error probability of any block

8. (8.3) Let $D_{ij}^2 \triangleq |\mathbf{s}_i - \mathbf{s}_j|^2$ be the distance between two signals in vector space. Define d_m as the minimum D_{ij}, $i \neq j$, and $\bar{D}$ as the average D_{ij}, $\bar{D}^2 = (\sum_i^M \sum_j^M D_{ij}^2)/M^2$. (a) Show that for M signals on a sphere of radius $\sqrt{E}$,

$$\bar{D}^2 = 2E - 2|\mathbf{s}|^2$$

where $\mathbf{s}$ is the centroid of the signal set. (The vector $\mathbf{s}$ has for its qth component $\sum_{i=1}^M s_{iq}/M$, where s_{ij} is the jth component of vector $\mathbf{s}_i$). (b) Under what condition on the set is $\bar{D}^2$ maximized? (c) Show that $\bar{D}^2 \geq d_m^2 (M - 1/M)$. (d) Use (c) to prove $d_m^2 \leq 2EM/M - 1$ for any signal set.

9. (8.3) Prove that in two-dimensional space, an equal energy simplex signal set corresponds to the three vertices of a triangle inscribed within a circle.

10. (8.4) In a binary system a block of k information bits is encoded into n encoder bits by an (n, k) code that has the capability of detecting up to d transmitted bit errors. Assuming that words detected in error are not used, what is the PWE of the original k information bits?

11. (8.4) Substitute (8.4.9) into (8.4.6) and prove (8.4.10) by taking the limit as $M \to \infty$. First prove that

$$\lim_{M \to \infty} [(M-1) \log_e [1 - \tfrac{1}{2} \text{Erfc}(x + \beta \log_2 M)]] = \begin{cases} -\infty, & \text{if } \beta < \log_e 2 \\ 0, & \text{if } \beta > \log_e 2 \end{cases}$$

12. (8.4) If a constant voltage c is added to every decoder output sample y_i in block decoding, the same PWE will occur since the c will not affect which is maximum. Consider an equally correlated signal set $\{s_i(t)\}$ with correlation γ and energy E. Let c have the value $c = \int_T^{T+\delta} An(t)\, dt + A^2 \delta$ for some δ. (a) Show that $(y_i + c)$ would be the decoder output for a signal set

$$s_i'(t) = \begin{cases} s_i(t) & 0 \leq t \leq T \\ A & T \leq t \leq T + \delta \end{cases}$$

(b) Show this new signal set, defined over $(0, T+\delta)$ has correlation $\gamma_{ii} = E + A^2\delta$ and crosscorrelation $\gamma_{ij} = E\gamma + A^2\delta$. (c) Show that for the choice $A^2\delta = -E\gamma$ the new set is an orthogonal set with energy $E(1-\gamma)$. (d) What can we conclude about the PWE of the two signal sets $\{s_i(t)\}$ and $\{s_i'(t)\}$?

13. (8.5) Derive the MSK result in (8.5.17) by expanding (8.5.15) and (8.5.16). Identify the phase angle ψ_{i1} and ψ_{i2}.

14. (8.6) A noncoherent decoder for MFSK can be constructed as follows. During each word time $(0, T)$ we time sample the decoder input waveform, $x(t)$ generating the N voltage samples $x(i/2B)$. We then compute the discrete Fourier transform function

$$\sum_{i=1}^{N} x\left(\frac{i}{2B}\right) e^{-j(2\pi i q/2BT)}$$

for different q. Explain how and why this can noncoherently decode the MFSK signal set.

15. (8.7) A pulse is received with one of M amplitudes $\{s_i\}$ and width T, along with additive white Gaussian noise. The noisy signal is integrated for T sec producing a voltage y. (a) Show that [prob pulse amplitude $= s_i | y$] $\geq$ prob [pulse amplitude $= s_j | y$] if $(s_i y - s_i^2 T) \geq (s_j y - s_j^2 T)$. (b) Show that this is equivalent to determining if y is closer to $s_i T$ or to $s_j T$. (c) Use these facts to define a maximum likelihood decoding procedure for ASK signaling.

16. (8.9) In Figure 8.19b all succeeding columns of the trellis have the same diagram. From this point on, therefore, the trellis can be represented by a *state diagram*, in which the nodes are the states and the interconnecting trellis branches are the state connections. Draw the state diagram for Figure 8.19b.

CHAPTER 9

FREQUENCY ACQUISITION, PHASE REFERENCING, AND TIMING

In our earlier discussions of RF carrier processing we tacitly assumed that the received RF frequency was known fairly accurately at the receiver. This allowed the RF front end filter to be properly tuned to the carrier and allowed RF mixing with a local oscillator to position the received carrier directly in the proper IF frequency band. The advantage of this is that the IF filter bandwidth can be made as narrow as possible, reducing the total noise involved in the demodulation processing. In many forms of practical systems, however, frequency instabilities and Doppler effects during transmission may prevent knowledge of the RF frequency location when communication first begins, and system requirements may prevent manual local frequency tuning. This is particularly true in deep space communications with moving satellites and in short burst communications to high speed aircraft. Note that even a 10% uncertainty in an RF frequency at *S*-band translates to an uncertainty of several hundred megahertz—several times larger than most IF bandwidths and much too large to allow rapid manual local frequency searching. For this reason, certain types of systems must perform an automatic *frequency acquisition* and *tracking* operation prior to any communication transmission. By frequency acquisition we refer to the operation of internally generating a local frequency identical to, or a fixed distance away from, the actual frequency of the received carrier. By frequency tracking we refer to the operation of continually maintaining this local frequency in synchronism with the received frequency, even though the latter may be undergoing variation during communication. Once acquisition has been successful and tracking initiated, communication can begin, and our analysis in Chapters 2–4 is then applicable.

In addition to the mandatory frequency synchronization of the RF and IF stages, there may be an additional requirement for *phase referencing* in the receiver. By phase referencing we refer to the general operation of providing a local carrier frequency in exact, or almost exact, phase synchronism with a received carrier (or subcarrier) at the same frequency. We have shown in earlier analysis several applications where phase synchronism allows specific performance advantages. Phase referencing may be desired at the RF, IF, or subcarrier levels, depending on system implementation. Phase referencing may be obtained simultaneously for all three stages from a single phase referencing subsystem or may be obtained individually at each stage by separate subsystems.

9.1. Frequency Acquisition

In communication systems using RF-IF conversion it is necessary that the received and local carrier frequencies be within the IF bandwidth of each other. If the frequencies are too widely separated for the IF and no provision is made to reduce the separation, no IF processing will take place. The operation of aligning these frequencies is called frequency acquisition. Frequency acquisition can be achieved by manually adjusting the local frequency or can be achieved by an automatic frequency acquisition loop (FAL).

A typical FAL block diagram is shown in Figure 9.1. The loop uses a single oscillator, called a *master* oscillator, and multiplies or divides its frequency to the RF and IF. The RF is used to mix the received carrier from the antenna to the IF stage, which is then mixed with the IF to a tracking frequency. The master oscillator is then translated to the tracking frequency, where it is phase detected with the received carrier in a phase lock loop (PLL). When in lock, the PLL forces the master oscillator to operate at the required frequency for maintaining the RF, IF, and tracking frequencies. This is referred to as a *double conversion* FAL, or simply called a "long" loop. When the FAL alone is used for acquiring the RF frequency, the operation is called *quiescent acquisition*. However, in some instances the master oscillator and received frequency may be so far apart that no quiescent acquisition can take place. In this case a *slewing* signal must be applied to the loop oscillator to force the frequencies into near alignment, from which the automatic tracking can take place. This slewing

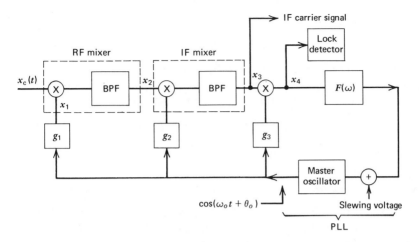

Figure 9.1. RF frequency acquisition loop block diagram.

operation is called *frequency searching*. With frequency searching often an auxiliary circuit called a *lock detector* is provided for recognizing when the frequencies are close to being aligned. The lock detector disengages the slewing operation and allows the PLL tracking to be initiated. In typical systems the lock detector monitors the PLL error voltage and uses a threshold test for lock-in recognition.

To design an acquisition loop properly, it is necessary to understand its basic quiescent operation. To determine these, we write the FAL loop equations throughout the system, where signals refer to points in Figure 9.1. Let the received RF signal be written as

$$x_c(t) = A \sin [\omega_c t + \theta(t) + \psi(t)] + n(t) \tag{9.1.1}$$

as in Chapter 3. Here A is the RF carrier amplitude, $\theta(t)$ represents phase or frequency modulation, and $n(t)$ is the RF noise. We assume only an angle modulated carrier, but the result extends to AM if we interpret (9.1.1) as the carrier component only. The function $\psi(t)$ again accounts for the extraneous phase or frequency variation and is now important to our discussion. If the received carrier has a frequency offset from the desired RF frequency ω_c, it must be represented by $\psi(t)$. Thus if the frequency offset is Ω rps, then in (9.1.1)

$$\psi(t) = \Omega t \tag{9.1.2}$$

If the frequency is varying or if oscillator phase noise is present, the analytical expression for $\psi(t)$ must be suitably modified. We denote the master oscillator signal as

$$x_0(t) = \cos\left[\omega_o t + \theta_o(t)\right] \tag{9.1.3}$$

where the amplitude level is absorbed into the mixer gain. The master oscillator output is frequency translated (multiplied or divided) by the frequency factor g_1 to the RF, producing the RF mixer signal

$$x_1(t) = \cos\left[g_1 \omega_o t + g_1 \theta_o(t)\right] \tag{9.1.4}$$

according to our discussion in Section 3.8. The output of the RF mixer is then

$$x_2(t) = K_{m_1} A \sin\left[(\omega_c - g_1 \omega_o)t + \theta + \psi - g_1 \theta_o\right] + n_{IF}(t) \tag{9.1.5}$$

where K_{m_1} is the mixer gain, $n_{IF}(t)$ is the mixed IF noise, and the t has been dropped from the phase functions. Recall that if the RF bandwidth is much wider than the bandwidth of θ_o in (9.1.3), the noise $n_{IF}(t)$ is simply a frequency shifted version of the RF bandpass noise. Simultaneously, the master oscillator is frequency translated by g_2 to provide the mixing signal for converting the IF carrier to the PLL tracking frequency. The loop input is then

$$x_3(t) = K_{m_1} K_{m_2} A \sin\left[[\omega_c - (g_1 + g_2)\omega_o]t + \theta + \psi - (g_1 + g_2)\theta_o\right] + n_3(t) \tag{9.1.6}$$

If we now desire the PLL to force the master oscillator to track the carrier in (9.1.6), it is necessary that $g_3 \omega_o = [\omega_c - (g_1 + g_2)\omega_o]$, or

$$g_1 + g_2 + g_3 = \frac{\omega_c}{\omega_o} \tag{9.1.7}$$

In this case the PLL has a mixer output given by

$$x_4(t) = K_m A \sin\left[\theta + \psi - (g_1 + g_2 + g_3)\theta_o\right] + n_m(t) \tag{9.1.8}$$

where $K_m = K_{m_1} K_{m_2} K_{m_3}$ is the mixer gain product and $n_m(t)$ is the low frequency mixer noise process described in Section 3.8. Inserting (9.1.7) yields

$$x_4(t) = K_m A \sin\left[\theta + \psi - \left(\frac{\omega_c}{\omega_0}\right)\theta_o\right] + n_m(t) \tag{9.1.9}$$

Comparison of (9.1.9) with the basic PLL system equation [cf. (4.6.4)] shows that this tracking phase error function is identical to that generated in the equivalent system shown in Figure 9.2. The latter is a modified PLL equivalent system containing the actual PLL low pass filter, the total FAL mixer gains, and an effective phase multiplication by the ratio ω_c/ω_o. The

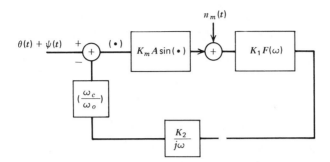

Figure 9.2. The equivalent FAL system.

loop is nonlinear in that it involves the sine of the phase tracking error. Thus the double conversion RF acquisition loop of Figure 9.1 is equivalent in its tracking capability to a modified, nonlinear PLL. Note that the loop does not depend on the actual values of the frequency translation parameters g_1, g_2, and g_3. This means the FAL can be generalized to other implementations, corresponding to different combinations of these parameters. For example, letting $g_3 = 1$ corresponds to phase tracking directly at the master oscillator frequency ω_o. Letting $g_2 = 0$ corresponds to eliminating the intermediate IF frequency mixing and converting the system to a single convertion FAL. Letting $g_1 = 0$ produces an FAL that frequency locks to the IF instead of the RF. In all cases the performance is governed by the loop model of Figure 9.2. The selection of the master oscillator frequency ω_o is generally based on hardware considerations. The lower its value, the less it must be deviated for RF frequency acquisition, but the larger the frequency multiplication to RF that must be provided.

To evaluate acquisition behavior, we need only study the equivalent system. Note that the effective loop input is the RF carrier phase variation $\theta(t) + \psi(t)$. The RF phase variation tracking error generated in the FAL is then defined as

$$\theta_e(t) \triangleq \psi(t) - \left(\frac{\omega_c}{\omega_o}\right)\theta_o(t) \qquad (9.1.10)$$

The corresponding RF frequency error in radians per second is then

$$\omega_e(t) \triangleq \frac{d\theta_e}{dt} = \frac{d}{dt}\left[\psi(t) - \left(\frac{\omega_c}{\omega_o}\right)\theta_o(t)\right] \qquad (9.1.11)$$

In our earlier analysis of tracking loops we assumed the phase error $\theta_e(t)$ was small, which linearized the loop. During frequency acquisition, however, we cannot initially impose this assumption. In fact, we are primarily interested in the quiescent locking procedure specifically, when the frequency separation is large. In this case we must treat the FAL as a true nonlinear system, and evaluate its dynamical behavior from the basic nonlinear differential equation describing its loop. The equation can be easily generated from (9.1.11), using the fact that

$$\frac{d\theta_o}{dt} = K_1 K_2 F\{AK_m \sin(\theta + \theta_e) + n_m(t)\}$$

$$= AK_m K_1 K_2 F\{\sin(\theta + \theta_e)\} + K_1 K_2 F\{n_m(t)\} \qquad (9.1.12)$$

Here $F\{x(t)\}$ represents the output time function of the loop filter $F(\omega)$ when its input is $x(t)$, K_2 is the sensitivity gain of the master oscillator, and K_1 is the gain of the loop filter. When (9.1.12) is inserted into (9.1.11) we obtain the basic system differential equation

$$\frac{d\theta_e}{dt} + AKF\{\sin(\theta + \theta_e)\} = \frac{d\psi}{dt} - K_1 K_2 \left(\frac{\omega_c}{\omega_o}\right) F\{n_m(t)\} \qquad (9.1.13)$$

where K is now the total component gain around the entire loop,

$$K = K_m K_1 K_2 \left(\frac{\omega_c}{\omega_o}\right)$$

$$= K_{m_1} K_{m_2} K_{m_3} K_1 K_2 \left(\frac{\omega_c}{\omega_o}\right) \qquad (9.1.14)$$

where it is to be understood that a single conversion FAL would have the appropriate mixer gain (K_{m_1} or K_{m_2}) deleted. Frequency acquisition therefore reduces to simply an investigation of the nonlinear equation in (9.1.13). The forcing function of the system is the extraneous phase variation $\psi(t)$, relative to the desired RF carrier frequency, and the filtered mixer noise. Note the manner in which the carrier modulation influences behavior, effectively appearing as a variable coefficient in the system equation. To handle the modulation term analytically it is convenient to expand the sine term as

$$\sin(\theta + \theta_e) = \sin(\theta_e(t))\cos(\theta(t)) - \cos(\theta_e(t))\sin(\theta(t)) \qquad (9.1.15)$$

If the $\theta(t)$ corresponds to a modulating subcarrier with amplitude Δ and frequency ω_s outside the bandwidth of the loop filter, it is best to expand (9.1.15) using Bessel functions, as in (5.4.5). The low pass filtering operation of F in (9.1.13) then passes only the zero order term, yielding

$$AKF\{\sin(\theta_e + \theta)\} \cong AKJ_0(\Delta)F\{\sin\theta_e\} \qquad (9.1.16)$$

In this case the modulation merely suppresses the carrier amplitude A, effectively changing its value to

$$A_c = AJ_0(\Delta) \qquad (9.1.17)$$

This is simply the amplitude of the carrier component of (9.1.1), and we see that only this amplitude enters the system differential equation. This reflects the fact that only the carrier component of the RF signal is being tracked by the loop. The system equation in (9.1.13) becomes

$$\frac{d\theta_e}{dt} + A_c KF\{\sin(\theta_e)\} = \frac{d\psi}{dt} - \left(\frac{K}{K_m}\right)F\{n_m(t)\} \qquad (9.1.18)$$

If the modulation $\theta(t)$ is composed of a sum of N modulating subcarriers, with deviations Δ_i, respectively, then (9.1.17) is instead

$$A_c = A \prod_{i=1}^{N} J_0(\Delta_i) \qquad (9.1.19)$$

If the RF carrier modulation was a unit level binary waveform (e.g., a binary code or digital PCM waveform), modulated directly onto the carrier with phase deviation $\pm\Delta_c$, then (9.1.15) becomes, from (2.3.7),

$$r(t)\sin \Delta_c \cos \theta_e(t) - (\cos \Delta_c)\sin(\theta_e(t)) \qquad (9.1.20)$$

Since the first term will typically have time variations outside the loop filter bandwidth, only the second term affects the loop equation. This means A is effectively converted to the value

$$A_c = A \cos \Delta_c \qquad (9.1.21)$$

This again corresponds to the suppressed carrier amplitude due to the binary waveform. Lastly, if $\theta(t)$ is a random modulating waveform, then $\cos \theta(t)$ and $\sin \theta(t)$ in (9.1.15) must be treated as a random process, requiring an analysis similar to Section 2.3. Taking only the constant portion of the $\cos \theta(t)$ term, and assuming $\theta_e(t)$ is small, we have

$$\sin[\theta(t) + \theta_e(t)] \approx |\Psi_\theta(1)|\sin \theta_e(t) - \sin \theta(t) \qquad (9.1.22)$$

where $\Psi_\theta(\omega)$ is the first order characteristic function of $\theta(t)$. This means we can write

$$A_c = A|\Psi_\theta(1)| \qquad (9.1.23)$$

and the second term in (9.1.22) can be incorporated with the right side of (9.1.18), forming a new effective noise term

$$n_m(t) + (A_c K)\sin \theta(t) \qquad (9.1.24)$$

In this case the random carrier modulation has suppressed the carrier component, as we know from (2.3.16), while adding an effective noise interference to the desired loop tracking operation.

In summary, we see that any modulation superimposed on the carrier during frequency acquisition acts to reduce the RF carrier amplitude A to

Table 9.1. Summary of Effective Carrier Amplitudes

Modulation Format	A_c
No modulation	A
Subcarrier, deviation Δ	$AJ_0(\Delta)$
K subcarriers, deviation Δ_i	$A \prod_{i=1}^{K} J_0(\Delta_i)$
Binary waveform, deviation Δ_c	$A(\cos \Delta_c)$
Binary waveform (Δ_c) plus K subcarriers, (Δ_i)	$A(\cos \Delta_c) \prod_{i=1}^{K} J_0(\Delta_i)$
Random, char. func. $\Psi_\theta(\omega)$	$A\|\Psi_\theta(1)\|$

A = RF carrier amplitude.
$A^2/2$ = RF carrier power.

an effective value of A_c in the loop dynamical equation. The various forms of A_c are summarized for convenience in Table 9.1. These values of A_c are merely the amplitudes of the RF carrier component, and the discussion here has shown that only this component is being used for frequency acquisition.

9.2. Dynamical Analysis of Frequency Acquisition

We wish to examine the frequency acquisition behavior described by (9.1.13). Initially we concentrate only on the properties of the actual frequency synchronization itself and neglect the effect of the added noise $n_m(t)$. We initially assume no modulation term $\theta(t)$, contending that such modulation will be placed outside the filter bandwidth or that no modulation will be transmitted during the frequency acquisition operation. We assume the received and local RF frequencies differ by Ω rps, as in (9.1.2), so that

$$\frac{d\psi}{dt} = \Omega \tag{9.2.1}$$

Under these conditions, quiescent frequency acquisition is described by the differential equation

$$\frac{d\theta_e(t)}{dt} + A_c KF\{\sin \theta_e(t)\} = \Omega \tag{9.2.2}$$

Let us consider first the case of a simple first order loop, where the filter is effectively removed, that is, $F\{\sin \theta_e\} = \sin \theta_e$. In this case (9.2.2) becomes

$$\frac{d\theta_e(t)}{dt} + A_cK \sin \theta_e(t) = \Omega \qquad (9.2.3)$$

The frequency error of the FAL, as given in (9.1.11), is then given specifically by

$$\omega_e(t) = \Omega - A_cK \sin (\theta_e(t)) \qquad (9.2.4)$$

To determine frequency behavior, we must solve (9.2.3), for $\theta_e(t)$ and insert the solution in (9.2.4). We immediately note a basic property of the solution. If the frequency error $\omega_e(t)$ is to be driven to zero (i.e., the frequency acquisition is to occur), then it is necessary that $\theta_e(t)$ approach a final value θ_{e0} such that $\Omega - A_cK \sin \theta_{e0} = 0$, or

$$\theta_{e0} = \sin^{-1}\left(\frac{\Omega}{A_cK}\right) + i2\pi, \qquad i = 0, 1, 2 \ldots \qquad (9.2.5)$$

This, however, has a solution only if $\Omega/A_cK \le 1$. Since the loop noise bandwidth of a first order loop is given by $B_L \triangleq A_cK/4$ (see Table 4.1) we can restate this condition as

$$\Omega \le 4B_L \qquad (9.2.6)$$

Thus a first order frequency tracking loop eventually acquires an RF frequency offset Ω rps only if (9.2.6) is satisfied. This requires that Ω be within the loop noise bandwidth. If the frequency offset is too large, no steady state acquisition is achieved. If (9.2.6) is satisfied, the system achieves a steady state of zero frequency error and an RF phase error given (9.2.5). This means the RF frequencies have been brought into synchronism and the local RF carrier tracks the received RF carrier with the phase error θ_{e0}. It is interesting to compare this result with our earlier conclusion in Table 5.1 concerning linear PLL. There we concluded that a first order loop always tracks a frequency offset Ω rps with a phase error of Ω/A_cK rad. We now see that this earlier result is valid only if $\Omega/A_cK \ll 1$, which, of course, is the condition that the loop be linear.

Actual loop behavior is best illustrated by examining the relation between the frequency variable ω_e and the phase error θ_e, as given by (9.2.4). A plot of the equation $\omega_e = \Omega - A_cK \sin \theta_e$ is shown in Figure 9.3, indicating the behavior of the instantaneous value of ω_e as a function θ_e. In nonlinear analysis the plot is called a *phase plane* diagram, and equivalently illustrates how θ_e and its derivative are related. The arrows on the curve indicate directions of phase variation with time, since they correspond to positive and negative derivatives of $\theta_e(t)$. At any instant of

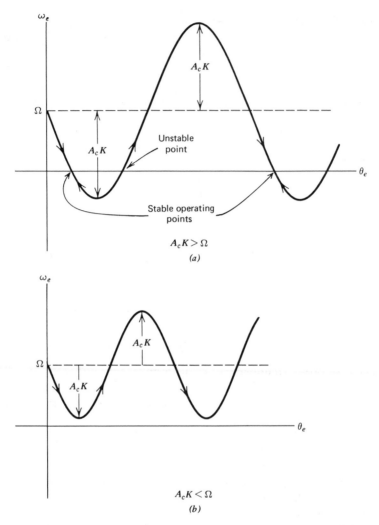

Figure 9.3. Phase plane diagram for first order frequency tracking loop. (a) $A_c K \geq \Omega$, (b) $A_c K \leq \Omega$.

time, $\theta_e(t)$ and the corresponding value of $\omega_e(t)$ can be located on the diagram. As $\theta_e(t)$ traces out its solution with time [obtained by solving (9.2.3)] along the abscissa, following the arrow directions, $\omega_e(t)$ traces out its behavior along the ordinate. If (9.2.6) is satisfied, the phase plane plot crosses the $\omega_e = 0$ axis at the steady state values of (9.2.5). In this case the solution trajectory begins at a point on the curve corresponding to the initial condition of $\theta_e(t)$ and moves according to the arrows to the nearest solution point of (9.2.5). Note that not all solution values of (9.2.5) become steady state values for θ_{e0}, since the arrows force operation away from certain solution points. (The latter points are called *unstable* solution points.) When (9.2.6) is not satisfied, as shown in Figure 9.3b, no $\omega_e = 0$ solution points exist, and the solution for $\theta_e(t)$ slides continually to the right, increasing with time. This means the phase error is undergoing continuous 2π phase changes, and the loop is said to be *slipping cycles*. The error frequency $\omega_e(t)$ therefore hovers about Ω, implying that the FAL is not approaching frequency acquisition.

An important parameter in frequency acquisition is the time required to achieve synchronization, the latter called the *acquisition time* of the system. To determine acquisition time for nonlinear systems, it is necessary to know how fast we move along the phase plane trajectories. For our first order system this requires a specific solution for $\theta_e(t)$ in (9.2.3). The equation is a classical first order nonlinear differential equation, often called the *"pendulum equation,"* since it describes the angle motion of a hanging pendulum. When (9.2.6) is satisfied, solutions exist in the form of log functions (Problem 9.3.). Although the solution approaches the steady state only asymptotically, the phase error traces through a cycle in Figure 9.3a in a time period of approximately several multiples of $A_c K$ [1, 2]. Thus first order loop acquisition time is about $5/B_L$ sec.

In general, acquisition loops have some type of loop filtering. Let us extend our discussion to the case where

$$F(\omega) = \frac{j\omega\tau_2 + 1}{j\omega\tau_1 + 1} \qquad (9.2.7a)$$

and

$$\tau_1 = \frac{A_c K}{\omega_n^2} \qquad \tau_2 = \frac{2\zeta}{\omega_n} - \frac{1}{A_c K} \qquad (9.2.7b)$$

corresponding to a second order loop model and a second order differential equation in (9.2.2). As $\tau_1 \to \infty$, (9.2.7) approaches an integrator, and the loop behaves as the double integrating loop discussed in Table 4.2. To determine the system equation we convert (9.2.7) to a differential operator in which the filter output is related to its input by

$$\left(1 + \tau_1 \frac{d}{dt}\right) \text{output} = \left(1 + \tau_2 \frac{d}{dt}\right) \text{input} \qquad (9.2.8)$$

If we operate on both sides of (9.2.2) with $(1 + \tau_1 d/dt)$ and insert (9.2.8), we have

$$\left(1 + \tau_1 \frac{d}{dt}\right)\frac{d\theta_e}{dt} + A_c K\left(1 + \tau_2 \frac{d}{dt}\right)\sin\theta_e = \left(1 + \tau_1 \frac{d}{dt}\right)\Omega \qquad (9.2.9)$$

Expanding and taking derivatives yields the second order system equation

$$\tau_1 \frac{d^2\theta_e}{dt^2} + (1 + A_c K \tau_2 \cos \theta_e) \frac{d\theta_e}{dt} + A_c K \sin \theta_e = \Omega \qquad (9.2.10)$$

The preceding equation must be solved for $\theta_e(t)$, and the frequency error is then obtained by differentiating $\theta_e(t)$. We immediately see from (9.2.10) that if $\omega_e(t) = 0$ is to be a solution, it is necessary that $\theta_e(t)$ have a steady state value θ_{e0} such that $A_c K \sin \theta_e = \Omega$, or

$$\theta_{e0} = \sin^{-1}\left(\frac{\Omega}{A_c K}\right) + i2\pi, \qquad i = 0, 1, 2, \ldots \qquad (9.2.11)$$

No such solution can exist unless†

$$\Omega \le A_c K \qquad (9.2.12)$$

We again see a limitation to the maximum frequency offset Ω that can possibly be acquired by quiescent frequency acquisition. This maximum frequency offset depends on the loop gain, and increasing K extends this acquisition range. Since acquisition loops are typically designed with fixed natural frequencies ω_n, we can rewrite (9.2.12) from (9.2.7b),

$$\Omega \le \tau_1 \omega_n^2 \qquad (9.2.13)$$

We now see that for $\tau_1 \to \infty$, the frequency range becomes infinite, and the FAL has the capability of acquiring any RF frequency. To further examine second order loop acquisition let us consider specifically the case where $\tau_1 \gg 1$. By dividing (9.2.10) by τ_1 and letting $\tau_1 \to \infty$, the system equation becomes

$$\frac{d^2\theta_e}{dt^2} + (2\zeta\omega_n \cos \theta_e) \frac{d\theta_e}{dt} + \omega_n^2 \sin \theta_e = 0 \qquad (9.2.14)$$

To determine the frequency-phase behavior (phase plane trajectories), we note from (9.2.14) that since $\omega_e = d\theta_e/dt$,

$$\frac{d\omega_e}{d\theta_e} = \frac{d\omega/dt}{d\theta_e/dt} = -(2\zeta\omega_n) \cos \theta_e - \frac{\omega_n^2 \sin \theta_e}{\omega_e} \qquad (9.2.15)$$

The functional behavior of ω_e with θ_e is obtained from exact solution of (9.2.15). Note that for large values of ω_e (9.2.15) has the approximate behavior

$$\omega_e \approx -2\zeta\omega_n \sin \theta_e + \text{constant} \qquad (9.2.16)$$

---

† Although a solution cannot exist if (9.2.12) is not satisfied, acquisition may not necessarily occur if it is. It has been shown [6] that acquisition actually requires $\Omega \le A_c K [4\zeta\omega_n/K]^{1/2}$. The latter is referred to as the *pull in range* of the acquisition loop.

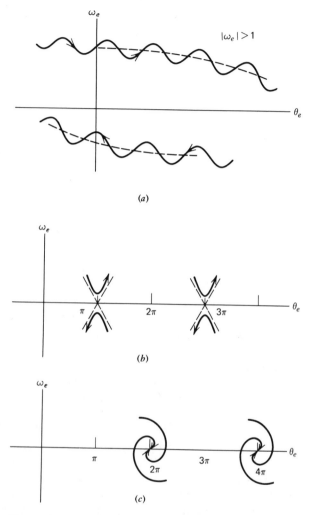

Figure 9.4. Phase plane trajectories for second order loop. (a) $\omega_e \gg 1$, (b) $\omega_e \approx 0$, unstable saddle point, (c) $\omega_e \approx 0$, spiral node.

and therefore varies sinusoidally with θ_e (Figure 9.4a), following the upper trajectory to the right and the lower to the left. To determine the amount of frequency change that will occur, we multiply both sides of (9.2.15) by $\omega_e d\theta_e$ and integrate over an integer number of cycles of θ_e, yielding

$$\omega_e^2(\theta_{e_2}) - \omega_e^2(\theta_{e_1}) = -2\zeta\omega_n \int_{\theta_{e_1}}^{\theta_{e_2}} \omega_e(\theta) \cos\theta \, d\theta \qquad (9.2.17)$$

since the last term in (9.2.15) will integrate to zero. Here θ_{e_1}, θ_{e_2} are the initial and final phase angles. From integrating by parts, we can establish

$$\int_{\theta_{e_1}}^{\theta_{e_2}} \omega_e(\theta) \cos \theta \, d\theta = \int_{\theta_{e_1}}^{\theta_{e_2}} \sin \theta \, d\omega_e(\theta) \tag{9.2.18}$$

and from (9.2.15)

$$\int_{\theta_{e_1}}^{\theta_{e_2}} \sin \theta \, d\omega_e = -2\zeta\omega_n \int_{\theta_{e_1}}^{\theta_{e_2}} \cos \theta \sin \theta \, d\theta - \omega_n^2 \int_{\theta_{e_1}}^{\theta_{e_2}} \frac{\sin^2 \theta}{\omega_e(\theta)} d\theta \tag{9.2.19}$$

Since the first term on the right is zero, we use (9.2.17), (9.2.18), and (9.2.19) to conclude

$$\omega_e^2(\theta_{e_2}) - \omega_e^2(\theta_{e_1}) = -2\zeta\omega_n^3 \int_{\theta_{e_1}}^{\theta_{e_2}} \frac{\sin^2 \theta}{\omega_e(\theta)} d\theta \tag{9.2.20}$$

Although (9.2.20) does not easily produce a solution, we do note the following. For any $\omega_e > 0$, the right-hand side is negative, which means ω_e must decrease as θ_e increases. For $\omega_e < 0$, ω_e becomes less negative as θ_e decreases. Hence as ω_e follows the trajectories of Figure 9.4a, it must decrease in magnitude, and the phase plane trajectories must continually converge toward the abscissa, as shown.

When ω_e approaches zero, its solution behavior is governed by the *singular points* [3, 4] of (9.2.15). (Singular points are the phase plane values where $d\omega_e/d\theta_e = 0/0$.) Singular points of (9.2.15) occur for $\omega_e = 0$ and

$$\theta_{e0} = i\pi, \qquad i = 0, 1, 2, \ldots \tag{9.2.21}$$

To determine solutions in the vicinity of such points, (9.2.15) can be linearized to form the equation $d\omega_e/d\theta_e \approx \omega_n^2(\theta_e - \theta_{e0})/\omega_e$. From this it can be established [3] that for i odd, an unstable saddle point (Figure 9.4b) exists, whereas for i even a stable spiral node (Figure 9.4c) exists. Thus as the frequency error is reduced and the system approaches the singular points of (9.2.21), the points for i odd represent unstable points from which solution trajectories are reflected away, whereas for i even solution trajectories spiral in. The steady state solutions for the system therefore correspond to $\omega_e = 0$ and $\theta_{e0} = 0, 2\pi, 4\pi$, and so on. Thus frequency acquisition is achieved, starting from an initial large frequency offset Ω rps, by following the oscillating trajectories of Figure 9.4a in which the frequencies are slowly pulling together. During this time the phase is continually changing by 2π rad and the loop is slipping cycles. As the trajectories approach the abscissa, the cycle slipping terminates and the loop spirals into its stable position.

The time to acquire with this system is determined by the amount of cycle slipping that occurs and the time needed to achieve the final phase state after the frequency difference has been reduced to zero. The latter phase pull in time is on the order of several loop time constants and is therefore approximately $2/\omega_n$ sec, where ω_n is the loop natural frequency. The time needed to reduce the frequency error can be obtained from (9.2.20). Setting $\theta_{e_1} = 0$, $\omega_e(0) = \Omega$, and $\omega_e(\theta_{e_2}) = 0$ shows that

$$\Omega^2 = 2\zeta\omega_n^3 \int_0^{\theta_{e_2}} \frac{\sin^2\theta}{\omega_e(\theta)} \, d\theta \qquad (9.2.22)$$

where θ_{e_2} is the accumulated phase shift during the frequency acquisition. To approximate the integral, we replace $\sin^2\theta$ by its maximum value and

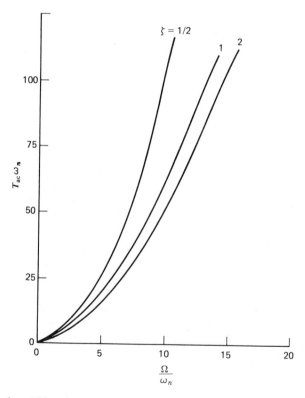

Figure 9.5. Acquisition time versus initial frequency offset Ω (ω_n = natural frequency, ζ = loop damping factor).

recall $\omega_e = d\theta_e/dt$ so that

$$\Omega^2 \cong 2\zeta\omega_n^3 \int_0^{\theta_{e_2}} \frac{d\theta}{d\theta/dt} = 2\zeta\omega_n^3 \int_0^{T_{\text{ac}}} dt$$

$$= 2\zeta\omega_n^3 T_{\text{ac}} \qquad (9.2.23)$$

where T_{ac} is the frequency acquisition time (time to accumulate the θ_{e_2} phase shift). Hence we have

$$T_{\text{ac}} \cong \frac{\Omega^2}{2\zeta\omega_n^3} \qquad (9.2.24)$$

We see that although the loop can theoretically acquire any frequency offset Ω, we see that the quiescent frequency acquisition time increases as Ω^2. Equation (9.2.24) is plotted in Figure 9.5 for several values of damping factor ζ. Note that if the initial offset is significantly outside the loop bandwidth, an appreciably long time may occur before acquisition is achieved. We also see that the acquisition time varies inversely as the cube of the loop natural frequency, suggesting that loop bandwidths should be made extremely large. Although this indeed is the preferable design, the actual bandwidth size must be traded against the noise effects within the loop, which we have temporarily neglected in our current study.

9.3. Frequency Searching with Oscillator Slewing

If the initial frequency uncertainty Ω is large, we have found that quiescent frequency acquisition may require a long time or may not be accomplished at all. In these cases, a slewing voltage must be inserted at the oscillator input (Figure 9.1) to aid the acquisition operation. This slewing voltage is in the form of a linearly increasing voltage such that if the FAL loop were opened, its RF frequency would increase as βt Hz. The parameter β is called the RF slewing rate and has units of rps/sec. In addition, β may be positive or negative, depending on the direction in which we slew. We are interested in determining the effect of a slewing voltage on the frequency acquisition operation.

We first note that inserting a slewing voltage with rate β in the loop is identical to inserting a phase signal $\beta t^2/2$ in the RF output phase $(\omega_c/\omega_o)\theta_o(t)$. This in turn adds a signal $-\beta t^2/2$ to the phase tracking error, which is identical to adding the same term to the forcing function $\psi(t)$ in Figure 9.2. This alters the system differential equation in (9.2.2) by adding the same term, $-\beta t^2/2$, to the right-hand side. It then becomes

$$\frac{d\theta_e}{dt} + A_c KF\{\sin\theta_e\} = \Omega - \beta t \qquad (9.3.1)$$

The slewing term βt adds to or subtracts from the original value of Ω, depending on the direction of the slewing signal. For a second order loop with $\tau_1 \gg 1$, the system equation now becomes

$$\frac{d^2\theta_e}{dt^2} + (2\zeta\omega_n \cos\theta_e)\frac{d\theta_e}{dt} + \omega_n^2 \sin\theta_e = -\beta t \qquad (9.3.2)$$

If we recompute (9.2.20) with the slewing inserted, and integrate over an integer number n of phase error cycles, we find

$$\omega_e(n2\pi) - \omega_e(0) = 2\zeta\omega_n \int_0^{2n\pi} \frac{-\beta - \omega_n^2 \sin^2\theta}{\omega_e(\theta)} d\theta \qquad (9.3.3)$$

We see that if the frequency error is positive, it will not decrease during n phase cycles if β is negative (in the opposite direction of $\omega_e(0)$) and if $|\beta| > \omega_n^2$. That is, if the slewing rate is too fast and in the wrong direction, the frequency error is continually increased and the FAL cannot acquire the RF carrier frequency in spite of two loop integrations. If $|\beta| < \omega_n^2$, there exists the possibility of frequency acquisition even though we are slewing in the wrong direction. In this case the loop dynamics may be sufficient to acquire the RF carrier by internally compensating for the slewing signal.

When the slewing is in the proper direction, the frequency error will be reduced even without the loop dynamics. In this case the original frequency uncertainty Ω, no matter how large, will eventually be driven toward zero by the slewing operation. The important dynamical question concerns the behavior of the loop as the instantaneous frequency error passes through zero and the slewing continues to be applied. To determine this, we resort to our linear theory and examine linear loop behavior as the phase input forcing function has the form $\beta t^2 / 2$. The linear system equivalent to (9.3.2), obtained by setting $\cos\theta_e = 1$ and $\sin\theta_e = \theta_e$, will have the solution

$$\theta_e(t) = \left(\frac{\beta}{K}\right)t + \frac{\beta}{\omega_n^2} \qquad (9.3.4)$$

This implies that an offset RF frequency error of β/K rad/sec will exist in the FAL. To reduce this to zero it is necessary to remove the slewing signal (i.e., set $\beta = 0$) as the loop passes through the zero error position. This removal is provided by the lock detector subsystem in Figure 9.1. Although we do not pursue lock detector design here (see Problem 9.4), we point out that the response time of the lock detector should be fast enough to recognize FAL lockup as it occurs. Hence the lock detector should have a response time approximately equal to the acquisition loop bandwidth.

The preceding statements concerning the frequency search behavior, as the FAL passes through frequency lock, were based on the assumption that the loop had time to operate linearly. Since the loop will have a nonzero response time in its linear mode, it is necessary that the slewing rate β not be too fast for the system. That is, the slewing signal must cause the frequency error to be within the loop bandwidth long enough for the loop dynamics to acquire the RF carrier. If RF slewing is at the rate of β rps/sec, the FAL frequency error spends approximately ω_n/β sec in its loop bandwidth. If we assume a response time of $2/\omega_n$ sec is required, then the slewing rate must be such that $\omega_n/\beta \geq 2/\omega_n$, or

$$\beta \leq \frac{\omega_n^2}{2} \, \text{rad}/\text{sec}^2 \qquad (9.3.5)$$

Hence RF slewing rates are limited by the loop natural frequency.† Faster slewing rates may cause the FAL to pass through the RF carrier frequency without acquisition. When slewing at the maximum rate in the correct direction, the time required to reduce an initial offset of Ω to zero is

$$T_{ac} = \frac{\Omega}{\beta} = \frac{\Omega}{(\omega_n^2/2)} \qquad (9.3.6)$$

Compare this result with (9.2.24) for quiescent acquisition. For $\zeta = \frac{1}{2}$, we find quiescent acquisition is actually preferable to frequency searching if $\Omega \leq 2\omega_n$. That is, frequency searching is not advantageous if the expected frequency uncertainty is well within the FAL bandwidth. Again we see the obvious trade-off of improved dynamical performance of large loop bandwidths (no slewing necessary) versus the increased loop noise.

Our discussion of frequency searching has been based purely on the response capabilities of a noiseless FAL. When noise is added to our model in Figure 9.2, it circulates within the loop, causing random phase effects. This random phase interferes with the normal tracking operations of the loop and increases the chance of the loop not acquiring during the slewing operation. Hence the addition of noise reduces the maximum allowable slewing rate for attaining a given acquisition probability. A rigorous analysis relating allowable rates and the amount of loop noise is somewhat difficult, although simulation, laboratory testing, and approximate analysis have been performed. A result often quoted [5] for the maximum slewing rate β_{max} is the empirical formula,

$$\beta_{max} = \frac{\omega_n^2}{2}\left[1 - \frac{1}{CNR_L}\right], \qquad CNR_L \geq 1 \qquad (9.3.7)$$

† Slewing may also be constrained by hardware considerations that prevent "pulling" an oscillator frequency too quickly. However, remember that β refers to the RF slewing rate, which benefits from the frequency multiplication of the acquisition system.

where $\text{CNR}_L = A_c^2/2N_0B_L$ represents the RF carrier to noise ratio in the loop noise bandwidth. As the loop CNR decreases, the maximum slewing rate must be decreased to maintain a desirable acquisition probability. This of course leads directly to an increase in the acquisition time of (9.3.6).

9.4. RF Phase Referencing

When the RF or IF frequencies are brought into frequency alignment by a frequency acquisition system, they are also brought into a fixed steady state phase position (θ_{e_0} in our earlier equations). When this final phase position is known exactly (modulo 2π), it can theoretically be easily compensated, and the final phase error can be made zero as well. Thus phase synchronization is theoretically achieved in these cases along with the desired frequency alignment. This means that the FAL master oscillator is operating at precisely the correct frequency and phase so as to be translated to the RF or IF exactly in phase sync (or a fixed known phase offset) from the received RF carrier. This is referred to as receiver *phase synchronization*, and we say the receiver has been made *phase coherent* with the transmitter. If the final phase error state is unknown [i.e., it may depend on the initial frequency offset as in (9.2.5), which is unknown to the receiver], then phase synchronization is not truly achieved unless the residual phase error is small enough to be negligible.

Having a phase synchronized master oscillator means that any other frequency derived from the RF carrier is also phase synchronized. If, for example, the transmitter RF frequency was divided down to provide the IF and subcarrier frequencies, then a phase synchronized receiver oscillator can also generate these frequencies with phase synchronization. Since a phase synchronized receiver carrier can perform phase referencing operations, we see that the operations of a RF FAL are also linked to possible carrier processing in the IF and subcarrier stages.

An important problem is to continue to maintain phase synchronization once it has been achieved from the acquisition operation. This requires that the loop oscillator continually maintain its phase position relative to the carrier being tracked, in spite of variations that may occur in the phase (or frequency) error. There are two important indicators of the receiver's phase referencing performance. The first is the loop phase error, which determines how well the loop is maintaining phase coherence. This error is usually measured by its mean squared loop phase error, as developed back in Chapter 4. The second indicator is the phase noise appearing on the loop VCO output. Since this phase noise is superimposed on the VCO frequency, it indicates the amount of phase noise that the referencing loop will interject into the receiver operations. The reference noise is indicated by displaying the phase noise spectrum that is effectively superimposed on the VCO by the various sources. The conversion of this spectrum to a meaningful frequency and phase noise jitter is then obtained as with any noisy oscillator (see Appendix C). Note that the loop phase error and the loop reference noise refer to phase effects at two different points in the equivalent system diagram, although the two are obviously related through the loop equations.

The primary factors that tend to destroy the receiver phase synchronization are (1) dynamical tracking errors, as the received carrier phase tends to vary and the FAL oscillator attempts to follow, (2) front end and frequency multiplier noise that enters the loop; (3) instabilities in both the transmitter and receiver oscillators that cause inherent phase fluctuations; and (4) modulation inserted on the carrier during communications operations. For the most part, the effect of these disturbances can best be estimated from linear analysis. We assume a linear tracking loop initially in phase and frequency synchronization and insert the various disturbances as an effective loop phase input at the appropriate place in the loop. Such a model is shown in Figure 9.6. Since the loop is linear, the effect of each factor can be determined separately, then combined to determine the overall effect on the phase referencing operation. Let us first consider the contributions to loop tracking mean squared phase error of the various sources.

Dynamic Tracking Errors. The effect of loop phase errors due to tracking dynamics has been discussed in Section 5.5 and considered earlier in this

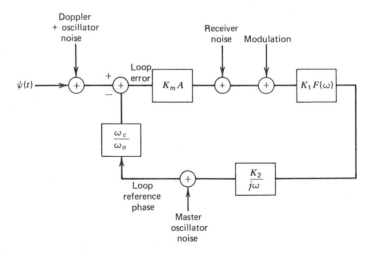

Figure 9.6. RF phase locking equivalent loop with typical tracking disturbances.

chapter. The primary tracking phase errors are due to sudden (step) changes in the RF frequency and to linear variations of this frequency. A linearly varying frequency can be caused by either a drifting of the transmitter or receiver oscillators, or by relative motion between the two. (In the latter case the rate of frequency change is referred to as the *Doppler rate*.) For a second order linear loop the phase error solution caused by a frequency step of Ω rps and a Doppler rate of $\hat{\Omega}$ rps/sec is given by

$$\theta_e(t) = \left(\frac{\hat{\Omega}}{A_c K}\right)t + \frac{\hat{\Omega}}{\omega_n^2} + \frac{\Omega}{A_c K} \tag{9.4.1}$$

Note that the instantaneous phase error now increases with time and will eventually become unbounded if the frequency drift is continually applied. This simply indicates that the second order loops with the filter in (9.7.2) cannot track linearly varying frequencies. If the linear frequency variation is due only to oscillator drift, its rate $\hat{\Omega}$ will typically be quite small, and the first term in (9.4.1) may be negligible over most communication periods. However, if the variation is due to transmitter or receiver acceleration, it may be necessary to compensate to prevent the FAL from breaking lock. Compensation can occur by injecting oscillator slewing to cancel the Doppler rate or by using transmitter frequency corrections. Note that the constant phase error caused by the Doppler term Ω can always be reduced by increasing the loop gain K.

Receiver Noise. Receiver noise, resulting from front end and multiplier noise, enters the tracking loop and causes random errors within the loop. This random error has the variance computed earlier in (4.7.10). For white noise and a linear loop model this becomes

$$\sigma_n^2 = \frac{N_0 B_L}{P_c} \tag{9.4.2}$$

where $P_c = A_c^2/2$ is the power of the carrier component being used for referencing and N_0 is the combined spectral level of all the equivalent receiver noise. The loop bandwidth B_L is the noise bandwidth of the overall RF tracking loop in Figure 9.6. This means B_L must be computed from the RF loop gain function

$$H(\omega) = \frac{A_c(K_m K_1 K_2 \omega_c/\omega_0)F(\omega)/j\omega}{1 + [A_c(K_m K_1 K_2 \omega_c/\omega_0)F(\omega)/j\omega]} \tag{9.4.3}$$

Since the function includes the RF frequency multiplication factor, B_L represents the loop bandwidth presented to the RF carrier.

To determine precisely the effect of receiver noise in an operating tracking loop, we require knowledge of the actual probability density of the random phase error produced by the noise. Since the input noise is Gaussian, linear theory predicts that the phase error due to the noise will also be Gaussian with the variance given in (9.4.2). More accurately, however, the tracking loop is truly nonlinear, described by the differential equation in (9.1.13), and a more rigorous analysis of noise effects requires further study of the basic nonlinear equation. Results of such nonlinear studies are included in References 6–9, and a thorough summary of noisy nonlinear phase synchronization theory is presented by Lindsey [1]. Although a rigorous investigation is beyond our scope here, we point out several results associated with nonlinear loops and white Gaussian input noise. The steady state probability density of the loop phase error for a first order nonlinear loop has been found to have the form

$$p(\theta_e) = \frac{e^{\rho \cos \theta_e}}{2\pi I_0(\rho)}, \qquad |\theta_e| \leq \pi \tag{9.4.4}$$

where $I_0(\rho)$ is the zero order Bessel function of argument ρ, and

$$\rho = \frac{P_c}{N_0 B_L} \tag{9.4.5}$$

is the effective loop CNR. This solution is plotted in Figure 9.7 for several values of ρ. Note that the density indeed approaches a Gaussian shape with variance $1/\rho$ for large ρ, as predicted by linear theory, but tends to spread as ρ is decreased (see Problem 9.8). For higher order loops it has been shown that (9.4.4) serves a good approximation for ρ values larger than about 5, as long as the B_L used to define ρ is taken to be the actual loop noise bandwidth [9]. The mean squared value of the phase error in (9.4.4) is plotted in Figure 9.8 and compared to the mean square phase error $\sigma_n^2 = 1/\rho$ in (9.4.2) predicted by linear theory. We see that linear theory accurately predicts mean squared phase error performance for $\rho \geq 5$. For smaller values the system departs from its linear performance, and the more accurate density in (9.4.4) must be used.

In practice, we also find that tracking loop noise actually causes other effects than simply generating the mean squared phase error in (9.4.2). If we recall our phase plane trajectories in Section 9.2, we see that stable phase lock points are actually dispersed along the phase error axis. The frequency tracking dynamics in the absence of noise caused us to slide into one of these stable points and remain there. In an actual system, however, external disturbances such as noise may intermittently force us out of a stable state and into a new phase stable position. (This would correspond,

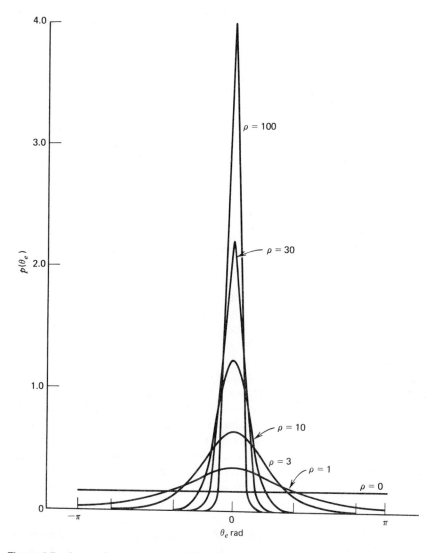

Figure 9.7. Loop phase error probability density—first order nonlinear loop, steady state theory.

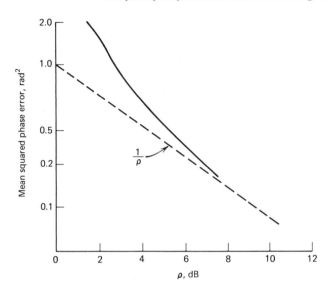

Figure 9.8. Loop mean squared phase error versus loop CNR ($\rho = \text{CNR}_L$).

for example, to sliding over the sinusoidal hump in Figure 9.3a and pulling into the adjacent phase lock position.) During this transient the phase error changes by multiples of 2π, corresponding to *cycle slipping*, and we say the loop has "lost lock" temporarily. As the phase error slides into the new stable position, the loop regains phase lock. Although the total time out of lock may be quite short, phase referencing operations using the loop oscillator are temporarily interrupted. In addition, loss of lock is exhibited throughout the receiver processing as phase impulses, or phase "clicks," superimposed on oscillator waveforms. If these clicks occur too often, interference in subsequent receiver operations may become significant.

An important parameter associated with cycle slipping is the mean time over which loss of lock occurs, that is, the expected time between cycle slips. Simulation and experimental studies [11, 12, 15] have attempted to predict mean cycle slipping time in tracking loops. Figure 9.9 shows a plot of estimated mean slip times for a second order loop, as a function of loop CNR ρ in (9.4.5) for several values of loop damping factor ζ. Theoretical studies of nonlinear loops using diffusion theory [10, 11], and more recently, renewal theory [13, 14], have verified these estimates. Note that the mean time between cycle slips decreases rapidly as ρ is decreased, and such performance may be the ultimate limitation on operating CNR. Note also the increase in slip time with damping factor, with an order of magnitude increase occurring as ζ increases to infinity. In the latter case, the

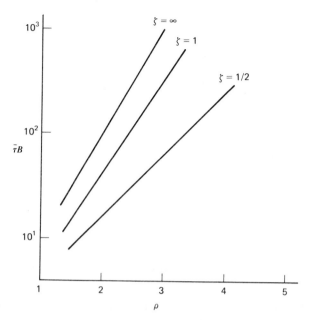

Figure 9.9. Average time between loop cycle slips $\bar{\tau}$ versus loop CNR (ζ = loop damping factor).

second order loop is performing practically as a first order loop and the lock points are of a more stable nature. For $\zeta = \frac{1}{2}$, the behavior in Figure 9.9 is adequately approximated by the equation

$$\bar{\tau}B_L = \frac{\pi}{4}e^{2\rho_c} \tag{9.4.6}$$

where $\bar{\tau}$ is the mean slip time and B_L is the loop noise bandwidth. Mean slip times can be used to predict the probability of a cycle slip occurring in a given time period τ. Studies [13] have confirmed that this probability $P_{cs}(\tau)$ is given quite accurately by the relation

$$P_{cs} \cong 1 - e^{-(\tau/\bar{\tau})} \tag{9.4.7}$$

Equation (9.4.7) is plotted in Figure 9.10 for several values of ρ.

Oscillator Phase Noise.   Undesired random phase variations on either the transmitter or receiver oscillators enter the tracking loop and disturb phase synchronization. Noise on the transmitter oscillator superimposes a random phase variation on the carrier extraneous phase $\psi(t)$ in (9.1.1), which therefore enters the tracking loop as an input forcing function (Figure 9.6).

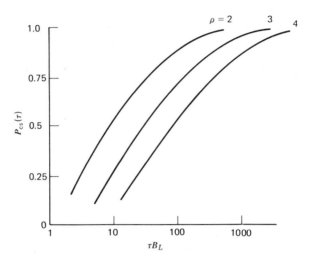

Figure 9.10. Probability of a cycle slipping in τ sec (ρ = loop CNR).

If we assume the input RF carrier oscillator phase noise has spectral density $S_\psi(\omega)$, then the mean squared loop phase error contribution that it produces is

$$\sigma_\psi^2 = \frac{1}{2\pi} \int_{-\infty}^{\infty} S_\psi(\omega)|1 - H(\omega)|^2 \, d\omega \qquad (9.4.8)$$

where $H(\omega)$ is again the tracking loop transfer function in (9.4.3). Oscillator phase noise is typically characterized by the onesided flicker noise spectrum (Appendix C, Equation C.4.7)

$$\hat{S}_\psi(\omega) = \frac{S_3}{\omega^3} + \frac{S_2}{\omega^2} + S_0 \qquad (9.4.9)$$

Here the first term is referred to as frequency *flicker noise*, the second term is referred to as *white frequency noise*, and S_0 is the *white phase noise* term, the latter assumed to exist to some upper frequency ω_u. For a second order loop with damping factor ζ and loop natural frequency ω_n, $H(\omega)$ *in* (9.4.3) *becomes*

$$H(\omega) = \frac{1 + (2\zeta/\omega_n)j\omega}{1 + 2\zeta/\omega_n)j\omega - (\omega/\omega_n)^2} \qquad (9.4.10)$$

When (9.4.9) and (9.4.10) are used in (9.4.8), the mean squared loop phase error contribution is

$$\sigma_\psi^2 = \frac{1}{2\pi} \int_{-0}^{\infty} \left[\frac{S_3}{\omega^3} + \frac{S_2}{\omega^2} + S_0 \right] \left[\frac{(\omega/\omega_n)^4}{[1 - (\omega/\omega_n)^2]^2 + 4\zeta^2(\omega/\omega_n)^2} \right] d\omega$$
$$(9.4.11)$$

For the case $\zeta = 1/\sqrt{2}$, the preceding integrates to [Section A.2, Integrals],

$$\sigma_\psi^2 = \left(\frac{S_0}{2\pi}\right)\omega_u + \frac{0.94S_2}{B_L} + \frac{0.03S_3}{B_L^2} \qquad (9.4.12)$$

where $B_L = 0.53\omega_n$ is the loop noise bandwidth, obtained from Table 4.1. Hence transmitter oscillator noise effects are reduced by using as large a loop bandwidth as possible. Since the individual terms vary differently with B_L, it is important to characterize properly the phase noise spectrum, that is, determine whether the flat or reciprocal frequency portion is more significant.

Phase noise on the loop VCO oscillator enters the equivalent system at a different point (Figure 9.6). If this local oscillator noise has spectrum $S_{os}(\omega)$, it produces the mean squared loop phase error

$$\sigma_{eo}^2 = \left(\frac{\omega_c}{\omega_0}\right)^2 \frac{1}{2\pi} \int_{-\infty}^{\infty} S_{os}(\omega)|1 - H(\omega)|^2 \, d\omega \qquad (9.4.13)$$

Of obvious importance is the fact that the integral is multiplied up by the frequency multiplication factor of the loop in determining the loop error. Thus local oscillator noise is extremely important in phase tracking loops with large frequency translations, and selection of "quiet" master oscillators is mandatory for good phase tracking.

Modulation. When communicating, modulation impressed on the carrier being tracked affects the transponder phase referencing operation. In addition to the suppression of the carrier power that it may cause, the modulation enters the tracking loop as an equivalent phase noise interference, as shown in Figure 9.6. Its mean squared error contribution can therefore be determined by replacing the noise spectrum in (9.4.2) with the corresponding phase modulation spectrum. Thus a phase modulation spectrum $S_m(\omega)$ produces the mean squared phase error

$$\sigma_{em}^2 = \frac{1}{2\pi} \int_{-\infty}^{\infty} S_m(\omega)|H(\omega)|^2 \, d\omega \qquad (9.4.14)$$

If the modulation is on a subcarrier located outside the loop bandwidth, the integrated effect in (9.4.14) is negligible. If the modulation was applied directly to the RF carrier, the portion of it entering the loop bandwidth will cause phase tracking errors. Typically, the bandwidth of such modulation is much larger than that of the loop, and the carrier modulation spectrum can be modeled as a flat spectrum in (9.4.14). In this case, σ_{em}^2 can be evaluated as in (9.4.2) with N_0/P_c replaced by the phase modulation spectral level. Also of particular concern is the presence of modulation spectral frequency lines that appear directly in the loop bandwidth. In the latter case it is necessary to insure that $|H(\omega)|$ is suitably small at the line frequency to reduce this component to a negligible phase error contribution.

The phase referencing noise spectrum is that generated on the VCO output by the previously stated sources. To determine the resulting spectrum we must determine the phase transfer function from each source input to the VCO output (see Figure 9.6). These transfer functions are summarized in Table 9.2 for each source, along with the resulting phase noise spectrum they produce. Note that only the frequency components of the input phase, noise, and modulation that fall in the loop bandwidth contribute to the reference phase noise spectrum, whereas the out-of-band frequencies of the loop VCO spectrum add to the reference noise. In addition, all loop input spectra are effectively reduced by the square of the frequency multiplication factor (ω_c/ω_o) in determining the reference phase noise.

In summary, we see that accurate receiver phase referencing is hindered by several primary causes. Which of these sources is involved generally dictates system implementation, since the minimization of each effect often requires contrasting design procedures for loop bandwidths. For example, dynamical and oscillator jitter phase errors are reduced by relatively fast loops, with large bandwidths and low damping factors, whereas noise and modulation effects require narrow bandwidths. Thus phase referencing loop design invariably reduces to careful trade-off analyses of phase errors.

Table 9.2. Reference Phase Spectral Contributions

Source of Phase Noise	Transfer Function to Reference VCO Output	Reference Phase Noise Spectrum Contribution
Input phase spectrum, $S_{\theta_1}(\omega)$	$\dfrac{H(\omega)}{r}$	$\dfrac{S_{\theta_1}(\omega)}{r^2}|H(\omega)|^2$
Receiver noise, N_0	$\dfrac{H(\omega)}{A_c r}$	$\dfrac{N_0}{A_c^2 r^2}|H(\omega)|^2$
Loop VCO, $S_{os}(\omega)$	$1 - H(\omega)$	$S_{os}(\omega)|1 - H(\omega)|^2$
Modulation, $S_m(\omega)$	$\dfrac{H(\omega)}{r}$	$\dfrac{S_m(\omega)}{r^2}|H(\omega)|^2$

$H(\omega) = $ loop gain function.
$r = \omega_c/\omega_0 = $ RF frequency/reference frequency.

In addition, phase referencing must be properly interfaced with the frequency acquisition operation that it initially performs. In Section 9.2 we found that frequency acquisition time, whether using quiescent or searching operations, is reduced with increasing bandwidths, which is contrary to the noise reduction requirements of phase referencing. Thus FAL design also involves careful trade-off of the acquisition prerequisite as well as the required phase synchronization. In this regard several types of hybrid systems have been suggested, in which wideband RF loops are used for initial frequency acquisition, then switched to narrowband operation for accurate phase referencing.

9.5. Subcarrier Phase Referencing

In demodulating or decoding subcarriers, it is advantageous to use phase referenced operations. This requires that the receiver baseband stages have available a subcarrier phase synchronized with the received subcarrier. If this received subcarrier was generated at the transmitter by dividing down the RF carrier then a phase synchronized receiver subcarrier can always be obtained by dividing down accordingly the phase referenced RF carrier from the FAL. That is, the loop master oscillator can be used for subcarrier generation as well. This has the advantage that dividing down the RF frequency also divides down the phase error. Hence if the RF is phase synchronized with a total mean squared phase error of σ_e^2, a subcarrier at ω_s derived from the RF will have a mean squared phase error of $\sigma_e^2(\omega_s/\omega_c)^2$. Thus the subcarrier will be much more accurately phase synchronized than the RF carriers. The primary difficulty with deriving subcarrier referencing in this way is the need to achieve large accurate division factors without exceedingly complicated circuitry. Furthermore, many systems do not use RF acquisition, or do not have RF-subcarrier related frequencies. For these reasons subcarrier phase referencing must often be done separately at the receiver baseband stages.

Subcarrier phase referencing can be accomplished by a tracking loop subsystem at the receiver baseband (Figure 9.11) just as in the RF stage. The tracking loop attempts to phase lock to the subcarrier frequency component of ω_s immersed within the recovered baseband signal. The system is therefore identical to that of Figure 9.1, with ω_0 set to ω_s, $g_3 = 1$, and the RF and IF stages removed. The equivalent tracking loop model is given by Figure 9.6 with the frequency multiplication factor equal to one. The ensuing phase error analysis in Section 9.4 is therefore immediately applicable with proper conversion to the subcarrier frequency. For example, the frequency offsets and drifts that the loop must track are those of

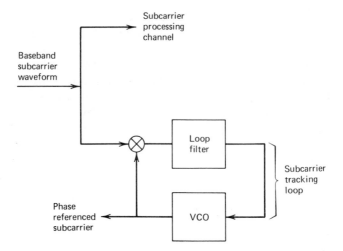

Figure 9.11. Subcarrier phase referencing subsystem.

the subcarrier. [If they are caused by Doppler effects they must be divided down from the RF uncertainties by the frequency ratios, as predicted by (2.7.1).] The transponder input noise is that of the demodulated baseband, and oscillator jitter and modulaton refers to that on the subcarrier. With these modifications the phase error terms of Section 9.4 can be evaluated and the loop designed for maximum subcarrier referencing accuracy.

The subcarrier being tracked has an amplitude A_s, which depends on the manner in which the subcarrier was demodulated from the RF carrier. The subcarrier component A_{sc} used in the subcarrier tracking loop is dependent on the degree of suppression applied to it. If the subcarrier is phase modulated onto the RF carrier and is RF demodulated by RF carrier tracking, then A_s depends on the suppression effects of all other RF modulation occurring simultaneously. Table 9.3 summarizes the form of A_s for different types of RF modulation alternatives. If, in addition, the subcarrier was itself modulated, then A_s is further suppressed to the value A_{sc} used for subcarrier tracking. This modification is also included in Table 9.3.

Invariably the subcarrier will have modulation on it, and its suppression effect on the subcarrier component being tracked is important, since it leads directly to further reduction of loop CNR. Of particular importance is the case of digital phase modulation by a binary sequence with phase deviation Δ_c. As shown in Table 9.3, the subcarrier component will be proportional to $\cos \Delta_c$. When $\Delta_c = \pi/2$ and the binary sequence is a PCM digital signal, the subcarrier corresponds to an antipodal PSK binary

Table 9.3. Subcarrier Amplitudes

Carrier Modulation	A_s	Subcarrier Modulation	A_{sc}
Single subcarrier, dev. Δ	$AJ_1(\Delta)$	sine wave, Δ	$A_s J_0(\Delta)$
Subcarrier (Δ) plus K other subcarriers (Δ_i)	$AJ_1(\Delta) \prod\limits_{i=1}^{K} J_0(\Delta_i)$	K sine waves (Δ_i) binary waveform	$A_s \prod\limits_{i=1}^{K} J_i(\Delta_i)$ $A_s \cos \Delta_c$
Subcarrier (Δ) plus binary waveform (Δ_c)	$AJ_1(\Delta) \cos (\Delta_c)$	random $\Psi_\theta(\omega)$	$A_s|\Psi_\theta(1)|$
Subcarrier (Δ) plus random modulation ($\Psi_\theta(\omega)$)	$AJ_1(\Delta)|\Psi_\theta(1)|$		
Subcarrier (Δ) plus K other subcarriers (Δ_i) plus binary waveform Δ_c plus random modulation $\Psi_\theta(\omega)$	$AJ_1(\Delta)\left[\prod\limits_{i=1}^{K} J_0(\Delta_i)\right] \cos \Delta_c |\Psi_\theta(1)|$		

A = RF carrier amplitude.

waveform. In this case the subcarrier component used by the loop in Figure 9.11 for subcarrier referencing is eliminated. However, an accurate phase synchronized subcarrier reference is nevertheless mandatory for coherent PSK decoding. Thus for one of the most important applications of subcarrier referencing, phase synchronization cannot be achieved by a standard carrier tracking system, and the reference must be derived in some other way from the received subcarrier. Phase referencing from a modulated subcarrier with no subcarrier component is referred to as *suppressed carrier* referencing, or simply *carrier extraction*.

There are two standard methods for achieving suppressed subcarrier phase referencing, both using modified tracking loops. One method is the use of a *squaring loop*, in which we precede a subcarrier PLL with a nonlinear, memoryless, squaring device (Figure 9.12a). The input to the squarer is the noisy PSK subcarrier signal

$$x(t) = A_s \sin\left[\omega_s t + \frac{\pi}{2} m(t) + \psi(t)\right] + n_b(t) \tag{9.5.1}$$

where $n_b(t)$ is the baseband noise, $m(t) = \pm 1$ is the PCM waveform, and

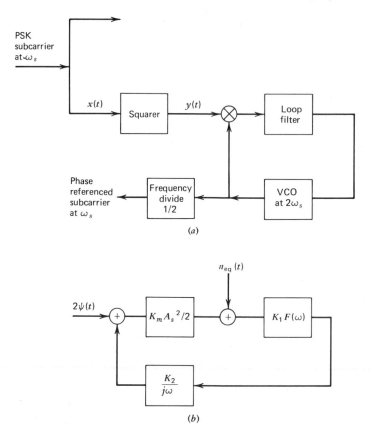

Figure 9.12. Squaring subcarrier referencing subsystem. (a) Block diagram, (b) equivalent loop model.

$\psi(t)$ is the extraneous phase variation that must be tracked for coherent phase referencing. The output of the squarer is then

$$y(t) = x^2(t) = A_s^2 \sin^2 \left[\omega_s t + \frac{\pi}{2} m(t) + \psi(t) \right]$$

$$+ 2n_b(t)A_s \sin \left[\omega_s t + \frac{\pi}{2} m(t) + \psi(t) \right] + n_b^2(t)$$

$$= \frac{A_s^2}{2} + \frac{A_s^2}{2} \sin \left[2\omega_s t + 2\psi(t) \right] + n_b^2(t)$$

$$+ 2n_b(t)A_s \sin \left[\omega_s t + \frac{\pi}{2} m(t) + \psi(t) \right] \qquad (9.5.2)$$

The output contains a constant term, a carrier component at $2\omega_s$ with no modulation and with the phase variation multiplied by 2, and two interference terms involving the noise. A PLL with a VCO at $2\omega_s$ can now track the carrier component in (9.5.2). If it successfully performs the tracking, then its VCO output is given by $\cos(2\omega_s t + \theta_2(t))$, where $\theta_2(t)$ is a close replica of $2\psi(t)$. By now frequency dividing the VCO output by 2 we generate the reference carrier at ω_s

$$\cos\left(\omega_s t + \frac{\theta_2(t)}{2}\right) \tag{9.5.3}$$

where now the phase $\theta_2(t)/2$ is approximately phase locked to $\psi(t)$ in (9.5.1). Note that the role of the squarer is effectively to remove the digital modulation while frequency translating the subcarrier to $2\omega_s$ for phase locking.

The ability to achieve accurate phase referencing with squaring systems can be determined from the equivalent loop tracking diagram in Figure 9.12b. The loop is effectively driven by $2\psi(t)$ and therefore has twice the extraneous phase variation of the subcarrier itself. The equivalent noise $n_{eq}(t)$ entering the loop is simply the loop mixer output noise produced by the squarer noise terms [i.e., the last two terms in (9.5.2)]. These correspond to the squared subcarrier filtered noise and the signal-noise cross product term. To evaluate the spectrum of $n_{eq}(t)$ we need only determine the spectrum of the squarer noise in the vicinity of $2\omega_s$. This can be determined by recalling that the squared Gaussian noise term has a spectrum given by twice the convolution of $S_{nb}(\omega)$ with itself [16], and the spectrum of the cross product term is given by the convolution of $S_{nb}(\omega)$ with $2S_s(\omega)$, where $S_{nb}(\omega)$, $S_s(\omega)$ are the spectrum of the filtered noise and subcarrier signals in (9.5.1). These spectra are sketched in Figure 9.13. In (a) is the baseband noise, and (b) shows the spectrum of its square. Figure 9.13c shows a typical subcarrier spectrum, and its convolution with the noise of (a) is shown in (d). Figure 9.13e then shows the spectrum of $n_{eq}(t)$, obtained by shifting and superimposing the portion of (b) and (d) around $2\omega_s$. Note that the spectra are multiplied and spread by the squaring (convolution) operation. In the vicinity of zero, $n_{eq}(t)$ has approximately a flat spectrum with level

$$S_{ne}(0) \approx \frac{N_{b0}^2 B_{sc}}{2} + \frac{N_{b0} A_s^2}{2} \tag{9.5.4}$$

Since the noise bandwidth B_L of the tracking loop is generally much smaller than that of the subcarrier bandwidth B_{sc}, the mean squared noise value entering the loop bandwidth is therefore approximately $2S_{ne}(0)B_L$.

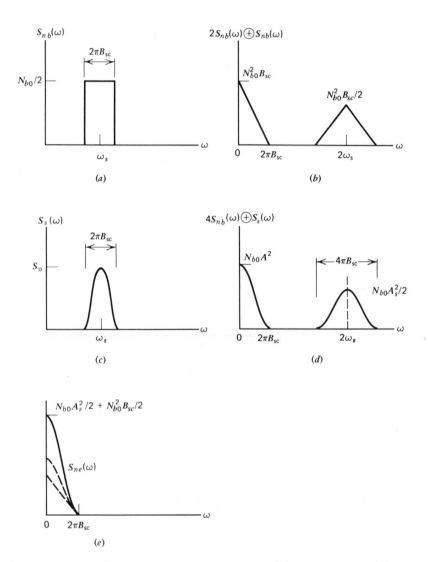

Figure 9.13. Noise spectral densities due to squaring. (a) Baseband noise, (b) noise squared, (c) subcarrier, (d) subcarrier convolved with noise, (e) $n_{eq}(t)$ spectrum, $S_{ne}(\omega)$.

The resulting subcarrier reference phase variance, after phase dividing by 2, is then

$$\sigma_e^2 = \frac{1}{4}\left[\frac{2S_{ne}(0)B_L}{A_s^4/8}\right]$$

$$= \frac{2N_{b0}B_L}{A_s^2} + \frac{2N_{b0}'^2 B_{sc}B_L}{A_s^4}$$

$$= \left(\frac{N_{b0}B_L}{A_s^2/2}\right)\mu \tag{9.5.5}$$

where

$$\mu \triangleq 1 + \frac{N_{b0}B_{sc}}{A_s^2} \tag{9.5.6}$$

The parameter μ is called the *loop squaring factor* and accounts for the increases in error caused by the squaring operation. Since $\mu \geq 1$, squaring loop errors are larger than those for standard tracking loops, the amount of the increase dependent on the subcarrier CNR. In effect, μ exhibits the fact that the loop noise changes from a squared noise term to a signal-noise cross product as the subcarrier CNR increases. From (9.4.5), we can define the parameter

$$\hat{\rho} = \frac{\rho}{\mu} = \frac{A_s^2/2N_{b0}B_L}{\mu} \tag{9.5.7}$$

as the effective loop CNR of the squaring loop, in that its reciprocal generates the loop tracking error, as in (9.4.2). In this way, $1/\mu$ can be interpreted as a CNR suppression factor. Note that the dc term in (9.5.2) is given by the subcarrier power, $A_s^2/2$. Hence a low pass filter that extracts this term at the squarer output provides a convenient measure of subcarrier power for signal monitoring or automatic gain control.

A problem that occurs in phase referencing with a squaring loop is the phase ambiguity that must be resolved before the phase reference can be used for PSK decoding. This ambiguity is caused by the one-half frequency division of the VCO output. The resulting divided down reference may be 180° out of phase with the desired reference (i.e., both the desired reference and its negative correspond to the same double frequency waveform at the VCO). This phase ambiguity must be resolved prior to PSK decoding, since the negative reference will decode all bits exactly opposite to their true polarity. Resolving this ambiguity is usually accomplished by first decoding a known binary word.

A second way to achieve suppressed carrier synchronization is by the use of the *Costas* [17], or *quadrature*, loop shown in Figure 9.14a. The

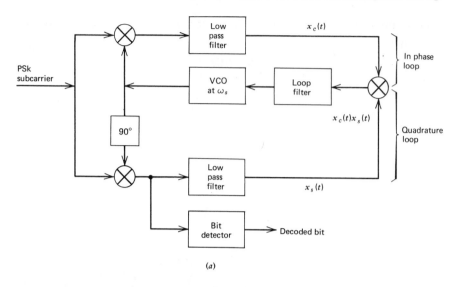

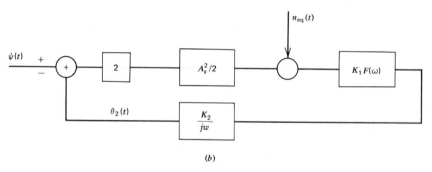

Figure 9.14. Costas phase referencing loop. (*a*) Block diagram, (*b*) equivalent loop model.

system involves two parallel tracking loops operating simultaneously from the same VCO. One loop, called the *in-phase* loop, uses the VCO directly for tracking, and the second (*quadrature* loop) uses a 90° shifted VCO. The mixer outputs are each multiplied, filtered, and used to control the VCO. The low pass filters in each arm must be wide enough to pass the subcarrier modulation without distortion. If the input to the Costas loop is the subcarrier signal in (9.5.1), the in-phase mixer generates the signal

$$x_c(t) = A_s \cos\left[\frac{\pi}{2}m(t) + \psi(t) - \theta_2(t)\right] + n_{mc}(t) \qquad (9.5.8)$$

while the quadrature mixer simultaneously generates

$$x_s(t) = A_s \sin\left[\frac{\pi}{2}m(t) + \psi(t) - \theta_2(t)\right] + n_{ms}(t) \qquad (9.5.9)$$

The output of the multiplier is then

$$x_c(t)x_s(t) = \frac{A_s^2}{2}\sin\left[\pi m(t) + 2\psi(t) - 2\theta_2(t)\right] + n_{eq}(t)$$

$$= \frac{A_s^2}{2}\sin\left[2\psi(t) - 2\theta_2(t)\right] + n_{eq}(t) \qquad (9.5.10)$$

where

$$n_{eq}(t) = n_{mc}(t)n_{ms}(t) + n_{mc}(t)\sin\left[\frac{\pi}{2}m(t) + \psi - \theta_2\right]$$

$$+ n_{ms}(t)\cos\left[\frac{\pi}{2}m(t) + \psi - \theta_2\right] \qquad (9.5.11)$$

The second equality in (9.5.10) follows, since $m(t) = \pm 1$. When the phase error, $\psi(t) - \theta_2(t)$, is small, the loop has the equivalent linear model of Figure 9.14b. The Costas loop therefore tracks the phase variation $\psi(t)$ with the VCO without interference from the subcarrier modulation $m(t)$. The quadrature tracking loops have effectively removed the modulation, allowing the loop filter and VCO to accomplish the subcarrier synchronization. The mixer noises $n_{mc}(t)$ and $n_{ms}(t)$ in (9.5.11) are base-band versions of the subcarrier noise in (9.5.1). When this latter noise has the spectrum in Figure 9.13a, it can be shown (Problem 9.10) that $n_{eq}(t)$ in (9.5.11) has the identical spectrum as in Figure 9.13e. The frequency division by the factor of two can be accomplished directly in the loop gain, since the error term, $\psi - \theta_2$, must be derived from the argument of the sine term in (9.5.10). Thus the equivalent Costas quadrature loop is identical to the equivalent squaring loop system of Figure 9.12b and generates the same mean squared subcarrier phase error given in (9.5.5).

An advantage of the Costas loop is that the PSK decoding is partially accomplished right within the loop. If the loop is tracking well so that $\theta_2(t) = \psi(t)$, we see the quadrature mixer output in (9.5.9) is

$$x_s(t) \cong A_s \sin\left[\frac{\pi}{2}m(t)\right] + n_{ms}(t)$$

$$= A_s m(t) + n_{ms}(t) \qquad (9.5.12)$$

This result is identical to that produced at the multiplier output in the coherent PSK decoder of Figure 7.9. The addition of a bit integrator

(Figure 9.14a) therefore completes the PSK decoder. Hence both decoding and phase referencing can be accomplished directly from the Costas loop. For this reason the quadrature arm is often called the *decisioning* arm and the in-phase arm the *tracking* arm. One can think of the multiplier of the system as allowing the bit polarity of the decisioning loop to correct the phase error orientation of the tracking loop, thereby removing the modulation. Often Costas loops are designed with the decisioning arm filter followed by a limiter. At high SNR, the limiter output will have a sign during each bit interval that is identical to the present data bit polarity. This limiter output then multiplies the loop error and removes the loop modulation. In effect, the data bit sign is used to aid the tracking loop, and such modified systems are often called *data aided carrier extraction* loops. Note that when tracking perfectly, the output of the in-phase mixer always has a *dc* signal proportional to the subcarrier amplitude A_s, regardless of the PSK bit, making it convenient for amplitude level measurements and automatic gain control

When digital data is modulated on the subcarrier with QPSK (quadraphase) encoding, carrier extraction must be accomplished by a slightly different means than with PSK. This is due to the fact that the modulated subcarrier, although still containing a suppressed carrier component, contains one of four possible phases, rather than two, during each bit interval. To extract the QPSK suppressed carrier, a fourth power, instead of a squaring, system must be used, as shown in Figure 9.15a. The QPSK subcarrier is passed into the fourth power device, and a phase lock loop can be locked to the fourth harmonic. To see this, we write the subcarrier QPSK signal as

$$x(t) = A_s \cos (\omega_s t + \theta(t) + \psi) \tag{9.5.13}$$

where $\theta(t)$ is one of the phase angles $n\pi/2$, $n = 1, 2, 3, 4$, during each bit time. When raised to the fourth power, the fourth harmonic contains the carrier signal component

$$(A_s^4/16) \cos [4(\omega_s t + \theta(t) + \psi)] = (A_s^4/16) \cos (4\omega_s t + 4\psi) \tag{9.5.14}$$

and therefore contains an unmodulated subcarrier component at $4\omega_s$. The subsequent tracking loop at $4\omega_s$ in Figure 9.15a tracks the phase variation $4\psi(t)$. Frequency division of the loop VCO reference output by a factor of 4 produces a phase locked subcarrier reference that is phase coherent with the suppressed QPSK subcarrier.† Noise effects in the fourth power system

---

† A phase ambiguity will again occur from this frequency division. In this case we have a fourth order ambiguity since frequency division by a factor of four can produce any of the four phase angles $i\pi/4$, $i = 1, 2, 3, 4$, on the divided reference frequency.

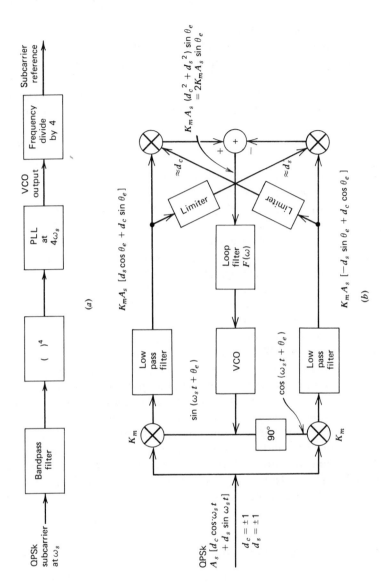

Figure 9.15. QPSK carrier extraction systems. (*a*) Fourth power loop. (*b*) Costas crossover loop.

431

are now more complicated than for the squaring loop since the multiplication terms involving the QPSK subcarrier and noise crossproducts are present in all orders up to the fourth power. It has been shown [19] that if the subcarrier bandpass filter in Figure 9.15a has the value $B_{sc} = 2/T$, where T is the QPSK bit time, then the fourth power loop generates the mean squared phase error at the subcarrier frequency of approximately

$$\sigma^2 = \frac{2N_0 B_L}{A_s^2}\left[1 + 3.8\left(\frac{1}{CNR_{sc}}\right) + 4.2\left(\frac{1}{CNR_{sc}}\right)^2 + \left(\frac{1}{CNR_{sc}}\right)^3\right] \quad (9.5.15)$$

where B_L is the PLL noise bandwidth and $CNR_{sc} = A_s^2/2N_0 B_{sc}$ is the CNR at the output of the subcarrier bandpass filter. The bracketed term now plays the same role for the fourth power loop as the squaring factor μ in (9.5.5) for the squaring loop. We emphasize that (9.5.15) refers to the phase error between the received subcarrier and the divided-down reference frequency at ω_s, and the phase error in the loop at $4\omega_s$ has a mean squared value that is $4^2 = 16$ times larger.

Carrier extraction for QPSK signals can also be derived from a modified form of the Costas loop involving loop crossover arms, as shown in Figure 9.15b. The signal waveforms are shown throughout the figure. Note that each filter output contains data bits from both quadrature carriers that effectively compose the QPSK subcarrier waveform. However, the sign of the output of the arm filters, produced by the limiters, are used to crossover and mix with the opposite arm signal. If the loop phase error is small, the output of the limiters correspond to the cosine term of the arm filter outputs, and therefore have the sign of the data bits of the quadrature components, as shown. Hence, the limiters effectively demodulate the QPSK quadrature bits, and the crossover produces a common phase error term which is canceled after subtraction, leaving a remainder error term proportional to $\sin \theta_e$. The latter signal can then be used to generate an error signal to phase control the loop VCO, thereby closing the QPSK Costas loop.

9.6. The Effect of Imperfect Phase Referencing on Bit Detection

Now that we have discussed phase referencing and have developed properties of the phase error process $\theta_e(t)$ occurring in a typical loop, we can reinvestigate PSK decoding. Recall that a coherent PSK decoder, using a subcarrier reference waveform with phase error $\theta_e(t)$, will decode with the bit error probability in (7.2.13),

$$PE(\theta_e) = \tfrac{1}{2}\,\text{Erfc}\left[\left(\frac{E}{N_{b0}}\right)^{1/2}\frac{1}{T}\int_0^T \cos\left(\theta_e(t)\right)dt\right] \quad (9.6.1)$$

The phase error function $\theta_e(t)$ is that generated in subcarrier reference loops of Sections 9.4 and 9.5. As such, $\theta_e(t)$ evolves as a time process having a bandwidth approximately equal to that of the loop itself. We wish to determine the effect of this error process on the bit error performance described by (9.6.1).

We first consider the case where the PSK bit rate $1/T$ is much larger than the noise bandwidth of the loop, which is the typical situation confronted in practice. Under these conditions the error process $\theta_e(t)$ changes very slowly over a bit period T and may be considered a constant in (9.6.1). This means we can approximate (9.6.1) with

$$PE(\theta_e) = \tfrac{1}{2} \operatorname{Erfc}\left[\left(\frac{E}{N_{b0}}\right)^{1/2} \cos(\theta_e)\right] \tag{9.6.2}$$

where θ_e is the phase error value during the bit interval. Noting that the preceding is a conditional probability for a specific θ_e, we determine the resulting average eror probability by averaging over the randomness in the phase error θ_e. Hence

$$PE = \int_{-\infty}^{\infty} PE(\theta_e) p(\theta_e)\, d\theta_e \tag{9.6.3}$$

To compute the average, we use for the density of θ_e the steady state error density in (9.4.4). Substitution yields

$$PE = \frac{1}{2\pi} \int_{-\pi}^{\pi} \tfrac{1}{2} \operatorname{Erfc}\left[\left(\frac{E}{N_{b0}}\right)^{1/2} \cos\theta_e\right] \frac{e^{\rho \cos\theta_e}}{I_0(\rho)}\, d\theta_e \tag{9.6.4}$$

where ρ is the CNR in the reference loop bandwidth. Equation (9.6.4) is shown plotted in Figure 9.16 for several values of ρ. As $\rho \to \infty$, the phase error density $p(\theta_e)$ approaches a delta function at $\theta_e = 0$ (see Figure 9.7) and (9.6.4) approaches the idealized antipodal error probability of Figure 7.7. However, as ρ is reduced, PE is increased at each value of E/N_{b0}, because of imperfect phase referencing used in the decoding. Note that the decoding performance may actually be significantly worse than that predicted by the ideal performance if the CNR used for referencing is not adequate. Thus practical PSK decoder design requires a careful interface of the detection and referencing subsystem performance.

It is important to note that each curve in Figure 9.16 tends to flatten out for large values of E/N_{b0}. This implies that imperfect phase referencing produces an irreducible detection error that cannot be overcome by simply increasing bit energy E. This can be seen by noting that if $|\theta_e| > \pi/2$, the cosine term in (9.6.2) changes sign and bit error will most likely be made. This implies

$$PE(\theta_e) \to 1, \quad \text{if } |\theta_e| > \frac{\pi}{2} \tag{9.6.5}$$

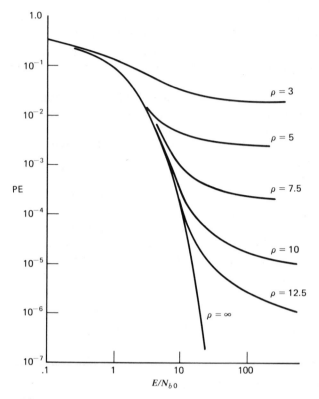

Figure 9.16. PSK error probabilities with nonideal phase referencing ($\rho = \text{CNR}_L$, $B_L T \ll 1$).

Therefore even if $E/N_{b0} \rightarrow \infty$, so that $\text{PE}(\theta_e) \rightarrow 0$ for $|\theta_e| \leq \pi/2$, we still have

$$\text{PE} = \int_{|\theta_e| > (\pi/2)} (1)p(\theta_e)\,d\theta_e$$

$$= \text{Prob}\left[|\theta_e| > \frac{\pi}{2}\right] \qquad (9.6.6)$$

This value represents the asymptotic irreducible error probability. Thus PSK decoding performance is ultimately limited by the capabilities of the referencing system alone. If the subcarrier referencing is obtained by tracking the subcarrier carrier component, then the loop CNR p will depend on the subcarrier component amplitude A_{sc} (Table 9.3) while E/N_{b0} depends on the total subcarrier amplitude A_s. If the subcarrier referencing is suppressed carrier, as with squaring or Costas loops, then the referencing and decoding is accomplished from the same waveform. In this case the parameters E/N_{b0} and ρ in (9.6.4) are directly related via the subcarrier power $A_s^2/2$. In fact,

$$\frac{E}{N_{b0}} = \frac{(A_s^2/2)T}{N_{b0}} = \frac{(A_s^2/2)}{N_{b0}B_L}(TB_L) = \rho(TB_L) \qquad (9.6.7)$$

Since we have assumed $TB_L \ll 1$, $\rho \gg E/N_{b0}$, and the referencing CNR is generally many times larger than the decoding bit energy to noise level. For these reasons suppressed carrier decoders exhibit little performance reduction caused by imperfect phase referencing when operating with reasonably high bit energy levels.

When the PSK bit rate is less than the referencing loop bandwidth, $B_L T \geq 1$ and the integral in (9.6.1) cannot be neglected. In this case, $\theta_e(t)$ varies rapidly during the bit interval, and we can interpret the integral as a T sec time average of the time function $\cos \theta_e(t)$. Denoting this as $\overline{\cos \theta_e(t)}$, we argue that if T is long enough, this time average will be given by the steady state statistical average of $\cos \theta_e$. Hence

$$\overline{\cos \theta_e(t)} = \int_{-\pi}^{\pi} \cos \theta_e \left(\frac{e^{\rho \cos \theta_e}}{2\pi I_0(\rho)}\right) d\theta_e$$

$$= \frac{I_1(\rho)}{I_0(\rho)} \qquad (9.6.8)$$

where $I_0(\rho)$ and $I_1(\rho)$ are the zero and first order Bessel functions. The resulting bit error probability for the low bit rate case is therefore given instead by

$$PE = \tfrac{1}{2} \text{Erfc}\left[\left(\frac{E}{N_{b0}}\right)^{1/2} \overline{\cos \theta_e}\right]$$

$$= \tfrac{1}{2} \text{Erfc}\left[\left(\frac{E}{N_{b0}}\right)^{1/2} \left(\frac{I_1(\rho)}{I_0(\rho)}\right)\right] \qquad (9.6.9)$$

Equation (9.6.9) can be interpreted as the error probability of a coherent decoder in which the bit energy E is replaced by the modified energy

$$\hat{E} = E\left[\frac{I_1(\rho)}{I_0(\rho)}\right]^2 \qquad (9.6.10)$$

The bracketed factor is shown in Figure 9.17 as a function of ρ. Again, the degradation in performance caused by imperfect phase referencing systems is evident. To design a binary PSK system to operate at a prescribed PE would require first determining the necessary energy value E for the perfectly referenced case, then modifying by the suppression

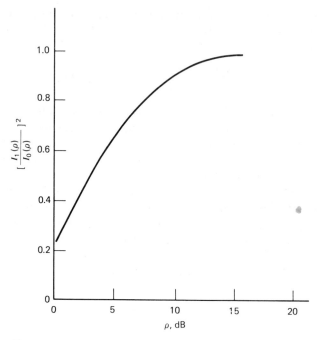

Figure 9.17. Energy degradation with ρ ($\rho = \text{CNR}_L$, $B_L T \gg 1$).

effect in Figure 9.17 to determine the required bit energy E. We may note that for a fixed value of ρ, (9.6.9) continually decreases with increasing bit energy E, so that PSK operation with low bit rates does not exhibit the irreducible error of Figure 9.16.

When the PSK phase referencing is provided by squaring or Costas loop, the loop CNR must be replaced by the modified $\hat{\rho}$ in (9.5.7). Equation (9.6.9) becomes

$$\text{PE} = \tfrac{1}{2}\,\text{Erfc}\,[\Lambda] \tag{9.6.11}$$

where

$$\Lambda = \left(\frac{E}{N_{b0}}\right)^{1/2}\left(\frac{I_1(\hat{\rho})}{I_0(\hat{\rho})}\right) \tag{9.6.12}$$

and

$$\hat{\rho} = \frac{\rho}{\mu} = \frac{A_s^2/2}{N_{b0}B_L}\left[\frac{1}{1+[N_{b0}B_{sc}/(A_s^2/2)]}\right] \tag{9.6.13}$$

When the subcarrier bandwidth B_{sc} is taken as $1/T$, (9.6.13) can be rewritten as

$$\hat{\rho} = \frac{(E/N_{b0})}{TB_L}\left[\frac{1}{1+(N_{b0}/E)}\right] \tag{9.6.14}$$

with $E/N_{b0} = A_s^2 T/2N_{b0}$. The parameter Λ^2 in (9.6.12) is plotted in Figure 9.18 as a function of E/N_{b0} for several values of $TB_L \geq 1$. The variation from the dashed line shows the degradation to the subcarrier bit energy resulting from the combined squaring and nonideal phase referencing operations. Note the increased suppression with the parameter $B_L T$, exhibiting the disadvantage of using low bit rates relative to the tracking bandwidths. These results show again the importance of properly interfacing the decoding and synchronization subsystems in digital communication design.

To determine analytically the effect of phase referencing errors on block coded systems requires similar procedures, but generally is much more complicated. For an MFSK orthogonal word signaling set, the phase referencing error causes offset phase decoding in each decoder channel.

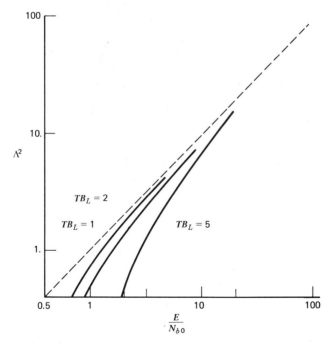

Figure 9.18. Energy degradation with E/N_{b0} for suppressed carrier referencing system ($B_L T \geq 1$).

This modifies the expression for PWE in (8.4.6) to

$$\text{PWE} = 1 - \int_{-\infty}^{\infty} \frac{e^{-y^2}}{\sqrt{2\pi}} \left[1 - \tfrac{1}{2} \text{Erfc} \left(y + \left(\frac{E}{N_{b0}} \right)^{1/2} \cos \theta_e \right) \right]^{M-1} dy$$

$$(9.6.15)$$

where θ_e is the phase error. For a particular value of θ_e, PWE for a specific bit energy E and block size k can be determined by reading off the value in Figure 8.6 at the value $E \cos^2 \theta_e$. Hence phase referencing errors degrade block coded performance just as in the binary case. To determine average performance with respect to a probability density on θ_e requires averaging (9.6.15). When the density in (9.4.4) is assumed, and the loop CNR ρ is related to E/N_{b0} via (9.6.7), the average of of (9.6.15) has been numeri-

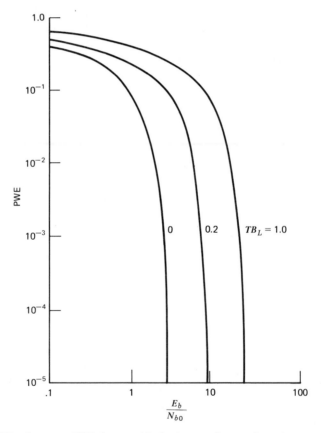

Figure 9.19. Average PWE for nonideal phase referenced systems—orthogonal signaling ($M = 64$ block encoding, E_b = energy per bit).

cally computed by Lindsey [18]. His result for the case $M = 64$ is shown in Figure 9.19. Note the improvement as TB_L is decreased (i.e., as the effective loop CNR used in the referencing system is increased). More complete studies of block coded phase referencing effects can be found in Lindsey and Simon [19].

9.7. Receiver Timing

In addition to the frequency acquisition and phase referencing, a communication receiver may have an inherent requirement to be time synchronized with the received waveform. In digital decoding, for example, we found it necessary to maintain decoder timing in order to identify the beginning and ending of bit and word intervals. Timing synchronization is also necessary for performing range measurements at the receiver [19, 20] and for properly aligning processing operations when dealing with a network of receivers. Also, a receiver that is properly synchronized with a transmitter can execute sequences of commands at prescribed points in time. Time synchronization in a receiver is generally achieved by timing circuitry, often operating in conjunction with the frequency and phase referencing operations. Thus timing circuitry is part of the general synchronization block depicted in our earlier diagrams.

The basic receiver timing problem can be described as one of having to generate receiver timing markers that are precisely synchronized to a sequence of timing reference points immersed within the received carrier waveform. Once the markers are synchronized, they can be used to control receiver operations (i.e., start and stop decoders, activate switches, etc.). We emphasize that it is the time of occurrence of the reference points at the receiver that is important. A receiver may well know that a timing point will be generated at the transmitter at two o'clock, say, but uncertainties in marker generation times and transmission delays prevent it from knowing the exact receiver arrival time. Theoretically, once a single timing point is known at the receiver, all subsequent time reference points are determined (e.g., the receiver can start its own stopwatch at that point). Practically, however, the future time referencing will eventually lose its accuracy (the stopwatch slows down) and the timing points must be retransmitted for timing update. This is often done periodically so that the transmitted time reference points generally appear as a periodic train of such points to which the receiver must synchronize its markers. The repetition period of the points should be as long as possible in order to simplify the timing operation but must be short enough to update receiver timing properly. The latter must occur often enough to take into account the rate of delay variations in the channel, and the memory capability of the receiver (i.e., the length of time for which the receiver can accurately keep track of the last time reference position). The receiver timing subsystem therefore has the task of continually synchronizing to these periodic time locations as they arrive. Since we can envision each timing point as a "tick" of a clock, the basic timing operation is equivalent to maintaining a receiver "clock" in synchronism with a clock immersed within the received waveform.

These are two standard procedures for establishing the timing operation. One is to use a separate subcarrier channel on the RF carrier to convey the time reference points to the receiver; the other is to have the receiver extract the time synchronization directly from the carrier modulation. The former is referred to as *separate channel timing*, and the latter is called *modulation derived timing*. Separate channel timing requires excess power (and sometimes bandwidth) to support the timing subcarrier, whereas modulation derived timing requires more complicated receiver processing. In this section we concentrate on separate subcarrier timing and refer modulation derived timing to Section 9.8.

The simplest method to achieve timing with a separate subcarrier is to use a baseband subcarrier whose frequency is selected to match exactly the desired period of time referencing points (or conversely, the reference period is selected to match the subcarrier frequency). This timing subcarrier must be located away from any baseband carrier modulation already present to avoid interference, as shown in Figure 9.20a. The system can then use the subcarrier tone itself as the timing waveform, using the location of the positive going zero crossings as the time reference points (Figure 9.20b). Receiver timing can be established by filtering off the timing subcarrier from the demodulated baseband and using a PLL to track the subcarrier tone. If the loop VCO tracked subcarrier with zero error, the VCO output would correspond to a noiseless version of the timing subcarrier. The timing marker sequence can then be physically generated by using the VCO output to trigger a pulse generated on its positive zero crossings (Figure 9.20c). With this method, timing system design reduces to the standard phase synchronization problem. The various sources of loop phase error now become sources of timing error, since a phase shift of the VCO output is equivalent to a time shift in its zero crossings. If the loop has a phase error of θ_e radians, the timing error is

$$\epsilon = \left(\frac{\theta_e}{2\pi}\right)\left(\frac{1}{f_t}\right) = \frac{\theta_e}{\omega_t} \sec \qquad (9.7.1)$$

where f_t is the timing subcarrier frequency, and $\omega_t = 2\pi f_t$. The corresponding mean squared timing error caused by a mean squared tracking phase

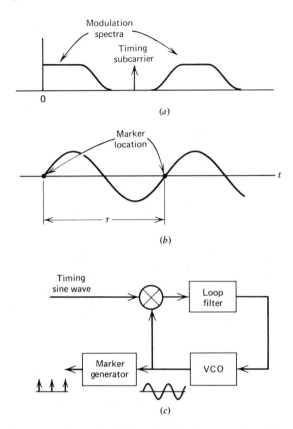

Figure 9.20. Separate subcarrier timing. (*a*) Spectral location, (*b*) subcarrier zero crossings as marker positions, (*c*) subcarrier tracking loop.

error of $\sigma_e^2 \, \text{rad}^2$, follows as

$$\overline{\epsilon^2} = \frac{\sigma_e^2}{\omega_t^2} \tag{9.7.2}$$

Equation (9.7.2) indicates that timing accuracy improves as the frequency of the timing subcarrier increases.† This implies that we should maintain timing with as high a subcarrier as possible. However, if τ is the desired period between time referencing points and if we increase the timing subcarrier frequency f_t beyond the desired fundamental timing frequency

---

† This statement may not always be true, since the loop phase error may increase with the timing frequency. This is especially true if a phase error results from Doppler tracking, since the Doppler offset will be proportional to frequency.

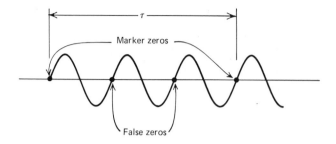

Figure 9.21. Subcarrier timing $(f_t > 1/\tau)$.

$1/\tau$, we increase the number of positive zero crossings in each τ sec interval (Figure 9.21). Even if the loop VCO was perfectly phase locked to the received sine wave, it would not know which zero crossings to use for generating the timing markers in Figure 9.20c. (It knows the markers must be separated by τ sec but does not know at which zero crossing to initiate the markers.) This can be seen analytically by writing the received and receiver sine waves with respect to the marker reference of the receiver. The received subcarrier is then

$$A_r \sin\left[\omega_t(t+\delta)+\theta_r\right] \tag{9.7.3}$$

where θ_r is its phase and δ is the time difference between the received subcarrier reference zero crossing and that used by the local PLL sine wave, as shown in Figure 9.22. The PLL signal is then

$$\cos\left(\omega_t t + \theta_0\right) \tag{9.7.4}$$

The tracking loop mixer output signal then follows as

$$A_r \sin\left(\omega_t\delta + \theta_e\right) \tag{9.7.5}$$

where $\theta_e = \theta_r - \theta_0$ is the phase error. The loop will be in phase lock if the mixer output in (9.7.5) is driven to zero. This occurs if $\theta_e = 0$ (phase synchronism) and if $\omega_t\delta = n2\pi$, or

$$\delta = \frac{n}{f_t}, \qquad n = 0, 1, 2, \ldots, \tau f_t \tag{9.7.6}$$

Each of the preceding values of δ corresponds to a legitimate stable phase lock position and shows that the PLL can achieve phase synchronism without the markers being aligned. This indicates an uncertainty of knowing the marker location even after phase lockup, and an initial ambiguity exists in starting the receiver marker generation. We see that there will be $\tau/(1/f_t) = f_t\tau$ candidate zero crossings that can possibly correspond to the

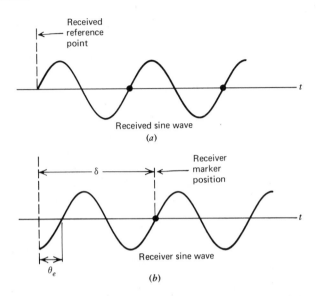

Figure 9.22. Received and receiver subcarriers with offset time reference points.

true marker position. We have therefore the contradictory requirements that increasing the timing subcarrier frequency f_t reduces mean squared timing error when in lock, (9.7.2), but increases the number of location ambiguities in (9.7.6). This ambiguity must be resolved before timing can begin (i.e., the receiver clock can be started). Once the ambiguity is correctly resolved, the timing markers will be continually generated with only the timing error of ϵ in (9.7.1) caused by the imperfect phase tracking. This leads to an eventual timing operation with the accuracy predicted by $\overline{\epsilon^2}$ in (9.7.2), as long as the PLL remains in lock. The operation of resolving the initial position ambiguity is called timing *acquisition*. The operation of continually driving the time marker generator in phase synchronism with the received timing points is called *clocking*. The timing subcarrier being used is therefore called the *clocking signal*. Hence timing with separate subcarriers reduces to simply a clocking operation preceded by an acquisition operation. When clocking is accomplished by a PLL, we see that timing is nothing more than phase referencing with respect to a specified time reference point provided by acquisition. If the clocking is temporarily lost (the PLL loses phase lock), acquisition may have to be reinitiated in order to resume timing.

Acquisition is generally accomplished with the aid of an auxiliary signal, called an *acquisition signal*, that is used in conjunction with the clocking signal. One method is to use two subcarriers for timing, one at the fundamental frequency $f_0 = 1/\tau$ as the acquisition signal and the other at a higher, phase coherent, harmonic subcarrier frequency as a clocking signal. The acquisition subcarrier has no ambiguities, and a PLL phase locked to this fundamental will generate timing markers with no uncertainties. The receiver then will switch to the higher frequency tracking at the phase locked harmonic subcarrier for the clocking. The timing markers generated from the acquisition fundamental are then used to acquire the clocking harmonic. Of all the ambiguous zero crossings of the harmonic, only that which will have occurred at the zero crossing of the fundamental is the true marker position. With acquisition completed, the higher subcarrier produces the clocking with the desired degree of timing accuracy. The power allocated to the timing operation must now be divided between the two subcarriers. This means that each phase locking operation, acquisition and clocking, must be accomplished with a fraction of the available timing power. Also, the baseband bandwidth allowed for the timing operation may have to be increased to handle both harmonics simultaneously. The harmonics must be spaced far enough apart so as to be outside the individual acquisition and clocking loop bandwidths, to prevent one frequency from interfering with the phase locking of the others.

Rather than use both an acquisition and clock subcarrier, another method is to modulate the acquisition fundamental onto the harmonic clock. A popular technique is to amplitude modulate with the fundamental, using standard AM demodulation, to obtain acquisition, then clocking on the AM carrier component. The modulated timing signal uses less overall bandwidth, essentially making use of the AM fundamental sidebands for acquisition. The power division is still required, with the power distribution now determined by the degree of amplitude modulation. To prevent acquisition (modulation) interference on the clock tracking, it is necessary that the clock loop bandwidth be much smaller than the fundamental acquisition frequency (i.e., the AM sidebands must be outside the loop bandwidth). If this cannot be avoided, the degree of acquisition modulation must be kept quite low to reduce interference, provided that transmitter timing power levels permit.

Acquisition signals need not be fundamental sine waves. Recently there has been interest in achieving time acquisition with the use of binary waveform signals called *acquisition codes*. These codes, being digital in nature, are convenient to generate and process and are well suited for PSK modulation on the clock. Let $q(t)$ represent a periodic binary pulse waveform having pulse width W, period τ, which is to serve as an acquisition code. Let us consider forming the timing waveform by PSK modulating the code $q(t)$ onto a clocking subcarrier at frequency f_t. The received version of

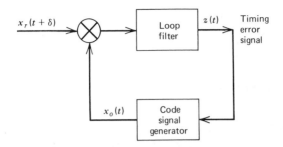

Figure 9.23. Code acquisition tracking loop.

this timing signal will then have the form

$$x(t+\delta) = A_s q(t+\delta) \sin\left[\omega_t(t+\delta) + \theta_r\right] \qquad (9.7.7)$$

where δ is the offset time shift that must be resolved. Note that the timing subcarrier waveform no longer corresponds to a single frequency as before, but instead involves a modulated subcarrier whose bandwidth has been spread to approximately 2/W Hz about the subcarrier frequency. Therefore a larger spectral width must be allocated for receiver timing subcarriers when acquisition codes are used. This required bandwidth is increased as the code symbols are made narrower. For receiver timing, we generate a receiver version of the same coded waveform and clock frequency, as

$$x_o(t) = q(t) \cos\left(\omega_t t + \theta_o\right) \qquad (9.7.8)$$

In the timing subsystem (Figure 9.23) we multiply the two periodic signals and use the dc component† for error control. This generates the multiplier error signal as

$$
\begin{aligned}
z(t) &= \frac{1}{\tau} \int_0^\tau x_r(t+\delta) x_o(t)\, dt \\
&= \frac{1}{\tau} \int_0^\tau A_s q(t+\delta) \sin\left[\omega_t(t+\delta) + \theta_r\right] q(t) \cos\left[\omega_t t + \theta_o\right] dt \\
&\approx \sin\left(\omega_t \delta + \theta_e\right) \left[\frac{1}{\tau} \int_0^\tau q(t) q(t+\delta)\, dt\right] \qquad (9.7.9)
\end{aligned}
$$

where $\theta_e = \theta_r - \theta_o$ is the clock phase error. The last approximation follows

---

† A true dc component appears in the output only if δ, θ_0, θ_r are not functions of t. Since these parameters may vary slowly in time, we should refer to a "slowly varying dc" component or a "low frequency" component. Basically, we mean that these parameters are essentially constant over τ sec time intervals.

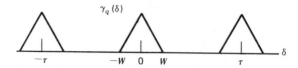

Figure 9.24. Example of code correlation function $\gamma_q(\delta)$.

from the assumption that θ_e and δ do not vary over τ sec intervals, and double frequency terms are neglected. Since the binary waveforms $q(t)$ are periodic, the integral in (9.7.9) is equivalent to the correlation function of $q(t)$, denoted

$$\gamma_q(\delta) = \frac{1}{\tau} \int_0^\tau q(t)q(t+\delta)\,dt \tag{9.7.10}$$

Hence the multiplier output in (9.7.9) is equivalently

$$z(t) = A_s \gamma_q(\delta) \sin\left[\omega_t \delta + \theta_e\right] \tag{9.7.11}$$

If the binary waveforms $q(t)$ were not present $[q(t) = 1,\ \gamma_q(\delta) = 1]$, $z(t)$ would be identical to the mixer term in (9.7.5), which had the ambiguities of (9.7.6). We now see that the insertion of the codes has modified the error signal to that in (9.7.11). Suppose now that the codes had the periodic correlation function that over a τ sec period has the form

$$\gamma_q(\delta) = \begin{cases} 1 - \dfrac{|\delta|}{W}, & |\delta| \le W, \\[2mm] 0 & W \le |\delta| \le \dfrac{\tau}{2} \end{cases} \tag{9.7.12}$$

shown in Figure 9.24. We then see that

$$z(t) = \begin{cases} A_s\left(1 - \dfrac{|\delta|}{W}\right)\sin\left[\omega_t \delta + \theta_e\right], & |\delta| \le W \\[2mm] 0, & W \le |\delta| \le \dfrac{\tau}{2} \end{cases} \tag{9.7.13}$$

Thus if $|\delta| \ge W$, no mixer signal is generated for timing error correction. The clocking loop is effectively opened and cannot phase lock to the clock signal. However, if $|\delta| \le W$, an error voltage is generated, which is then used to drive the clock loop into synchronism. The loop can therefore lock in at $\theta_e = 0$ and at any positive zero crossing point within $|\delta| \le W$; that is, at any ambiguity in $(-W, W)$. We see that if

$$W \le \frac{2}{f_t} = 2\left(\frac{2\pi}{\omega_t}\right) \tag{9.7.14}$$

then only one such zero crossing will occur (at $\delta = 0$). The latter corresponds to a perfectly acquired clock. With the correlation of (9.7.12), therefore, no clock tracking occurs unless $|\delta| < 2/f_t$, in which case the receiver clock pulls into phase synchronism with no ambiguities. Acquisition is thus achieved by using the code $q(t)$ to avoid false lock-in. When the timing offset is outside this range, mechanism must be available to change the receiver code time shift relative to the input, so as to alter the value of δ until pull-in does occur. To describe how this time shifting mechanism is implemented, it is first necessary to consider the basic structure of the code generators. We point out that if we select $W = 2/f_t$, as suggested by (9.7.14), then the number of pulses (*chips*) in the code is

$$n = \frac{\tau}{W} = \frac{\tau}{1/f_t} = \tau f_t \qquad (9.7.15)$$

Conversely, the number of code chips per period determines the usable clock frequency f_t. Thus the larger the code length (number of chips per period), the higher the clocking frequency that can be used with it and the more accurate the clocking after acquisition is completed.

If the clock frequency is operated at a higher multiple k of the code repetition rate $1/W$, then more than one ambiguity point is contained in the width of $\gamma_q(\delta)$ in (9.7.12). In particular, ambiguities will occur at $\delta = \alpha W/k$, $\alpha = 1, 2, \ldots, k$. Because of the correlation function, however, each such ambiguity point will correspond to a different peak amplitude for $z(t)$. Suppose a quadrature multiplier is operated in parallel with the tracking loop, as shown in Figure 9.25. This quadrature multiplication of

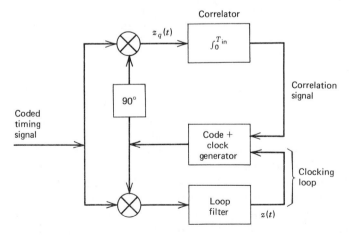

Figure 9.25. Code acquisition loop with correlation detection.

the input is achieved by using the quadrature version of the clock frequency, similar to that of the Costas referencing loops. Quadrature multiplication, using the same local code, will generate a mixer output signal

$$z_q(t) = A_s\left(1 - \frac{|\delta|}{W}\right)\cos\left(\omega_t\delta + \theta_e\right) \qquad (9.7.16)$$

When the loop mixer signal term is driven to zero at the ambiguity point $\delta = \alpha/f_t = \alpha W/k$, the quadrature multiplier will have the signal value

$$z_q(t) = A_s\left(1 - \frac{\alpha}{k}\right), \qquad \alpha \le k \qquad (9.7.17)$$

Thus each ambiguity point produces a different quadrature voltage. Theoretically, each voltage value is unique, and the ambiguities could be resolved by an accurate voltage measurement. The quadrature mixer is therefore providing an observation of the correlation condition within the loop. This is often referred to as a *correlation detector* and performs much like the lock detectors of the referencing loops in Figure 9.1. Practically, however, additive noise obscures the voltage values in (9.7.17), unless sufficiently long integration times (T_{in} in Figure 9.25) are used to distinguish their values.

The desired code correlation in (9.7.12) can be adequately approximated by the correlation function of codes generated from *digital feedback shift registers*. Such a code generator is shown in Figure 9.26a. Digital sequences are shifted through the register one bit at a time at a preselected shifting rate. The register output bits form the acquisition code by generating the binary waveforms in (7.5.3). At any shifting time, the contents of the register are binary combined and fed back to form the next input bit to the register. Once started with an initial bit sequence stored in the register, the device continually regenerates its own inputs, while shifting bits through the register to form the output code. The feedback logic and the initial bit sequence determine the structure of this output code. With properly designed logic, the output codes can be made to be periodic and the register acts like a free-running *code oscillator*. Codes generated in this manner are called *shift register codes*. The class of periodic shift register codes having the desired binary waveform correlation of Figure 9.26b is called *pseudorandom noise* (PRN) codes. PRN codes are known to have the longest period length associated with a given register size and are therefore also called *maximal length* shift register codes. If the register length is K bits, the PRN code will have a period of $n = 2^K - 1$ bits. The underlying mathematical structure of shift register codes and the techniques for

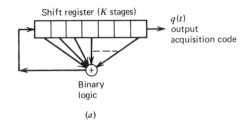

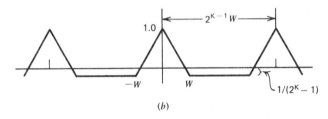

Figure 9.26. Shift register code generators. (*a*) Register diagram, (*b*) PRN code auto-correlation function.

determining proper feedback logic are well documented in the literature [20, 21] and will not be considered here.

A property of shift register codes well suited to timing acquisition is that time shifting of the code can be easily implemented by externally changing the bits in the register at any time shift. Changing these bits causes the output to jump to a new position in the code period, which then continues periodic generation from that point on. Thus, within one shift time, the output code can be forced to skip ahead in its cycle. When the desired code shift is only a relatively few chips (less than the register length), code shifting can be achieved by simply tapping the output at a different register stage.

A practical code acquisition subsystem is shown in Figure 9.27. A clock oscillator at frequency $f_t = n/\tau$ is used with a shift register code generator. If the register code has length n bits, then a code chip is being generated with each cycle of the clock, and the latter can be conveniently used for providing the register shifting rate. The output of the code generator is PSK onto the clock to form the receiver timing signal. This signal is multiplied with the incoming timing signal to generate the low frequency timing error control voltage, which is filtered and fed back to phase control the clock oscillator. If the codes are initially in near time lock ($\delta \le W$), the loop drives the clock and code generator into clock synchronism. The clock

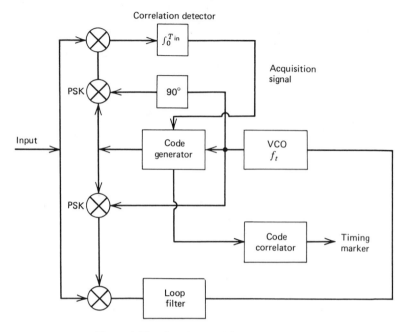

Figure 9.27. A code acquisition subsystem.

can then be used to generate the receiver timing markers in time synchronism with those of the received signal. However, we also note that the code correlation function in Figure 9.26b is much like a timing marker sequence itself. This suggests generating the receiver markers by physically computing the code correlation from the code generator output. As the code is generated, an instantaneous n bit digital correlation is made every register shift time with a stored version of the code. Only when the sequence of output code chips exactly matches the stored version (which only happens once during each code period) will a large correlation value occur. This can then be used as a timing marker. As long as the transmitter uses the same stored code version and the received and local codes are in alignment, the correlation markers will be generated in time synchronism. Note that since the code now generates the marker, the only role of the clock oscillator is to drive the code generator properly.

If the codes are initially offset by a larger difference, zero error signal is produced and the clock loop is inoperative. The status of the loop can be monitored by observing the voltage generated in a quadrature multiplier (correlation detector) operating in parallel with the loop. The correlation detector voltage is obtained by integrating (filtering) the quadrature multiplier output. The output of the correlation detector integrator has a mean (dc) value given by

$$z_q(t) = A_s \gamma_q(\delta) \cos(\omega_t \delta + \theta_e) \qquad (9.7.18)$$

If the codes are initially misaligned, $z_q(t) \approx 0$ and no detector voltage is noted. If the loop is in time lock, $\theta_e = 0$, $\delta = 0$, and $z_q(t) = A_s$. Hence by observing the quadrature integrated voltage, the acquisition condition of the loop is determined. If unacquired, a shifting signal can be applied to the code shift register to jump the code within its cycle, thereby altering the time offset between the incoming and receiver codes. This generates a new loop error signal that, after T_{in} sec of integration, can be monitored in the quadrature detector. If no acquisition is noted, the register is shifted again. This shifting is repeated, shifting the code one bit at a time, integrating for T_{in} sec, until δ is reduced to zero and the clock loop locks in. The latter is indicated by a sudden increase in correlation detector voltage. Thus acquisition with the use of PRN codes requires a shifting of δ values until clock and code synchronization is achieved. The key point is that false lock-in is theoretically avoided. Note also that once acquisition is achieved, clocking is accomplished with the full available PSK carrier amplitude A_s, and no power is wasted on an unused acquisition signal.

We immediately see a basic drawback to PRN code acquisition when long codes are used. We may have to shift through many code symbols before acquisition is achieved, each requiring T_{in} sec, so that total acquisition time may become quite lengthy. (On the average, approximately one-half the code symbols will be searched before acquiring.) For this reason, there has been interest in finding modifications that can reduce total acquisition time. One method is to use preacquisition detection of the code symbols. In this method the receiver first attempts to detect K consecutive code symbols on the timing subcarrier, using the detected symbols to load into the local shift register. If the symbols were detected correctly, the shift register is perfectly aligned with the incoming code and the subsequent symbols from the local register match the subsequent received symbols, since the latter were generated from the same register connection. However, the code symbol detection must be accomplished without carrier reference (the PSK code must be phase demodulated to a PCM waveform prior to symbol detection) and the subcarrier symbol energy must be suitably high to guarantee correct detection of all K symbols. If one is in error, the codes will not be aligned, and so indicated by the code correlator, and K new symbols must be detected. Another method to reduce acquisition time is to use parallel shift registers at the receiver, simultaneously searching separate portions of the code. The first one yielding a high correlation value is switched into the acquisition loop, whereas the others are disengaged.

9.8. Bit Timing

One of the most important applications of receiver timing is in bit and word decoding in digital systems. Timing markers are required at the bit or word rate of the transmitter, and the timing period τ corresponds to the bit interval T. Since the way in which timing markers are used in bit and word decoding is similar, we concentrate only on bit timing.

If a separate baseband subcarrier were to be used for bit timing alone, then any of our previously considered methods can be used. The transmitter simply separates the digital data and timing waveforms in the baseband format (e.g., by FDM), and the receiver baseband sybsystem processes the waveform, as in Figure 9.28. It merely separates the data and the timing information, carries out the timing acquisition and clocking operations previously discussed, and uses the generated timing markers to time the digital decoders. Since no data can be decoded until the acquisition is completed and the clock is locked in, timing parameters such as acquisition time become important in assessing the overall digital performance. Similarly, timing error variances will ultimately limit the decoding capability, as will be considered subsequently.

In many systems, however, the use of a separate subcarrier for the sole purpose of bit timing is avoided, because of the requirement of excess baseband power and bandwidth. The preference is to derive the bit timing directly from existing hardware, if possible, or to generate it using modulation derived timing. For example, in a PCM format, the timing markers are required at the beginning of each bit to start integration. If the receiver achieved RF phase reference, then by making the transmitted bit rate time synchronous with the RF frequency (i.e., the bit interval begins at an RF positive going zero crossing) bit clocking can be maintained from the

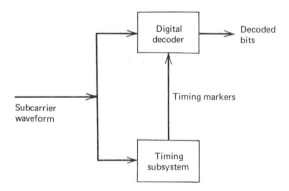

Figure 9.28. Bit timing with separate subcarrier timing.

receiver RF reference. We need only translate the RF, IF, or master oscillator frequency to the bit rate frequency and use the latter zero crossing for bit timing marker generation. The initial ambiguity can be resolved without the acquisition signal by making use of the decoding circuitry. We need only transmit a known bit sequence (say alternating ones and zeros) over the baseband digital channel and monitor the output voltage from the bit integrator. The integrator timing markers can then be shifted in time until a maximum bit correlation (either positive or negative) is observed on each bit. This signifies timing alignment, and the RF reference will maintain the proper clocking as long as it remains in lock. Note that we are basically using the known transmitted bits as an acquisition code and using the decoding integrator as a correlation monitor. Another technique is to make use of any inherent harmonic lines in the digitally modulated baseband spectrum of (7.1.9), treating the harmonic as a sine wave clock signal.

When no such reference carrier or harmonic is available, PCM timing must be obtained directly from the PCM waveform itself. Bit timing subsystems for performing this operation can be roughly classified into two basic types. One involves prefiltering of the data waveform to generate timing waveforms that can be used for time synchronization. The other uses some form of transition tracking of the PCM bit edges, in conjunction with data aided correction, to generate the timing. Both types of systems inherently involve modulation removal and harmonic generation, but the mechanics of each is accomplished in different ways.

Prefiltering systems (Figure 9.29) contain some form of digital waveform filtering (usually high pass filtering to accentuate the waveform transitions) placed in the timing subsystem branch of Figure 9.28. This filtering is then

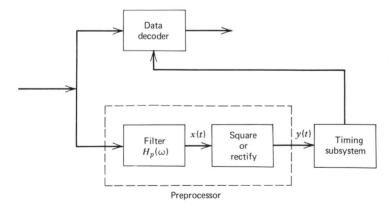

Figure 9.29. Preprocessor bit timing subsystem.

followed by rectification or squaring to remove the modulation and generate a timing waveform for the timing sybsystem that follows. If the binary waveform was composed of ideal, noiseless, PCM bit pulses, high pass filtering in the form of a differentiation would produce a sequence of positive and negative (according to the bit polarity) impulse functions at the bit transitions. Squaring rectifies these impulses into a periodic impulse train at the bit rate, with a power level dependent on the average number of impulses occurring [i.e., on the probability of a bit transition taking place (Problem 9.15)]. These impulses will be offet from their reference time by the transmission timing errors. Tracking of the bit rate harmonic in the timing subsystem by a PLL produces timing markers for the bit decoder that corrects for these offsets. When the baseband noise is included, a pure differentiation cannot be tolerated in the prefilter because of the excessive high frequency noise it produces. For this reason the filters used are usually bandlimiting to reduce this effect. If, in addition, the bit waveforms are not ideal pulses, the role of the preprocessing filter is no longer obvious, and it is not even evident that a timing harmonic can be generated. To examine this, let the baseband binary waveform during a bit interval $(0, T)$ be given as $\pm s(t)$, containing an inherent timing offset. Let the preprocessing filter $H_p(\omega)$ in Figure 9.29 be a bandlimiting filter with bandwidth $1/T$ Hz $(H_p(\omega)=0, |\omega| \geq 2\pi/T)$, converting the waveform $s(t)$ to a modified waveform $p(t)$ having transform

$$P(\omega) = H_p(\omega)F_s(\omega) \tag{9.8.1}$$

where $F_s(\omega)$ is the transform of $s(t)$. Since the binary waveform is now bandlimited, it will have intersymbol interference, and the signal output of the filter over $(0, T)$ must be written as in (7.7.2),

$$x(t) = \sum_{j=0}^{\infty} d_j p(t + jT) \tag{9.8.2}$$

where $d_j = \pm 1$ and represents the bit polarity j bits in the past. The signal component at the output of the squarer (rectifier) is then

$$y(t) = x^2(t) = \sum_{i=0}^{\infty} \sum_{j=0}^{\infty} d_i d_j p(t + jT) p(t + iT) \tag{9.8.3}$$

Since the data bits $\{d_j\}$ are independent and zero mean, $y(t)$ will have a mean signal component of

$$z(t) = \mathscr{E}[y(t)] = \sum_{j=0}^{\infty} (d_j)^2 p^2(t + jT) \tag{9.8.4}$$

This represents the signal waveform available over $(0, T)$ for timing. Since $d_j^2 = 1$, the data modulation is removed. Furthermore, since the summation term will be identical for every bit interval $(0, T)$, (9.8.4) corresponds to a periodic time function with period T sec. We see that indeed the pre-processing will always generate harmonic components from which any timing offset can be tracked. The Fourier transform of (9.8.4) can be obtained by applying the Poisson sum rule of Problem 6.4, yielding

$$Z(\omega) = \frac{2\pi}{T} \sum_{m=-\infty}^{\infty} P_2\left(\frac{2\pi m}{T}\right) \delta\left(\omega - \frac{2\pi m}{T}\right) \tag{9.8.5}$$

where $\delta(\omega)$ is the frequency delta function and

$$P_2(\omega) = P(\omega) \oplus P(\omega) \tag{9.8.6}$$

The harmonic at $m = 1$ (the bit frequency) therefore has the power

$$\frac{1}{2T}\left|P_2\left(\frac{2\pi}{T}\right)\right|^2 \tag{9.8.7}$$

This corresponds to the value of the convolution in (9.8.6) at $\omega = 2\pi/T$. Since $P(\omega)$ in (9.8.1) is bandlimited to $1/T$ Hz, this convolution value is identical to the result obtained by sliding the negative frequency part of $P(\omega)$ onto the positive frequency part and integrating (Figure 9.30a). This integration will be maximized only if $P(\omega)$ is symmetric about $1/2T$ Hz (Problem 9.14). Thus the role of the prefilter, when followed by a har-monic-tracking PLL for timing, is to shape the spectrum of $s(t)$ over $(0, 1/T)$ so as to be symmetrical about the one-half bit frequency, as shown in Figure 9.30b. The subsequent squaring then converts this to harmonic power at the bit frequency. Since $s(t)$ will typically have a low pass spectrum, $H_p(\omega)$ in (9.8.1) will tend to suppress the frequencies below $1/2T$ Hz, while amplifying those above it, similar to a bandlimited high pass characteristic. Deeper analysis into the preprocessor design problem would require investigation of the noise and interference developed in the system and its effect on the timing loop operation. The interested reader may pursue this matter in References 22–24.

A second way to achieve bit timing from a modulated PCM waveform is to utilize *transition tracking* on the bit edges, in conjunction with bit decisioning. Transition tracking is obtained by integrating over the bit transitions (bit edges) to generate a timing error voltage for feedback control. To see this, consider Figure 9.31a, which shows a periodic pulse sequence with a transition being integrated over a w sec integration inter-val. We see that if the integration is centered exactly on the pulse tran-sition, a zero integration value is produced, because of the equal positive and negative areas. If the integration is offset by δ sec, the integrator

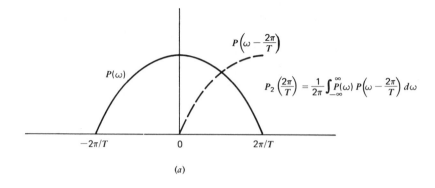

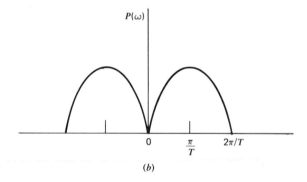

Figure 9.30. Preprocessor spectra. (a) $P_2(\omega)$ convolution, (b) symmetric version.

output generates the function $A_s g(\delta)$ shown in Figure 9.31b. This integrator output signal can then be used as an error correction voltage in a feedback loop, such as that shown in Figure 9.31c. The integrator output is filtered and fed back to control the timing (location) of the integration for the next transition. The timing adjustment can be achieved by controlling the phase of a VCO whose zero crossings drive the start of integration. Since the loop operates to hold the integrator centered on the transitions, timing markers can be generated from the VCO in synchronism with the pulse transitions for bit timing. Such a tracking loop is referred to as a *transition tracking loop*. Note that a timing error value is generated each pulse period, and the loop filter acts to smooth the error samples to give continuous loop control. In this sense, the transition tracking loop has the equivalent timing loop model shown in Figure 9.31d. Note the system is identical to the nonlinear sine wave phase tracking loop of Figure 9.2, except the nonlinear sine function is replaced by the nonlinear $g(\delta)$ function. Thus Figure 9.31d is the transition tracking equivalent to the phase

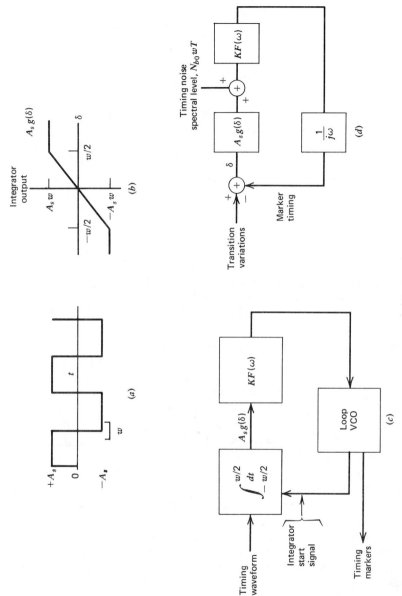

Figure 9.31. Transition tracking model. (*a*) Integrator diagram, (*b*) equivalent error signal, (*c*) tracking loop using error signal, (*d*) equivalent baseband model.

lock clocking loop. We see that $g(\delta)$ is linear in δ for $|\delta| \le w/2$, therefore the transition loop has an equivalent linear model, with the error detector having an effective gain G_e given by the slope of $g(\delta)$ at $\delta = 0$. Hence

$$G_e = \frac{A_s w}{w/2} = 2A_s \qquad (9.8.8)$$

This means transition tracking loops can be analyzed exactly as with phase lock loops. In particular, the transition loop will have a noise bandwidth B_L identical to that of a PLL with the same loop gain function. Tracking errors will be determined from the dynamic errors in trying to track the time variations of the pulse timing, and by the noise entering the loop, just as for a PLL. The timing mean squared error caused by the noise, expressed as a fraction of the pulse period T, is given by

$$\sigma_t^2 = \frac{(N_{b0} w T) 2 B_L}{4 A_s^2 T^2} \qquad (9.8.9)$$

where N_{b0} is the baseband spectral noise level. Note the direct dependence of the timing variance on the integration width w. However, it must be remembered that as w is made smaller, the linear range in Figure 9.31b is also decreased, making linear operation more difficult to retain.

Transition tracking of a PCM bit waveform, however, is slightly more complicated than with periodic pulse trains. This is due to the fact that a given bit of the sequence may have either polarity, which alters the direction of the bit transition. This means that the sign of the transition tracking loop error voltage depends on the direction of the transition (either from $+1$ to -1 or vice versa). This causes the smoothing of the loop filter over a sequence of random bits to average out the loop errors to zero for any δ. That is, no average loop correction voltage will be generated, because of the randomness of the bits. To compensate for this effect, bit transition rectification must be inserted. This rectification can be obtained by using error voltage inversion during a negative transition. By inserting a one bit delay in the loop input and by using the present bit decision, the inversion can be applied directly to the error voltage of the loop by multiplying by the bit polarity, as shown in Figure 9.32. The latter system is referred to as a *data aided transition tracking loop* [19, 25, 26]. The previous bit decisions are effectively used to remove the modulation (rectify the error waveform) so that unmodulated tracking can take place. In a sense, this is the bit timing equivalent of the suppressed carrier phase referencing systems. The bit delay can also be eliminated in modified versions of the system by simply hard limiting the baseband waveform in the upper arm and using the limiter output to invert the error. As long as the noise does not cause

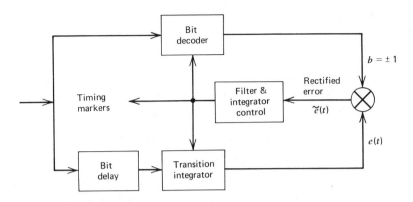

Figure 9.32. Data aided transition tracking loop.

the PCM waveform polarity to change during a bit time, the tracking error will be properly inverted, since the instantaneous limiter output will have the same polarity as the present bit.

Since the data aided loops use auxiliary bit decisions, the performance of the bit timing loop will be directly affected by the bit detection performance. However, since the bit timing is used directly for the bit decoding, the two operations are actually interrelated in a rather complicated way. To attempt to analyze this, we account for the role of the decision loop in Figure 9.32 as an effective multiplication of the transition tracking loop timing error $e(t)$ by a factor b depending on the bit decisions. Thus the loop timing error during a bit interval, when data aided tracking is used, is

$$\tilde{e}(t) = be(t) \tag{9.8.10}$$

where

$$b = \begin{cases} +1, & \text{if bit is correctly detected} \\ -1, & \text{if bit is incorrectly detected} \end{cases} \tag{9.8.11}$$

Since the bit is detected by a PCM binary test, b is actually a random variable. We can therefore write

$$b = \begin{cases} +1, & \text{with prob } 1 - \text{PE}(E/N_{b0}, e) \\ -1, & \text{with prob } \text{PE}(E/N_{b0}, e) \end{cases} \tag{9.8.12}$$

where $\text{PE}(E/N_{b0}, e)$ is the PCM bit error probability of the decoding channel. We have emphasized the fact that it depends on the timing error e (assumed to be constant during the bit interval) and on the bit energy to noise level ratio. If the baseband PCM waveform has amplitude A_s and bit interval T, then

$$\frac{E}{N_{b0}} = \frac{A_s^2 T}{N_{b0}} \tag{9.8.13}$$

When (9.8.10) is inserted into the transition tracking loop equivalent model of Figure 9.31d, the parameter b becomes part of the effective loop gain. The average loop error gain in (9.8.8) is then

$$G_e = 2A_s[1 - \text{PE}(E/N_{b0}, e)] - 2A_s[\text{PE}(E/N_{b0}, e)]$$

$$= 2A_s[1 - 2\text{PE}(E/N_{b0}, e)] \tag{9.8.14}$$

We can again interpret the bracket as an effective multiplication of the pulse amplitude A_s. When (9.8.14) is used in the equivalent system model, we can foresee that the subsequent loop will have a relatively complicated description, since the gain coefficients themselves depend on the error process $e(t)$. This complexity simply manifests the interaction of the timing and decoding operations. This system has been rigorously studied by Lindsey and Simon [19]. For our purposes, however, some indication of the resulting performance can be achieved by making use of a small error approximation. We assume $e \cong 0$ and delete the dependence of PE on e, so that PE corresponds to the perfectly timed results of Chapter 7. As such it depends only on E/N_{0b}. From (9.8.9) the fractional mean squared bit timing error is then

$$\sigma_t^2 = \frac{2N_{b0}B_L(w/T)}{4A_s^2[1 - 2\text{PE}(E/N_{b0})]^2}$$

$$= \frac{w/2T}{\rho[1 - 2\text{PE}(\rho T B_L)]^2} \tag{9.8.15}$$

where

$$\rho = \frac{A_s^2}{N_{b0}B_L} = \frac{E}{N_{b0}}\left(\frac{1}{TB_L}\right) \tag{9.8.16}$$

and B_L is the loop noise bandwidth. Equation (9.8.15) is plotted as a function of ρ in Figure 9.33 for several values of $B_L T$ and for $w/T = 0.25$. The curve $1/\rho$ corresponds to the tracking variance of a system using the same power level without modulation. The result shows the degradation caused by the data aided operation. At high values of ρ, the decoding successfully removes the modulation, and the system performs in a manner identical to that of a pure tracking loop. As ρ is decreased, the decoding errors cause a sharper increase in tracking variance, although care must be used in accepting results at large variance values, because of our small error approximation. The curves, nevertheless, are helpful for determining threshold values for setting ρ in system design.

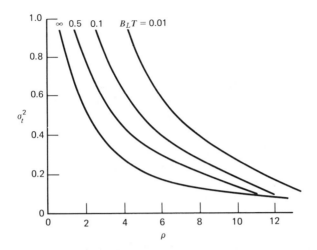

Figure 9.33. Mean squared timing error of data aided transition tracking loop, Equation (9.8.15). ($\rho = E/N_{bo}(1/TB_L)$, T = bit time, B_L = loop bandwidth, $w/T = 0.25$).

The variance curves of Figure 9.33 assume that bit transitions are always available for transition tracking. However, the use of bit edges for NRZ timing suffers from the fact that long sequences of identical bits produce no pulse transitions for clocking update, and the system must depend on "flywheel" action of the tracking loop oscillator during these periods. Effectively, the available harmonic tracking power in the transitions is reduced by the pulse absence, as in Problem 9.15. Insertion of intentional sync bit transitions may be necessary to guarantee the presence of a transition within a prescribed time interval. By using the Manchester coded PCM format in (7.2.8) a transition is guaranteed at the center of every bit interval, independent of the bit sequence, and the transition insertion is not necessary.

We have concentrated on only PCM bit timing. With PSK signals, bit timing can be obtained directly from the subcarrier reference used for the coherent decoding. Since the subcarrier frequency is generally many times the bit rate, the reference must be divided down to provide the clocking markers. This can be provided easily from either the squaring or Costas loop referencing system. Initial bit acquisition can again be obtained by monitoring the integrator voltage for maximal value. When the clocking is obtained directly from the reference loop oscillator in this way, the timing error is directly related to the loop phase error through (9.7.1). Hence the statistics of the clocking time error can be obtained from those of the phase error itself. Specifically, the timing error δ will have a probability density related to that of the phase error at frequency ω_t by

$$p(\delta) = \frac{p_\theta(\theta)}{1/\omega_t}\bigg|_{\theta = \omega_t \delta} = \omega_t p_\theta(\omega_t \delta) \tag{9.8.17}$$

and the timing error variance, normalized to the bit period, is

$$\sigma_t^2 = \frac{\sigma_\theta^2}{\omega_t^2 T^2} = \frac{(\sigma_\theta/2\pi)^2}{(f_t T)^2} \tag{9.8.18}$$

Thus the phase error of the tracking loop is reduced by the ratio of the bit period to the tracking frequency period. Since this ratio can be made quite large by using a PSK subcarrier frequency f_t much larger than the bit rate frequency $1/T$, timing error effects are usually negligible relative to phase error effects. (Recall that the detecting bit energy is degraded directly as $\cos^2 \theta_e$.)

To determine the actual effect of a given timing error on the bit error performance, we refer to our earlier results in Section 7.3. We found that a timing error of δ sec produced a bit error probability $PE(\delta)$ given in

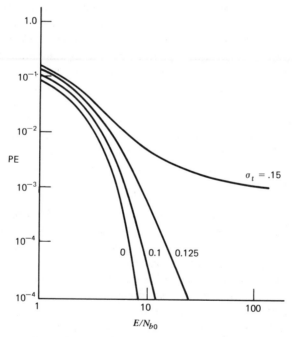

Figure 9.34. Bit error probabilities with timing errors. Here σ_t^2 is the timing variance of the timing subsystem.

(7.3.5). To compute the average error probability caused by a timing error statistically described by (9.8.17), we compute

$$PE = \int_{-\infty}^{\infty} PE(\delta)p(\delta)\,d\delta \qquad (9.8.19)$$

To evaluate this we assume the referencing error density of (9.4.4) in (9.8.17), normalized to produce a timing error variance of σ_t^2. The resulting integration in (9.8.19) produces the results in Figure 9.34 when $PE(\delta)$ has the PCM expression in (7.3.8). [This also corresponds to the PSK case in (7.3.12) if we neglect phase referencing errors.] We again see the presence of the degradation in error probability caused by the phase referencing discussed in Section 9.6. The latter is due to improper signal waveshape, whereas the timing degradation is due to improper integration intervals.

References

1. Lindsey, W. *Synchronization Systems in Communications and Control*, Prentice-Hall, Englewood Cliffs, N.J., 1972.
2. Adler, R. "A Study of Locking Phenomena in Oscillators," *Proc. IRE*, vol. 34, 1946, pp. 351–357.
3. Cunningham, J. *Introduction to Nonlinear Analysis*, McGraw-Hill, New York, 1958.
4. Hayashi, C. *Nonlinear Oscillations in Physical Systems*, McGraw-Hill, New York, 1964.
5. Frazier, J. and Page, J. "Phase Lock Loop Frequency Acquisition," *Trans. IRE*, vol. SET-8, September 1962, pp. 210–227.
6. Viterbi, A. *Principles of Coherent Communications*, McGraw-Hill, New York, 1966.
7. Lindsey, W. Nonlinear Analysis and Synthesis of Generalized Tracking System, *Proc. IEEE*, vol. 67, no. 10, October, 1969, pp. 1705–1722.
8. Tikhonov, V. "The Operation of Phase Automatic Frequency Control in the Presence of Noise," *Automation and Remote Control*, vol. 21, no. 3, 1960, pp. 209–214.
9. Charles, F. and Lindsey, W. "Some Analytical and Experimental Phase Locked Results for Low Signal to Noise Ratios," *Proc. IEEE*, vol. 54, September 1966, pp. 1152–1166.
10. Tauseworth, R. "Simplified Formula for Mean Slip Time of Phase Locked Loops with Steady State Phase Error," *IEEE Trans. Comm. Tech.*, vol. COM-20, June 1972, pp. 331–337.
11. Holmes, J. "First Slip Times Is Static Offset Phase Error for the First and Second Order Phase Locked Loop," *IEEE Trans. Comm. Tech.*, vol. COM-19, April 1971.
12. Schuckman, L. "Time to Cycle Slip in First and Second Order Phase Lock Loops," *Proc. Inter. Conf. Comm.*, San Francisco, June 1970, pp. 34–39.
13. Meyr, H. "Nonlinear Analysis of Corrective Tracking Systems Using Renewal Process Theory," *IEEE Trans. Comm. Tech.*, vol. COM-23, February 1975, pp. 192–203.
14. Lindsey, W. and Meyr, H. "Complete Statistical Description of Phase Error Process Generated by a Corrective Tracking System," *IEEE Trans. Info. Theory*, vol. IT-23, March 1977, pp. 194–203.

15. Tauseworth, R. "Theory and Practical Design of Phase Locked Receivers," Jet Propulsion Lab., Pasadena, Calif., TR No. 32–819, February 15, 1966, pp. 49–55.

16. Davenport, W. and Root, W. *An Introduction to the Theory of Random Signal and Noise,* McGraw-Hill, New York, 1958.

17. Costas, J. "Synchronous Communications," *Proc. IRE,* vol. 44, 1956, pp. 1713–1722.

18. Lindsey, W. "Design of Block Coded Communication Systems," *IEEE Trans. Comm. Tech.,* Vol. COM-15, August 1967, pp. 525–535.

19. Lindsey, W. and Simon, M. *Telecommunications Systems Engineering,* Prentice-Hall, Englewood Cliffs, N.J., 1974.

20. Golomb, S., Baumert, L., Easterling, M., Stiffler, J., and Viterbi, A. *Digital Communications,* Prentice-Hall, Englewood Cliffs, N.J., 1964.

21. Golomb, S. *Shift Register Sequences,* Holden Day, San Francisco, 1967.

22. Roza, E. "Analysis of Phase Locked Timing Extraction Circuits for Pulse Code Transmission," *IEEE Trans. Comm. Tech.,* vol. COM-22, September 1974.

23. Franks, L. and Bubrowski, J. "Statistical Properties of Timing Jitter in Timing Recovery Systems," *IEEE Trans. Comm. Tech.,* vol. COM-22, July 1974.

24. Takasaki, Y. "Timing Extraction in Baseband Pulse Transmission," *IEEE Trans. Comm. Tech.,* vol. COM-20, October 1972.

25. Lindsey, W. and Anderson, T. "Digital Data Transition Tracking Loops," *Proc. Inter. Tele. Conf.,* Los Angeles, October 1968, pp. 259–271.

26. Simon, M. "Optimization of the Performance of Digital Data Tracking Loop," *IEEE Trans. Comm. Tech.,* vol. COM-18, October 1970, pp. 686–690.

Problems

1. (9.1) Derive the second order differential equation that describes the system in (9.1.13) when the loop filter is a pure integrator (i.e., $F(\omega) = 1/j\omega$) and the received frequency is linearly increasing at rate β Hz/sec relative to the VCO frequency of the loop.

2. (9.1) Consider a tracking loop that replaces the frequency mixer by a general error detecting device $g(\theta_e)$. Derive the resulting system differential equation in the absence of modulation. Assume the remaining part of the system (filter and VCO) remain the same.

3. (9.2) Solve the first order system differential equation in (9.2.3). [You will not get an explicit solution for $\theta_e(t)$.] Consider the case where $\Omega/A_c k < 1$ and $\Omega/A_c k > 1$. Sketch the solution for $\theta_e(t)$ for both cases, paying careful attention to periodicity.

4. (9.2) Assume a frequency acquisition employs the lock detector subsystem in Figure P9.4. The output of the quadrature detector is $A \cos \theta_e(t) + n(t)$, where $n(t)$ is the RF mixer noise. This detector signal is low pass filtered and continually compared to a signal threshold. When

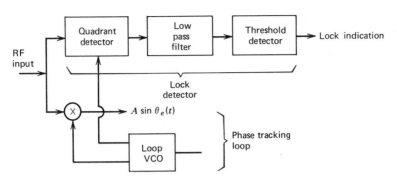

Figure P9.4

threshold is crossed, acquisition lock-up is signaled. (a) Explain how the detector works, considering both the case where $\theta_e(t) = \Omega t$, $\Omega \gg 1$ and where $\theta_e(t) \approx 0$. Explain the role of the low pass filter and how it should be designed. (b) If the nose $n(t)$ is Gaussian, determine the expressions for probabilities of correct lock detection and probability of false lock indication for a particular low pass filter $H(\omega)$.

5. (9.3) An RF frequency acquisition system must acquire a 5 GHz carrier in an uncertainty bandwidth of 100 MHz. The FAL uses a master oscillator at 10 MHz and a loop bandwidth of 10 Hz at the master oscillator frequency. (a) How long will the acquisition operation take, assuming we slew at the maximum rate allowed by (9.3.5) at the RF frequency? (b) How long will it take if we try to acquire at the master oscillator frequency?

6. (9.4) A 2 GHz carrier is received with an unknown frequency offset of up to $\pm$ one part million of the carrier frequency. It must be acquired in one millisec by a slewed acquisition loop using the maximum slewing rate in (9.3.5). How much RF carrier component power is required to maintain a mean squared RF tracking error of $0.01 \, \text{rad}^2$? The RF noise has a spectral level of $N_0 = -190 \, \text{dBW}$.

7. (9.4) It is known that the steady state probability density of the first order nonlinear tracking loop in Figure 7.2 must satisfy the differential equation

$$(\rho \sin \theta_e) p(\theta_e) + \frac{dp(\theta_e)}{d\theta_e} = C$$

where C is an unknown constant to be determined. Solve the equation to derive the density in (9.4.4).

8. (9.4) Consider the probability density in (9.4.4). (a) By approximating the $I_0(\rho)$ term for large ρ (Appendix A) and the cosine term for small θ_e, show that the density approaches a Gaussian density with variance $1/\rho$. (b) By using the Jacobi–Anger expansion equation (2.2.10) show that its mean squared value is given by

$$\sigma_e^2 = \frac{\pi^2}{3} + 4 \sum_{n=1}^{\infty} \frac{(-1)^n I_n(\rho)}{n^2 I_0(\rho)}$$

9. (9.5) Show that the squaring factor in (9.5.6) can be written in general form as

$$\mu = \left[1 + \frac{2}{A_s^2 N_{b0}} \int_{-\infty}^{\infty} R_n^2(\tau)\, d\tau \right]$$

where $R_n(\tau)$ is the autocorrelation function of the noise at the input to the squarer. (b) Evaluate this for the autocorrelation function $R_n(\tau) = (N_{b0}/4) B \exp(-B|\tau|)$.

10. (9.5) Determine the power spectrum of $n_{eq}(t)$ in (9.5.11). The processes $n_{mc}(t)$ and $n_{ms}(t)$ are uncorrelated Gaussian processes. Use autocorrelation analysis and apply convolution, using the spectral shapes in (9.14a) for the mixer noise. The autocorrelation of $n_{mc}(t)$, $n_{ms}(t)$ is one-half that of the mixer noise.

11. (9.5) Consider that the subcarrier signal in Figure 9.12 is passed through a hard limiter prior to subcarrier tracking. Rederive the expression for ρ in the tracking loop by taking into account the effect of the limiter on the original subcarrier waveform. (*Hint:* Follow the analysis in Section 5.5.)

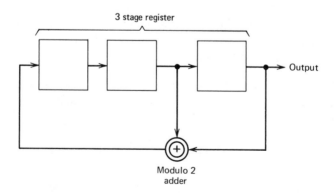

Figure P9.12

12. (9.7) Given the shift register code generator in Figure P9.12. Assume the register is in the all-zero state and begin by shifting a one into the register input. (a) Write the sequence of output bits that will occur. (b) Determine if the sequence is maximal length. (c) Derive its correlation function [use (8.3.8) and compute the correlation at each discrete shift]. (d) Show that changing a register bit at any shift time will jump the sequence to a new point in its cycle.

13. (9.7) Show that the power spectrum of a PRN binary waveform of period n bits and pulse width W is given by

$$S(\omega) = \left(\frac{n+1}{n^2}\right)\left[\frac{\sin(\omega W/2)}{(\omega W/2)}\right]^2 \sum_{\substack{i=-\infty \\ i \neq 0}}^{\infty} \delta\left(\omega - \frac{2\pi i}{nW}\right) + \frac{2\pi}{n^2}\delta(\omega)$$

14. (9.8) Show that $P_2(\omega)$ in Figure 9.30a is maximized by the spectrum in Figure 9.30b. (*Hint*: Apply the Schwartz inequality.)

15. (9.8) Given a periodic sequence of impulses in which there is a probability p that an impulse will occur in a particular periodic position (i.e., there is probability $1-p$ that a given impulse is missing). Show that the power at the impulse harmonic at the repetition rate frequency will depend on p. (*Hint*: Follow the spectral density development in Section 7.1.)

APPENDIX A

FOURIER TRANSFORMS AND IDENTITIES

A.1. The Fourier Transform

$$X(\omega)=\int_{-\infty}^{\infty} x(t)\, e^{-j\omega t}\, dt$$

The Inverse Fourier Transform

$$x(t)=\frac{1}{2\pi}\int_{-\infty}^{\infty} X(\omega)\, e^{j\omega t}\, d\omega$$

Time Convolution

$$x(t)\oplus y(t)=\int_{-\infty}^{\infty} x(t-\rho)y(\rho)\, d\rho$$

Frequency Convolution

$$X(\omega)\oplus Y(\omega)=\frac{1}{2\pi}\int_{-\infty}^{\infty} X(\omega-\rho)Y(\rho)\, d\rho$$

Transform Properties

linearity $\qquad \alpha x(t)+\beta y(t)\leftrightarrow\alpha X(\omega)+\beta Y(\omega)$

delay $\qquad\quad x(t-\alpha)\leftrightarrow e^{-j\omega\alpha}X(\omega)$

scale $\qquad\quad x(\alpha t)\leftrightarrow\dfrac{1}{|\alpha|}X\!\left(\dfrac{\omega}{\alpha}\right)$

conjugate	$x^*(t) \leftrightarrow X^*(-\omega)$
duality	$X(t) \leftrightarrow x(\omega)$
shift	$e^{j\alpha t} x(t) \leftrightarrow X(\omega - \alpha)$
differentiation	$\dfrac{d^n x(t)}{dt^n} \leftrightarrow (j\omega)^n X(\omega)$
integration	$\displaystyle\int_{-\infty}^{t} x(\rho)\, d\rho \leftrightarrow X(\omega)/j\omega$
multiplication	$x(t)y(t) \leftrightarrow X(\omega) \oplus Y(\omega)$
convolution	$x(t) \oplus y(t) \leftrightarrow X(\omega) Y(\omega)$

Useful Transforms

$x(t)$	$X(\omega)$
$e^{-bt}, \quad t > 0$	$\dfrac{1}{b + j\omega}$
$\cos(\omega_c t + \psi)$	$\pi[\delta(\omega - \omega_c)e^{j\psi} + \delta(\omega + \omega_c)e^{-j\psi}]$
$\sin(\omega_c t + \psi)$	$(\pi/j)[\delta(\omega - \omega_c)e^{j\psi} - \delta(\omega + \omega_c)e^{-j\psi}]$
1	$2\pi\delta(\omega)$
$\delta(t)$	1
$1, \quad t > 0$	$\dfrac{1}{j\omega}$
$\dfrac{1}{\sqrt{2\pi}} e^{-t^2/2}$	$e^{-\omega^2/2}$
$t, \quad t > 0$	$\left(\dfrac{1}{j\omega}\right)^2$
$t^2, \quad t > 0$	$\dfrac{2}{(j\omega)^3}$

A.2. Theorems, Identities, and Expansions

Parcevals

$$\int_{-\infty}^{\infty} x(t)y(t)\, dt = \frac{1}{2\pi} \int_{-\infty}^{\infty} X(\omega)Y^*(\omega)\, d\omega$$

$$\int_{-\infty}^{\infty} |x(t)|^2\, dt = \frac{1}{2\pi} \int_{-\infty}^{\infty} |X(\omega)|^2\, d\omega$$

Schwartz Inequality

$$\left| \text{Real} \int_{-\infty}^{\infty} x(t)y^*(t)\,dt \right| \le \left[\int_{-\infty}^{\infty} |x^2(t)|\,dt \int_{-\infty}^{\infty} |y^2(t)|\,dt \right]^{1/2}$$

equality holds if $x(t) = Cy(t)$.

Holder Inequality

$$\left| \sum_{i=0}^{\infty} a_i b_i \right| \le \left[\sum_{i=0}^{\infty} |a_i^2| \sum_{i=0}^{\infty} |b_i^2| \right]^{1/2}$$

equality if $a_i = Cb_i$.

Final Value

$$\lim_{t \to \infty} x(t) = \lim_{\omega \to 0} [j\omega X(\omega)]$$

Bessel Identities

$$J_n(x) = \left(\frac{x}{2}\right)^n \sum_{k=0}^{\infty} \frac{(-x^2/4)^k}{k!(n+k)!}$$

$$= \frac{j^{-n}}{\pi} \int_0^{\pi} e^{jx \cos \theta} \cos(n\theta)\,d\theta$$

$$J_n(xe^{jm\pi}) = e^{jnm\pi} J_n(x)$$

$$I_n(x) = j^n J_n(x/j)$$

$$= \frac{1}{\pi} \int_0^{\pi} e^{x \cos \theta} \cos(n\theta)\,d\theta$$

Useful Trigonometric Identities

$$\sin(\alpha \pm \beta) = \sin \alpha \cos \beta \pm \cos \alpha \sin \beta$$

$$\cos(\alpha \pm \beta) = \cos \alpha \cos \beta \mp \sin \alpha \sin \beta$$

$$\tan(\alpha \pm \beta) = \frac{\tan \alpha \pm \tan \beta}{1 \mp \tan \alpha \tan \beta}$$

$$\sin \alpha \sin \beta = \tfrac{1}{2} \cos(\alpha - \beta) - \tfrac{1}{2} \cos(\alpha + \beta)$$

$$\cos \alpha \cos \beta = \tfrac{1}{2} \cos(\alpha - \beta) + \tfrac{1}{2} \cos(\alpha + \beta)$$

$$\sin \alpha \cos \beta = \tfrac{1}{2} \sin(\alpha - \beta) + \tfrac{1}{2} \sin(\alpha + \beta)$$

$$e^{\pm j\theta} = \cos \theta \pm j \sin \theta$$

$$\cos\theta = \tfrac{1}{2}(e^{j\theta}+e^{-j\theta})$$

$$\sin\theta = (e^{j\theta}-e^{-j\theta})/2j$$

$$\sin^2\theta + \cos^2\theta = 1$$

$$\cos^2\theta - \sin^2\theta = \cos 2\theta$$

$$\cos^2\theta = \tfrac{1}{2}(1+\cos 2\theta)$$

$$\cos^3\theta = \tfrac{1}{4}(3\cos\theta + \cos 3\theta)$$

$$\sin^2\theta = \tfrac{1}{2}(1-\cos 2\theta)$$

$$\sin^3\theta = \tfrac{1}{4}(3\sin\theta - \sin 3\theta)$$

Expansions

$$(1+x)^n = 1 + nx + \frac{n(n-1)}{2!}x^2 + \cdots \qquad |nx| < 1$$

$$e^x = 1 + x + \frac{1}{2!}x^2 + \cdots$$

$$a^x = 1 + x\ln a + \frac{1}{2!}(x\ln a)^2 + \cdots$$

$$\ln(1+x) = x - \tfrac{1}{2}x^2 + \tfrac{1}{3}x^3 + \cdots$$

$$\sin x = x - \tfrac{1}{3!}x^3 + \tfrac{1}{5!}x^5 - \cdots$$

$$\cos x = 1 - \frac{1}{2!}x^2 + \frac{1}{4!}x^4 - \cdots$$

$$\tan x = x + \tfrac{1}{3}x^3 + \tfrac{2}{15}x^5 + \cdots$$

$$\sum_{m=0}^{M} x^m = \frac{(x^M - 1)}{(x-1)}$$

$$e^{a\cos b} = \sum_{i=0}^{\infty} \epsilon_i I_i(a)\cos(ib), \qquad \epsilon_0 = 1,\ \epsilon_i = 2,\ i \geq 1$$

$$\cos(x\sin\theta) = J_0(x) + 2\sum_{k=1}^{\infty} J_{2k}(x)\cos(2k\theta)$$

$$\sin(x\sin\theta) = 2\sum_{k=0}^{\infty} J_{2k+1}(x)\sin[(2k+1)\theta]$$

$$\cos(x\cos\theta) = J_0(x) + 2\sum_{k=0}^{\infty} (-1)^k J_{2k}(x)\cos(2k\theta)$$

$$\sin (x \cos \theta) = 2 \sum_{k=0}^{\infty} (-1)^k J_{2k+1}(x) \cos [2k+1)\theta]$$

$$\int_0^{\infty} \frac{dx}{1+x^n} = \frac{(\pi/n)}{\sin (\pi/n)}, \quad n>1 \qquad \int_0^{\infty} \frac{dx}{(a^2+x^2)^2} = \frac{\pi}{4a^3}$$

$$\int_0^{\infty} \frac{x^u \, dx}{1+x^n} = \left(\frac{\pi}{n}\right) \csc \left[\frac{(u+1)\pi}{n}\right] \qquad \int_0^b \frac{dx}{1+x^2} = \tan^{-1} (b)$$

A.3.   Calculus of Variations

We wish to minimize or maximize the integral

$$\Gamma[H(\omega)] = \int_{-\infty}^{\infty} f[H(\omega)] \, d\omega \tag{A.3.1}$$

with respect to the function $H(\omega)$, where $f(x)$ is a differentiable functional. To accomplish this, we use the *calculus of variations* [1, 2]. We let $H_o(\omega)$ be the desired solution to be determined. Assuming this solution exists, then any perturbation from this solution must produce a larger value of Γ (smaller, if $H_o(\omega)$ is a maximizing solution). Thus for any arbitrary function $\eta(\omega)$ and any $\epsilon > 0$, we must have

$$\Gamma[H_o(\omega) + \epsilon \eta(\omega)] \geq \Gamma[H_o(\omega)] \qquad \text{if a minimum}$$
$$\leq \Gamma[H_o(\omega)] \qquad \text{if a maximum} \tag{A.3.2}$$

This means $\Gamma[H_o(\omega) + \epsilon \eta(\omega)]$, considered as a function of ϵ, must have an extremal point at $\epsilon = 0$, no matter how $\eta(\omega)$ is selected. This means

$$\left. \frac{\partial \Gamma[H_o(\omega) + \epsilon \eta(\omega)]}{\partial \epsilon} \right|_{\epsilon=0} = 0 \tag{A.3.3}$$

or equivalently,

$$\left. \int_{-\infty}^{\infty} \frac{\partial f[H_o(\omega) + \epsilon \eta(\omega)]}{\partial \epsilon} \, d\omega \right|_{\epsilon=0} = 0 \tag{A.3.4}$$

The only way the integral can be zero for every choice of $\eta(\omega)$ is to have the integrand zero. Hence we require

$$\left. \frac{\partial f[H_o(\omega) + \epsilon \eta(\omega)]}{\partial \epsilon} \right|_{\epsilon=0} = 0 \tag{A.3.5}$$

for every $\eta(\omega)$. The preceding is a necessary condition that any solution $H_o(\omega)$ must satisfy. That is, if a solution exists, it must satisfy (A.3.5). The latter equation is a form of the *Euler–Lagrange* equation. Any such extremal solution found must then be tested in (A.3.1) to determine if it is a valid minimum (or maximum).

To minimize or maximize (A.3.1) subject to the constraint

$$\int_{-\infty}^{\infty} g[H_o(\omega)] \, d\omega = C \qquad (A.3.6)$$

we instead define

$$\Gamma[H(\omega)] = \int_{-\infty}^{\infty} f[H(\omega)] \, d\omega + \lambda \left[\int_{-\infty}^{\infty} g[H(\omega)] \, d\omega - C \right] \qquad (A.3.7)$$

$$= \int_{-\infty}^{\infty} [f[H(\omega)] + \lambda g[H(\omega)]] \, d\omega - \lambda C \qquad (A.3.8)$$

where λ is an arbitrary constant (called the Lagrange multiplier). Since extremal points of $\Gamma[H(\omega)]$ must be extremal points of the first integral in (A.3.7) [since the last terms must sum to zero for $H(\omega) = H_o(\omega)$], we can instead proceed to determine the extremal points of (A.3.8). Since the last term is a constant and will not affect the minimization, we seek the function $H(\omega)$ that minimizes (or maximizes) instead

$$\int_{-\infty}^{\infty} [f[H(\omega)] + \lambda g[H(\omega)]] \, d\omega \qquad (A.3.9)$$

Proceeding as in (A.3.1)–(A.3.5), we therefore attempt to find extremal solutions $H_o(\omega)$ to

$$\frac{\partial}{\partial \epsilon} [f[H_o(\omega) + \epsilon \eta(\omega)] + \lambda g[H_o(\omega) + \epsilon \eta(\omega)]] \Big|_{\epsilon=0} = 0 \qquad (A.3.10)$$

for arbitrary $\eta(\omega)$. Any solution $H_o(\omega)$ will necessarily contain the arbitrary parameter λ. To determine the proper value of λ, we substitute $H_o(\omega)$ back into (A.3.6) and find the necessary λ to satisfy the constraint.

References

1. Fomin, A. and Gelfang, A. *Calculus of Variations*, Prentice-Hall, Englewood Cliffs, N.J., 1965.
2. Thomas, J. *Statistical Communication Theory*, Wiley, New York, 1967, Appendix H.

APPENDIX B

RANDOM VARIABLES
AND RANDOM PROCESSES

In this appendix we briefly summarize the basic definitions and conclusions of probability theory, random variables, and random processes needed throughout the text. The prime objective is for this discussion to serve as a fundamental review while achieving the prerequisite level and introducing the symbolic notation used. A reader previously versed in these areas will find this appendix unnecessary. The reader who is somewhat familiar with the material or the reader who may not have kept up with these topics will find this appendix beneficial as a summary or refresher discussion.

B.1. Random Variables

Let x be a random variable, having the probability density $p_x(\xi)$, where $\int_{-\infty}^{\infty} p_x(\xi)\, d\xi = 1$. The probability that x lies in an interval (a, b) is given by

$$\text{Prob}\,[a \leq x \leq b] = \int_{a}^{b} p_x(\xi)\, d\xi \tag{B.1.1}$$

The *average (expected, mean) value* of any function of x, $f(x)$, is denoted by

$$\mathscr{E}[f(x)] = \int_{-\infty}^{\infty} f(\xi) p_x(\xi)\, d\xi \tag{B.1.2}$$

where $\mathscr{E}$ is called the expectation operator. The moments of x are defined by

$$m_n = \mathscr{E}[x^n] \tag{B.1.3}$$

The first several moments are the most important and are denoted

$$m_1 = mean \text{ of } x = \mathcal{E}[x] \tag{B.1.4a}$$

$$m_2 = mean \text{ squared value of } x = \mathcal{E}[x^2] \tag{B.1.4b}$$

Also important is the *variance* of x defined by

$$\sigma^2 = \mathcal{E}[(x - m_1)^2] = m_2 - m_1^2 \tag{B.1.5}$$

The *characteristic function* of x is defined as

$$\Psi_x(\omega) = \mathcal{E}[e^{j\omega x}] = \int_{-\infty}^{\infty} e^{j\omega\xi} p_x(\xi)\, d\xi \tag{B.1.6}$$

Note that $\Psi_x(\omega)$ is the Fourier transform of $p_x(\xi)$. Also, the moments of x can be determined directly from its characteristic function by

$$m_n = \frac{1}{j^n} \left[\frac{\partial^n \Psi_x(\omega)}{\partial \omega^n} \right]_{\omega=0} \tag{B.1.7}$$

If x is a random variable, then $y = g(x)$ is also a random variable obtained from x by the transformation g. The probability density of y is related to that of x via

$$p_y(\xi) = \left[\frac{p_x(x)}{|dg/dx|} \right]_{x=g^{-1}(\xi)} \tag{B.1.8}$$

where $g^{-1}(\xi)$ is the solution of $\xi = g(x)$ for x.

The most important random variable in engineering is the Gaussian random variable. A Gaussian random variable is one whose probability density is

$$p_x(\xi) = \frac{1}{\sqrt{2\pi}\sigma} e^{-(\xi-m)^2/2\sigma^2} \tag{B.1.9}$$

where m is its mean and σ^2 its variance. The density is sketched in Figure B.1. The probability that a Gaussian random variable will lie in the interval $(m - b, m + b)$ and about its mean m is then

$$\text{Prob}\,[(m-b) \le x \le (m+b)] = \int_{m-b}^{m+b} \frac{1}{\sqrt{2\pi}\sigma} \exp\left(-\frac{(\xi-m)^2}{2\sigma^2} \right) d\xi$$

$$= \frac{2}{\sqrt{\pi}} \int_0^{b/\sqrt{2}\sigma} e^{-u^2}\, du \tag{B.1.10}$$

The function

$$\text{Erf}\,(a) = \frac{2}{\sqrt{\pi}} \int_0^a e^{-u^2}\, du \tag{B.1.11}$$

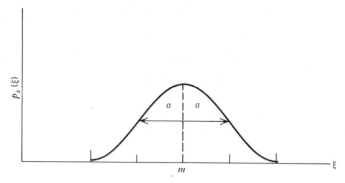

Figure B.1. The Gaussian probability density.

Table B.1. The Error Function Erf (a)

a	erf (a)	a	erf (a)
0.00	0.00000	1.05	0.86244
0.05	0.05637	1.10	0.88021
0.10	0.11246	1.15	0.89612
0.15	0.16800	1.20	0.91031
0.20	0.22270	1.25	0.92290
0.25	0.27633	1.30	0.93401
0.30	0.32863	1.35	0.94376
0.35	0.37938	1.40	0.95229
0.40	0.42839	1.45	0.95970
0.45	0.47548	1.50	0.96611
0.50	0.52050	1.55	0.97162
0.55	0.56332	1.60	0.97635
0.60	0.60386	1.65	0.98038
0.65	0.64203	1.70	0.98379
0.70	0.67780	1.75	0.98667
0.75	0.71116	1.80	0.98909
0.80	0.74210	1.85	0.99111
0.85	0.77067	1.90	0.99279
0.90	0.79691	1.95	0.99418
0.95	0.82089	2.00	0.99532
1.00	0.84270	2.50	0.99959
		3.00	0.99998

is called the *error function* and is tabulated in Table B.1. The *complementary error function* Erfc (a) is defined as

$$\text{Erfc } (a) = 1 - \text{Erf } (a) \tag{B.1.12}$$

The Erfc function is accurately approximated by

$$\text{Erfc } (a) \cong \frac{e^{-a^2}}{\sqrt{\pi}a} \tag{B.1.13}$$

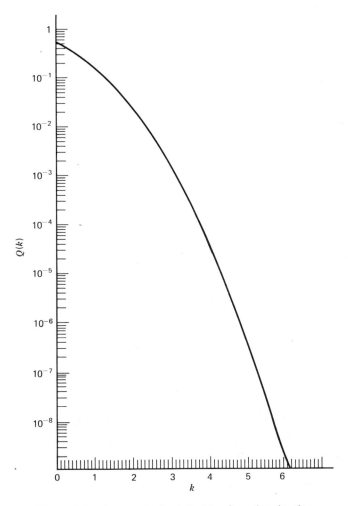

Figure B.2. Area under the tail of the Gaussian density.

when $a \geq 3$. Another important function related to the Gaussian density is the area under the tail of the curve, defined as

$$Q(k) = \frac{1}{\sqrt{2\pi}} \int_k^\infty e^{-u^2/2} \, du \qquad (B.1.14)$$

and is plotted in Figure B.2. This function is related to the Erfc function by

$$\text{Erfc}\,(a) = 2Q(\sqrt{2}a)$$

More extensive tabulations of these functions are available [1].

B.2. Joint Random Variables

Let x, y be a pair of real random variables having joint probability density $p_{xy}(x, y)$. Then

$$\text{Prob}\,[x \in I_x \quad \text{and} \quad y \in I_y] = \int_{I_y} \int_{I_x} p_{xy}(x, y) \, dx \, dy \qquad (B.2.1)$$

The individual densities $p_x(x)$ can be obtained by integrating the joint density,

$$p_x(x) = \int_{-\infty}^\infty p_{xy}(x, y) \, dy; \qquad p_y(y) = \int_{-\infty}^\infty p_{xy}(x, y) \, dx \qquad (B.2.2)$$

The joint expectation operator becomes

$$\mathscr{E}_{xy}[f(x, y)] = \int\int_{-\infty}^{\infty} f(x, y) p_{xy}(x, y) \, dx \, dy \qquad (B.2.3)$$

for any function f of x and y. The (ij) joint moments are then defined by

$$m_{ij} = \mathscr{E}_{xy}[x^i y^i] \qquad (B.2.4)$$

and the joint characteristic function is

$$\Psi_{xy}(\omega_1, \omega_2) = \mathscr{E}_{xy}[\exp\,(j\omega_1 x + j\omega_2 y)] \qquad (B.2.5)$$

Note that the joint characteristic function is the two dimensional Fourier transform of $p(x, y)$. An important moment in (B.2.4) is $m_{11} = \mathscr{E}(xy)$. The variables x and y are said to be *uncorrelated* if $m_{11} = \mathscr{E}(x)\mathscr{E}(y)$. The variables are said to be *independent* if the joint density factors as

$$p_{xy}(x, y) = p_x(x)p_y(y) \qquad (B.2.6)$$

The conditional density of x given $y = \hat{y}$ is denoted

$$p_{x|y}(x|\hat{y}) = \frac{p_{xy}(x, \hat{y})}{p_y(\hat{y})} \qquad (B.2.7)$$

and can be interpreted as the probability density of x when y has the value $\hat{y}$. The conditional expectation operator is then

$$\mathscr{E}_{x|y}[f(x)] = \int_{-\infty}^{\infty} f(x) p_{x|y}(x|y) \, dx \qquad (B.2.8)$$

and is a function of y. The joint expectation in (B.2.3) is seen to be equivalent to the sequence of expectations,

$$\mathscr{E}_{xy}[f(x, y)] = \mathscr{E}_y[\mathscr{E}_{x|y}[f(x, y)]] \qquad (B.2.9)$$

and can therefore be obtained by first conditional averaging, then averaging over the conditioning variable. Conditional moments and characteristic functions can be obtained as in (B.1.3) and (B.1.6) using conditional probabilities.

When dealing with sums of independent random variables, $z = x + y$, we note that the characteristic function of z becomes

$$\Psi_z(\omega) = \mathscr{E}_z[e^{j\omega z}] = \mathscr{E}_z[e^{j\omega(x+y)}] = \Psi_x(\omega)\Psi_y(\omega) \qquad (B.2.10)$$

and is therefore equal to the product of the characteristic function of each term. Being a Fourier transform, the probability density of z is then the convolution of the individual densities of x and y:

$$p_z(z) = \int_{-\infty}^{\infty} p_x(z - u) p_y(u) \, du \qquad (B.2.11)$$

If (u, v) are two new random variables formed from (x, y) by the transformations

$$u = g_1(x, y)$$
$$v = g_2(x, y) \qquad (B.2.12)$$

then the joint density of (u, v) is obtained from that of (x, y) by the density transformation

$$p_{uv}(u, v) = \frac{1}{|J|} p_{xy}(x, y)\Big|_{(x, y) = g^{-1}(u, v)} \qquad (B.2.13)$$

where J is the determinant of the Jacobian matrix

$$J = \det \begin{bmatrix} \dfrac{\partial g_1}{\partial x} & \dfrac{\partial g_1}{\partial y} \\[2mm] \dfrac{\partial g_2}{\partial x} & \dfrac{\partial g_2}{\partial y} \end{bmatrix} \qquad (B.2.14)$$

and $(x, y) = g^{-1}(u, v)$ is the inverse solution of the pair of equations in (B.2.12) for (x, y). That is, (B.2.12) is solved for x and y and the results are substituted into the right side of (B.2.13) to yield $p_{uv}(u, v)$.

A pair of random variables are *jointly Gaussian* variables if

$$p_{xy}(x, y) = \frac{1}{2\pi|\mu|^{1/2}}$$

$$\times \exp - \left[\frac{\mu_{02}(x - m_{10})^2 + \mu_{20}(y - m_{01})^2 - 2\mu_{11}(x - m_{10})(y - m_{01})}{2|\mu|} \right]$$

(B.2.15)

where $|\mu| = \mu_{20}\mu_{02} - \mu_{11}^2$, $\mu_{20} = \sigma_x^2$, $\mu_{02} = \sigma_y^2$, and $\mu_{11} = m_{11} - m_{10}m_{01}$. The corresponding joint characteristic function of (B.2.5) follows as

$$\Psi_{xy}(\omega_1, \omega_2) = \exp \tfrac{1}{2}[\mu_{20}\omega_1^2 + \mu_{02}\omega_2^2 + 2\mu_{11}\omega_1\omega_2] \exp j[m_{10}\omega_1 + m_{01}\omega_2]$$

(B.2.16)

B.3. Sequences of Random Variables

Let $\mathbf{x} = (x_1, x_2, \ldots, x_N)$ be a sequence (ordered set) of N real random variables. Many of the properties of $\mathbf{x}$ are then simply the N-dimensional extension of the properties of two random variables. In particular, the N-dimensional joint density $p_{\mathbf{x}}(x_1, x_2, \ldots, x_N)$ can be used to derive joint probabilities of being in specific intervals. The joint average over the sequences of any function $f(\mathbf{x})$ is given by

$$\mathscr{E}_{\mathbf{x}}[f(\mathbf{x})] = \int_X f(\mathbf{x})p_{\mathbf{x}}(\mathbf{x}) \, d\mathbf{x}$$

$$= \int_{-\infty}^{\infty} \cdots \int_{-\infty}^{\infty} f(x_1, x_2, \ldots, x_N)p_{\mathbf{x}}(x_1, x_2, \ldots, x_N) \, dx_2 \ldots dx_N$$

(B.3.1)

which we see is an N-fold integral over N-dimensional Euclidean space X.

A sequence $\mathbf{x}$ is said to be an *uncorrelated* sequence if every pair of components are uncorrelated. That is, (x_i, x_j) are uncorrelated for every $i \neq j$. The sequence is *independent* if components are pairwise independent.

The N-dimensional joint characteristic function is the N-dimensional extension of (B.2.5), and therefore becomes

$$\Psi_{\mathbf{x}}(\omega_1, \omega_2, \ldots, \omega_N) = \mathscr{E}_{\mathbf{x}}[\exp (j\omega_1 x_1 + j\omega_2 x_2 + \cdots + j\omega_N x_N)] \quad \text{(B.3.2)}$$

A Gaussian sequence has the joint density given by

$$p_{\mathbf{x}}(x_1, x_2, \ldots, x_N) = \frac{1}{(2\pi)^{N/2}\sqrt{|M|}} \exp\left[-\tfrac{1}{2}\chi^{\mathrm{Tr}}M^{-1}\chi\right] \qquad (B.3.3)$$

where

$$\chi = \begin{bmatrix} x_1 - m_{x_1} \\ x_2 - m_{x_2} \\ \vdots \\ x_N - m_{x_N} \end{bmatrix}_{N \times 1}, \qquad M = \begin{bmatrix} \mu_{11} & \mu_{12} & \cdots \\ \mu_{21} & \mu_{22} & \\ \vdots & & \ddots \\ & & & \mu_{NN} \end{bmatrix} \qquad (B.3.4)$$

and m_{x_i} is the mean of x_i, $\mu_{ij} = \mathscr{E}_{\mathbf{x}}[(x_i - m_{x_i})(x_j - m_{x_j})]$, and $|M|$ is the determinant of M. The matrix M is called the *covariance matrix* of the sequence. We see that if the sequence is uncorrelated, $\mu_{ij} = 0$, $i \neq j$, then the quadratic form in the exponent of (B.3.3) expands such that $p_{\mathbf{x}}(x_1 \cdots x_N)$ is a product of terms in each x_i. Hence uncorrelated Gaussian sequences are also independent sequences.

We often must sum the components of a random sequence. Define $z = x_1 + x_2 + \cdots + x_N$ as the sum variable. If the sequence is independent, the characteristic function of z is then

$$\Psi_z(\omega) = \prod_{i=1}^{N} \Psi_{x_i}(\omega) \qquad (B.3.5)$$

which is the product of the individual characteristic functions. The corresponding probability density becomes

$$p_z(z) = p_{x_1}(x_1) \otimes p_{x_2}(x_2) \otimes \cdots \otimes p_{x_N}(x_N) \qquad (B.3.6)$$

where $\otimes$ denotes convolution. The sum density is therefore obtained by an $(N-1)$-fold convolution of the individual densities. For a Gaussian sequence, z will always be a Gaussian random variable, whose mean is the sum of the means of each component. If the Gaussian sequence is independent, than z will have in addition a variance equal to the sum of the component variances. If the sequence is not Gaussian, then it has been shown that under relatively weak conditions, the density of z will converge to a Gaussian density as $N \to \infty$ (central limit theorem).

B.4. Random Processes

A random process $x(t)$ defines a random variable over a continuum of scalar points t. That is, at each scalar point t, $x(t)$ is a random variable.

Random processes are also called *stochastic processes*, and if t denotes time they are also referred to as random, or *noise, waveforms*.

If $x(t)$ is a random process, then a sequence of scalar ponts $(t_1, t_2, \ldots, t_N)$ defines a sequence of random variables. This sequence is then described by the Nth order joint probability density of the random variables involved, which in general depends on the location of the points $\{t_i\}$. A random process is said to be *completely described statistically* if all of its Nth order densities for all $N \to \infty$ are known.

A single point t, defines a single random variable, which we denote x_1. The probability density of that variable x_1, $p_{x_1}(\xi)$ is called a *first order density* of the process. From such a density, probabilities, averages, moments, and so on, can be computed for the process at that t_1. If the $p_{\bar{x}_1}(\xi)$ does not depend on the location of t_1, the process is said to be *first order stationary* and all the resulting averages and probabilities do not depend on t_1.

Two points, t_1 and t_2, define a pair of random variables, $x_1 \triangleq x(t_1)$ and $x_2 \triangleq x(t_2)$. Such a pair is statistically described by its joint second order density $p_{x_1 x_2}(\xi_1, \xi_2)$, which generally depends on the location of t_1 and t_2. From this joint density, joint probabilities, averages, and moments can be determined. An important joint moment of the process $x(t)$ is its *autocorrelation* function

$$R_x(t_1, t_2) = \mathscr{E}[x_1 x_2] \tag{B.4.1}$$

The average on the right is the joint correlation moment m_{11} in (B.2.4) for the random variable x_1 at t_1 and x_2 at t_2. Specifically,

$$\mathscr{E}[x_1 x_2] = \int_{-\infty}^{\infty} \int_{-\infty}^{\infty} \xi_1 \xi_2 p_{x_1 x_2}(\xi_1, \xi_2)\, d\xi_1\, d\xi_2 \tag{B.4.2}$$

Since this second order density will, in general, depend on both the points t_1 and t_2, the expectation in (B.4.2) will produce a function of both t_1 and t_2; that is, it will depend on the specific two points involved. If $p_{x_1 x_2}(\xi_1, \xi_2)$ does not depend on t_1 and t_2 but only on the separation $\tau = t_2 - t_1$ between the points, the process is said to be *second order stationary*. If the autocorrelation function in (B.4.1) depends only on τ, the process is said to be *wide sense stationary* (WSS). Note that a second order stationary process is always WSS, but the converse is not necessarily true. A WSS process has the basic property that correlation values associated with points of the process depend only on their separation and not on where the points are located. That is, all pairs of points separated by τ have the same correlation value.

For WSS processes the function in (B.4.1) can be written simply as $R_x(\tau)$. It is obvious that $R_x(\tau)$ is always a real function in τ, since it involves averages of real random variables. In addition, relabeling the points t_1 and t_2 as t_2 and t_1, respectively, does not alter their correlation value in (B.4.2). Hence it follows that

$$R_x(-\tau) = R_x(\tau) \tag{B.4.3}$$

and the autocorrelation function must be even in τ. Furthermore, since it is necessary that

$$\mathscr{E}[(x_1 \pm x_2)^2] \geq 0 \tag{B.4.4}$$

for any t_1 and t_2, we can easily expand out and show

$$\mathscr{E}[x_1^2] + \mathscr{E}[x_2^2] \pm 2\mathscr{E}[x_1 x_2] \geq 0 \tag{B.4.5}$$

Noting that $x_1 = x(t_1)$ and $x_2 = x(t_2)$, and letting $t_2 = t_1 + \tau$, allows us to write (B.4.5) as

$$\mathscr{E}[x^2(t_1)] + \mathscr{E}[x^2(t + \tau)] + 2\mathscr{E}[x(t_1)x(t_1 + \tau)] \geq 0 \tag{B.4.6}$$

For WSS processes this becomes

$$R_x(0) + R_x(0) \pm 2R_x(\tau) \geq 0 \tag{B.4.7}$$

for all τ, or

$$|R_x(\tau)| \leq R_x(0) \tag{B.4.8}$$

Thus the autocorrelation function of a WSS stationary process must take on its maximum magnitude value at $\tau = 0$. This means its slope must be zero at $\tau = 0$.

The *power spectral density* $S_x(\omega)$ of a random process $x(t)$ is formally defined as

$$S_x(\omega) = \lim_{T \to \infty} \frac{1}{2T} \mathscr{E}[|X_T(\omega)|^2] \tag{B.4.9}$$

where $X_T(\omega)$ is the Fourier transform of the random process $x(t)$ restricted to the range $(-T, T)$ and zero elsewhere. Since $X_T(\omega)$ defines an integral of a random process, it is itself a random variable at each ω. The expectation $\mathscr{E}$ in (B.4.9) averages over the probability density of this random variable. The function obtained at each ω in the limit at T is increased and the averaging is repeated is the process spectral density. When the process is wide sense stationary, we can expand as

$$S_x(\omega) = \lim_{T \to \infty} \frac{1}{2T} \mathscr{E}\left[\int_{-T}^{T} x(t) e^{-j\omega t} dt \int_{-\infty}^{\infty} x(\rho) e^{j\omega\rho} d\rho\right] \tag{B.4.10}$$

Writing the product of integrals as a multiple integral, and inverting the integration and averaging operations, yields

$$S_x(\omega) = \lim_{T \to \infty} \frac{1}{2T} \int_{-T}^{T} \int_{-T}^{T} \mathscr{E}[x(t)x(\rho)] e^{-j\omega(t-\rho)} dt\, d\rho \tag{B.4.11}$$

The average in the integrand is recognized as the process autocorrelation function and, because of the stationarity assumption, can be written as

$$\mathscr{E}[x(t)x(\rho)] = R_x(t-\rho) \tag{B.4.12}$$

With (B.4.12) inserted, (B.4.11) can be rewritten as (see Papoulis [2, p. 325])

$$S_x(\omega) = \underset{T\to\infty}{\text{Lim}} \int_{-T}^{T} \left(1 - \frac{|u|}{T}\right) R_x(u) e^{-j\omega u} \, du \tag{B.4.13}$$

In the limit this becomes

$$S_x(\omega) = \int_{-\infty}^{\infty} R_x(u) e^{-j\omega u} \, du \tag{B.4.14}$$

Thus the spectral density of a wide sense stationary process is simply the Fourier transform of its autocorrelation function. This means the process autocorrelation function and spectral density form a transform pair.

Higher order levels of stationarity can be similarly defined. A process is *Nth order stationary* if the joint probability density between the N random variables $x_1, \ldots, x_N$ at times $t_1, \ldots, t_N$ depends only on the time separations $t_2 - t_1$, $t_3 - t_2$, $t_4 - t_3 \ldots$ of the points, no matter where they are chosen. A process is *strictly stationary* if it is stationary in all orders.

B.5. Gaussian Random Processes

A *Gaussian random process* is a process in which the sequence of random variables defined at any N points $\{t_i\}$ is a Gaussian sequence. That is, at any point t, $x(t)$ is a Gaussian random variable. Hence the general Nth order density of a Gaussian process is given by the general Nth order Gaussian density in (B.3.3). Such a density depends only on the mean and mean squared values at each t and on the correlation $\mathscr{E}[x(t_i)x(t_j)]$ between two points. For wide sense stationary Gaussian processes, this correlation is given precisely by $R_x(t_j - t_i)$ for any pair of points t_i and t_j. That is, the matrix M in (B.3.4) will have the general (i, j) entry

$$u_{ij} = R_x(t_j - t_i) - m_i m_j \tag{B.5.1}$$

where m_i is the mean of $x(t_i)$. Since the matrix M is all that is needed to specify a general Nth order Gaussian density for any N, the autocorrelation function $R_x(\tau)$, evaluated at the proper point $\tau = t_j - t_i$, allows us to describe uniquely all joint probability densities of the process. Hence a wide sense stationary Gaussian process is completely described statistically by its mean function and autocorrelation. By the uniqueness of the Fourier transform, it is therefore equivalently described by its power spectral density.

B.6. Representing Gaussian Processes

In analyzing systems that contain Gaussian random processes, it is often necessary to have a convenient mathematical representation of the process. The representation must be statistically equivalent (have the same N order densities for all N) to the original process, so that all subsequent operations and transformations are identical to dealing with the process itself. From our discussion in the previous section, a WSS Gaussian process has the advantage that any representation is statistically equivalent if it has the same autocorrelation or power spectral density of the process itself. Hence we need only guarantee equivalence in this respect.

Consider a zero mean Gaussian process having the arbitrary two-sided power spectral density $S_n(\omega)$ shown in Figure B.3a. We let ω_c be an arbitrary frequency along the frequency axis. We can always write this

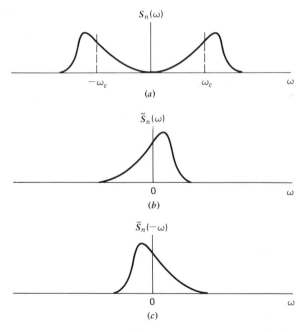

Figure B.3. Noise spectral densities. (a) Two-sided spectrum, (b) shifted version, (c) image of (b).

two-sided density as

$$S_n(\omega) = \tfrac{1}{2}\tilde{S}_n(\omega - \omega_c) + \tfrac{1}{2}\tilde{S}_n(-\omega - \omega_c) \tag{B.6.1}$$

where $\tilde{S}_n(\omega)$ is the shifted version of $S_n(\omega)$, obtained by shifting the one-sided spectrum (the latter obtained by folding the negative frequency part on to the positive frequency part) by an amount ω_c, as shown in Figure B.3b. Equation (B.6.1) is that necessary to regenerate $S_n(\omega)$ from $\tilde{S}_n(\omega)$. When written as in (B.6.1), the corresponding autocorrelation function of the process is then

$$
\begin{aligned}
R_n(\tau) &= \frac{1}{2\pi} \int_{-\infty}^{\infty} S_n(\omega) e^{j\omega\tau} d\omega \\
&= \frac{1}{2\pi} \int_{-\infty}^{\infty} \tfrac{1}{2}\tilde{S}_n(\omega - \omega_c) e^{j\omega\tau} d\omega + \frac{1}{2\pi} \int_{-\infty}^{\infty} \tfrac{1}{2}\tilde{S}_n(-\omega - \omega_c) e^{j\omega\tau} d\omega \\
&= \tfrac{1}{2}[\tilde{R}_n(\tau) e^{j\omega_c\tau} + \tilde{R}_n^*(\tau) e^{-j\omega_c\tau}] \\
&= \text{Real}\{\tilde{R}_n(\tau) e^{j\omega_c\tau}\}
\end{aligned}
\tag{B.6.2}
$$

where * denotes complex conjugate and

$$\tilde{R}_n(\tau) = \frac{1}{2\pi} \int_{-\infty}^{\infty} \tilde{S}_n(\omega) e^{j\omega\tau} d\omega \tag{B.6.3}$$

Here $\tilde{R}_n(\tau)$ is the transform of the shifted spectrum $\tilde{S}_n(\omega)$. We can further expand

$$\tilde{R}_n(\tau) = \tilde{R}_c(\tau) + j\tilde{R}_s(\tau) \tag{B.6.4}$$

where

$$\tilde{R}_c(\tau) = \frac{1}{2\pi} \int_{-\infty}^{\infty} \tilde{S}_n(\omega) \cos \omega\tau \, d\omega \tag{B.6.5a}$$

$$\tilde{R}_s(\tau) = \frac{1}{2\pi} \int_{-\infty}^{\infty} \tilde{S}_n(\omega) \sin \omega\tau \, d\omega \tag{B.6.5b}$$

When (B.6.5) is substituted into (B.6.2), we have

$$R_n(\tau) = \tilde{R}_c(\tau) \cos \omega_c\tau - \tilde{R}_s(\tau) \sin \omega_c\tau \tag{B.6.6}$$

The preceding represents a general expression for the autocorrelation function of a random process in terms of an arbitrary frequency ω_c. We would now like to find a representation of a random process that has the same autocorrelation as (B.6.6).

Let us consider writing the random process as

$$n(t) = n_c(t) \cos(\omega_c t + \psi) - n_s(t) \sin(\omega_c t + \psi) \tag{B.6.7}$$

where ψ is an arbitrary phase angle and $n_c(t)$ and $n_s(t)$ are each random processes. We first observe that if $n(t)$ on the left is to be a Gaussian process (a Gaussian variable at any t), the processes $n_c(t)$ and $n_s(t)$ must also be Gaussian processes (sums of Gaussian variables from Gaussian variables). In addition, let $n_c(t)$ and $n_s(t)$ have autocorrelation $R_c(\tau)$ and $R_s(\tau)$, respectively, and cross-correlation function $R_{cs}(\tau)$. The autocorrelation of $n(t)$ in (B.6.7) is then

$$
\begin{aligned}
R_n(\tau) &= \mathscr{E}[n(t)n(t+\tau)] \\
&= \mathscr{E}\big[[n_c(t)\cos(\omega_c t+\psi)-n_s(t)\sin(\omega_c t+\psi)] \\
&\quad \times[n_c(t+\tau)\cos(\omega_c(t+\tau)+\psi)-n_s(t+\tau)\sin(\omega_c(t+\tau)+\psi)] \\
&= R_c(\tau)\cos(\omega_c t+\psi)\cos(\omega_c(t+\tau)+\psi) \\
&\quad + R_s(\tau)\sin(\omega_c t+\psi)\sin(\omega_c(t+\tau)+\psi) \\
&\quad - R_{sc}(\tau)\sin(\omega_c t+\psi)\cos(\omega_c(t+\tau)+\psi) \\
&\quad - R_{cs}(\tau)\cos(\omega_c t+\psi)\sin(\omega_c(t+\tau)+\psi)
\end{aligned}
\tag{B.6.8}
$$

We now see that if

$$
\begin{aligned}
R_c(\tau) &= R_s(\tau) = \tilde{R}_c(\tau) \\
R_{cs}(\tau) &= \tilde{R}_s(\tau) = -R_{sc}(\tau)
\end{aligned}
\tag{B.6.9}
$$

then the autocorrelation in (B.6.9) of $n(t)$ in (B.6.7) is the same as that in (B.6.6) for any ψ. That is, $n(t)$ has the power spectral density $S_n(\omega)$. Note that (B.6.9) states that this occurs only if $n_c(t)$ and $n_s(t)$ have the proper autocorrelation ($\tilde{R}_c(\tau)$) and cross-correlation ($\tilde{R}_s(\tau)$). Hence we conclude that if $n_c(t)$ and $n_s(t)$ are Gaussian processes with correlation functions

$$
R_c(\tau) = R_s(\tau) = \frac{1}{2\pi}\int_{-\infty}^{\infty} \tilde{S}_n(\omega)\cos\omega\tau\,d\omega
\tag{B.6.10a}
$$

$$
R_{cs}(\tau) = \frac{1}{2\pi}\int_{-\infty}^{\infty} \tilde{S}_n(\omega)\sin\omega\tau\,d\omega
\tag{B.6.10b}
$$

then $n(t)$ in (B.6.7) is also Gaussian with the spectrum $S_n(\omega)$. Thus $n(t)$ can be made to represent any Gaussian process. If the processes involved are not Gaussian, we can only conclude that $n(t)$ will have the same autocorrelation function but will not be statistically equivalent.

The representation in (B.6.7) is called a *quadrature expansion* of the noise, $n_c(t)$ and $n_s(t)$ are called the *quadrature component* processes. Since (B.6.10a) is the component autocorrelation function, their spectral density follows as

$$S_c(\omega) = \int_{-\infty}^{\infty} \frac{1}{2\pi} \int_{-\infty}^{\infty} \tilde{S}_n(u) \cos(u\tau) \, du \, e^{-j\omega\tau} \, d\tau$$

$$= \frac{1}{2} \int_{-\infty}^{\infty} \tilde{S}_n(u)\delta(u-\omega) \, du + \frac{1}{2} \int_{-\infty}^{\infty} \tilde{S}_n(u)\delta(u+\omega) \, du$$

$$= \frac{\tilde{S}_n(\omega) + \tilde{S}_n(-\omega)}{2} \tag{B.6.11}$$

That is, $\tilde{S}_n(\omega)$ must be summed with its mirror image to obtain the component spectrum (Figure B.3c).

We note the quadrature components have the following resulting properties:

1. Since $R_{cs}(0) = 0$, $n_c(t)$ and $n_s(t)$ are always statistically uncorrelated (independent if Gaussian) at the same t.

2. If $\tilde{S}_n(\omega)$ is even in ω ($S_n(\omega)$ is symmetric about $\pm\omega_c$) then $R_{cs}(\tau) = 0$ for all τ. In this case the components become uncorrelated (independent if Gaussian) processes. This means that if the frequency ω_c (which has been arbitrary to this point) can be selected so that $S_n(\omega)$ is symmetric about it, the quadrature expansion in (B.6.7) will have statistically independent components. This requires $S_n(\omega)$ to be a bandpass spectrum and ω_c must be its center frequency. Furthermore, under these conditions, $S_c(\omega)$ in (B.6.11) reduces to $\tilde{S}_n(\omega)$, and the shifted (to zero) bandpass spectrum becomes the component spectrum. We often refer to the components here as being the "*low frequency equivalent*" of the original process. Since bandpass noise is common in communication systems, this quadrature expansion is extremely convenient for manipulation and interpretation, and is used extensively in Chapters 3 and 4 of the text.

3. Since the integral of the spectral density is the power, we see that

$$P_{n_c} = P_{n_s} = \frac{1}{2\pi} \int_{-\infty}^{\infty} \frac{\tilde{S}_n(\omega) + \tilde{S}_n(-\omega)}{2} \, d\omega$$

$$= \frac{1}{2\pi} \int_{-\infty}^{\infty} S_n(\omega) \, d\omega$$

$$= P_n \tag{B.6.12}$$

Hence $n_c(t)$ and $n_s(t)$ have the same power content as $n(t)$ itself. We emphasize that each does not have half the total power, as one may at first conclude.

4. If $S_n(\omega)$ is *narrowband* and symmetric about ω_c (i.e., bandwidth about ω_c is much smaller than ω_c), then the frequencies of the components will not overlap those of $n(t)$. The components $n_c(t)$ and $n_s(t)$ are then said to be "slowly varying" functions, and $n(t)$ can be interpreted as a "modulation" of the quadrature components onto the carrier at ω_c. This narrowband condition is typically indicative of communication receiver noise models.

It is often convenient to write the Gaussian noise process in (B.6.7) in terms of its complex expansion

$$n(t) = \text{Real} \{\alpha(t)\, e^{jv(t)}\, e^{j(\omega_c t + \psi)}\}$$
$$= \alpha(t) \cos[\omega_c t + \psi + v(t)] \tag{B.6.13}$$

where

$$\alpha(t) = [n_c^2(t) + n_s^2(t)]^{1/2}$$
$$v(t) = \tan^{-1}\left(\frac{n_s(t)}{n_c(t)}\right) \tag{B.6.14}$$

The random process $\alpha(t)$ is the *envelope* of the bandpass noise, whereas $v(t)$ is the *phase* process. The envelope process $\alpha(t)$ is a nonlinear function of the random quadrature components $n_c(t)$ and $n_s(t)$, and therefore is itself not a Gaussian process. Its probability density, at any t, is obtained by converting the Gaussian variables $n_c(t)$ and $n_s(t)$ through the transformation in (B.6.14). This envelope has been shown [2] to have the *Rayleigh* probability density at any t, given by

$$p_{\alpha_t}(\xi) = \frac{\xi}{P_n}\, e^{-\xi^2/2P_n}, \qquad \xi \geq 0 \tag{B.6.15}$$

We mention that if a sine wave $A \cos(\omega_c t + \psi)$ is added to $n(t)$, then the sum of the two waveforms has an envelope process

$$\alpha(t) = [(A + n_c(t))^2 + n_s^2(t)] \tag{B.6.16}$$

This envelope process, at any t, now has the probability density

$$p_{\alpha_t}(\xi) = \frac{\xi}{P_n}\, e^{-(\xi^2 + A^2)/2P_n}\, I_0(A\xi/P_n), \qquad \xi \geq 0 \tag{B.6.17}$$

The preceding is referred to as a *Rician* density.

References

1. Abramowitz, A. and Stegun, I. *Handbook of Mathematical Functions*, National Bureau of Standards, Washington, D.C., 1965, Chap. 26.
2. Papoulis, A. *Probability, Random Variables, and Random Processes*, McGraw-Hill, New York, 1965.

APPENDIX C

OSCILLATOR INSTABILITY AND PHASE NOISE

An ideal oscillator produces a pure sine wave carrier having a fixed frequency, constant phase, and constant amplitude. Oscillators used in practice, however, are nonideal and contain frequency instabilities and inherent phase and amplitude noise. These effects cause the oscillator parameters to vary from their design values, and therefore must be accounted for in system analysis. For example, variations in oscillator frequency implies a variation in the spacings between zero crossings of the oscillator output, which therefore disturb any system relying on its output for timing or phasing. Phase variations will appear as extraneous phase noise in any angle modulated communication system involving the oscillator. In this appendix we attempt to review the basic definitions, models, and measurement procedures for characterizing oscillator instability.

C.1. Oscillator Description

In describing nonideal oscillators, two basic parameters are generally of interest pertaining to its frequency output: (1) the instantaneous frequency of the oscillator and (2) the change in frequency over a specified period of time. The instantaneous frequency indicates exactly what frequency is being produced by the oscillator at any time, and therefore accounts for the instantaneous frequency difference from the true (design) frequency of the oscillator. The variation in oscillator frequency over a time period indicates how frequency stable the oscillator is (i.e., how the frequency at one time is related to the frequency at a later time).

The relative importance of each of these effects depends on the eventual use of the oscillator in an operating system. In a communication system utilizing accurate phase coherence, the instantaneous oscillator frequency is of primary importance. This is due to the fact that changes from the true design frequency appear as an effective phase noise in systems attempting to maintain phase coherence. In ranging and timing systems, frequency coherence at two different instants of time is of principal importance. Because of these diverse applications, oscillator analysis and specifications may differ in various types of systems.

The instantaneous frequency of an oscillator is usually characterized as having three basic components—a constant value, corresponding to the design frequency of the oscillator; a slowly varying drift component; and a randomly varying portion, as shown in Figure C.1. The oscillator drift term is due to device aging and is generally described by a polynomial variation with known coefficients, with the linear component dominating. The random variation is due to internal oscillator noise and appears as a frequency "jitter" on the desired oscillator frequency. Thus the instantaneous oscillator frequency can be written as

$$\omega(t) = \omega_c + \alpha t + \dot{\psi}(t) \tag{C.1.1}$$

where ω_c is the design frequency, α is the linear drift component in rps/sec, and $\dot{\psi}(t)$ is the random frequency jitter process. In general, the coefficient α is small, on the order of rps/hour, and therefore the drift

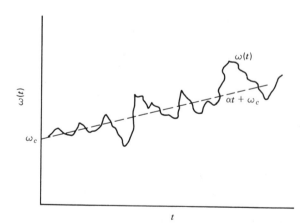

Figure C.1. Oscillator frequency variation.

effect is observable only over long periods of time. For this reason, frequency drift is often related to the "long term" oscillator behavior. On the other hand, the drift effect is negligible over short observation times, and the frequency noise term $\dot{\psi}(t)$ is of prime importance. For this reason, the latter is associated with the "short term" oscillator behavior.

The integral of $\omega(t)$ is called the *instantaneous phase* of the oscillator, $\theta(t)$. Hence

$$\theta(t) = \int_{-\infty}^{t} \omega(\rho)\, d\rho$$

$$= \omega_c t + \tfrac{1}{2}\alpha t^2 + \psi(t) \tag{C.1.2}$$

The function $\psi(t)$, the integral of the frequency noise $\dot{\psi}(t)$, is called the *phase noise* process, denoted by

$$\psi(t) = \int_{-\infty}^{t} \dot{\psi}(\rho)\, d\rho \tag{C.1.3}$$

Note that the phase noise, being an integral of a noise process, is inherently a nonstationary process. This becomes a basic point in oscillator analysis, and care must be used in defining meaningful measurements on oscillator phase processes.

The *mean frequency* over an interval $(t, t+T)$ of an oscillator frequency output $\omega(t)$ is defined as

$$\bar{\omega}(T) = \frac{1}{T} \int_{t}^{t+T} \omega(\rho)\, d\rho$$

$$= \omega_c + \alpha(t + \tfrac{1}{2}T) + m(T) \tag{C.1.4}$$

where

$$m(T) = \frac{1}{T} \int_{t}^{t+T} \dot{\psi}(\rho)\, d\rho \tag{C.1.5}$$

The mean frequency of an oscillator is therefore a measurement of the average frequency that would be observed over the time period $(t, t+T)$. If T is the reciprocal of the approximate bandwidth of the observing system, the $\bar{\omega}(T)$ represents the instantaneous frequency observed by the system with bandwidth $1/T$. For an ideal oscillator $\bar{\omega}(T) = \omega_c$, and the observed mean frequency is always the true frequency for any T. Note that the linear drift term adds a nonstationary term to $\bar{\omega}(T)$, which also appears as a linear drift of the mean frequency at the same rate. Thus frequency drift rates can always be determined from mean frequency measurements for any interval T. The term $m(T)$ in (C.1.4) adds a random component to the mean frequency. In general, $m(T)$ depends on t, but is usually modeled as a zero mean, stationary process.

The oscillator mean frequency can also be written directly in terms of the phase process of the oscillator. In terms of the oscillator phase $\theta(t)$ in (C.1.2), the mean frequency $\bar{\omega}(T)$ becomes

$$\bar{\omega}(T) = \frac{\theta(t+T) - \theta(t)}{T} \qquad (C.1.6)$$

The mean frequency of an oscillator can therefore be determined by computing the phase difference at the beginning and end of the observation interval. This phase difference is often referred to as the *accumulated phase* over T sec. Hence mean frequency can be equivalently obtained by measuring accumulated phase and dividing by the observation interval T. Using (C.1.3), (C.1.5) expands as in (C.1.4), where now

$$m(T) = \frac{\psi(t+T) - \psi(t)}{T} \qquad (C.1.7)$$

Thus the random component to the mean frequency can be determined from the accumulated phase of the phase noise process. Note that (C.1.6) gives an alternative, and perhaps more meaningful, definition to the notion of mean frequency $\bar{\omega}(T)$. If this mean frequency is normalized by the design frequency ω_c, then

$$\frac{\bar{\omega}(T)}{\omega_c} = \frac{\theta(t+T) - \theta(t)}{\omega_c T} \qquad (C.1.8)$$

The denominator is recognized as the total phase that would be accumulated in T sec by an *ideal* oscillator. Hence the normalized mean frequency of an oscillator is actually the ratio of the accumulated phase of that oscillator over T sec, divided by the phase that would be accumulated by an ideal oscillator. This interpretation aids in deriving an equivalence between mean frequency variation and the concept of phase stability for an oscillator.

C.2. Mean Frequency Jitter (Phase Stability)

For a typical oscillator the mean frequency differs from the true frequency by the frequency drifts and the frequency noise, as indicated in (C.1.4). Therefore the mean frequency is itself a process that evolves randomly in time. Its average value at any time is simply

$$\mathcal{E}[\bar{\omega}(T)] = \omega_c + \alpha(t + \tfrac{1}{2}T) \qquad (C.2.1)$$

which is simply the deterministic, drifted frequency. Variations of $\bar{\omega}(T)$ from this average frequency therefore represent random mean frequency changes, or *jitter*, on the oscillator output, and are therefore accounted for by $m(T)$ in (C.1.5). Thus mean frequency jitter is dictated by the integral of the frequency noise process $\dot{\psi}(t)$. The mean square frequency jitter, relative to the deterministic average value, then follows as

$$\sigma^2(T) = \mathscr{E}[m^2(T)] \qquad (C.2.2)$$

Here $\sigma(T)$ defines the rms mean frequency jitter observed over T sec, and indicates the expected variation of the oscillator mean frequency from its drifted value. Obviously, the value of $\sigma(T)$ will depend on the model for the statistics of the frequency noise process $\dot{\psi}(t)$. It is important to note that $\sigma(T)$ depends on T, and therefore will be a function of the averaging interval used to specify the mean frequency. We can also write the rms jitter in terms of the accumulated phase noise in (C.1.7). We denote

$$D_\psi(T) = \mathscr{E}[(\psi(t+T) - \psi(t))^2] \qquad (C.2.3)$$

as the *mean squared accumulated phase noise* over T sec. $D_\psi(T)$ is also called the *structure function* of the phase process $\psi(t)$ [1]. If we substitute in (C.2.2), then

$$\sigma(T) = \frac{(D_\psi(T))^{1/2}}{T} \qquad (C.2.4)$$

Thus rms mean frequency jitter is directly related to the accumulated phase noise. For this reason, rms mean frequency jitter is said to be an indication of the *phase stability* of the oscillator [2]. It is therefore common to speak of phase noise and phase stability, rather than frequency jitter, when describing short term conditions. However, it is important to recognize that it is accumulated phase noise that is basically of importance, and not the statistics of the phase noise itself.

The *phase stability function* of an oscillator is formally defined as [3]

$$\frac{\sigma(T)}{\omega_c} = \frac{(D_\psi(T)^{1/2}}{\omega_c T} \qquad (C.2.5)$$

and is therefore simply the ratio of the rms mean frequency jitter to the oscillator design frequency ω_c. As in (C.1.8), we can again interpret (C.2.5) as the ratio of the rms accumulated phase to the accumulated phase of an ideal oscillator. Specification of (C.2.5), at a particular value of T, therefore becomes a common way to characterize oscillator stability.

In summary, then, several basic measurement procedures are available for assessing the degree of oscillator mean frequency jitter, simply by referring to the preceding mathematical definitions. One can simply observe the oscillator frequency process through a system of appropriate bandwidth and carefully compute the variation from the design value.

Alternatively, one can compute the rms accumulated phase over T sec intervals and normalize by T.

C.3. Frequency Stability

In addition to the mean frequency jitter (phase stability), an oscillator is also described by its inherent *frequency stability*—the amount by which the frequency changes over a particular time interval. The difference between two mean frequency values, each measured over T sec intervals and separated by a τ sec time period, is formally given by

$$\Delta_\omega(\tau) = \frac{1}{T} \int_{t+\tau}^{t+\tau+T} \omega(\rho) \, d\rho - \frac{1}{T} \int_t^{t+T} \omega(\rho) \, d\rho \qquad (C.3.1)$$

where $\omega(t)$ is again the oscilllator frequency process. For an ideal oscillator, $\Delta_\omega(\tau)$ is identically zero, and the ideal oscillator is said to be completely frequency stable. For the nonideal oscillator $\Delta_\omega(\tau)$ is a measure of the mean frequency variation that occurs over τ sec. For the oscillator model of (C.1.1)

$$\Delta_\omega(\tau) = \alpha\tau + \Delta\dot{\psi}(\tau) \qquad (C.3.2)$$

where

$$\Delta\dot{\psi}(\tau) = \frac{1}{T} \int_{t+\tau}^{t+\tau+T} \dot{\psi}(\rho) \, d\rho - \frac{1}{T} \int_t^{t+T} \dot{\psi}(\rho) \, d\rho \qquad (C.3.3)$$

The first term in (C.3.2) is due to frequency drift and contributes a deterministic, constant part to the mean frequency variation. Note that this term is proportional to τ with the same rate coefficient as the frequency drift but does not depend on T. The second term in (C.3.2) adds a random variation to the frequency difference, because of the presence of the frequency noise. When the frequency noise $\dot{\psi}(t)$ is a zero mean process, as is usually assumed, $\Delta\dot{\psi}$ has a zero mean, and the average value of $\Delta_\omega(\tau)$ produces the drift term in (C.3.2). Thus

$$\mathscr{E}[\Delta_\omega(\tau)] = \alpha\tau \qquad (C.3.4)$$

This means oscillator drift rate can be measured by determining the average frequency variation over any interval τ and dividing by τ. The mean squared frequency variation, relative to the deterministic drift terms, that occurs in τ sec is then

$$D_{\dot{\psi}}(\tau) = \mathscr{E}[(\Delta\dot{\psi})^2] \qquad (C.3.5)$$

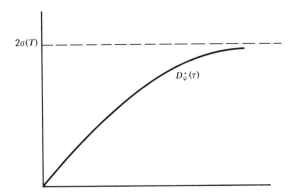

Figure C.2. Frequency noise structure function.

The behavior of $(D_{\dot\psi}(\tau))^{1/2}$, as a function of τ, therefore indicates the rms fluctuations that can be expected at the oscillator output. The function $D_{\dot\psi}(\tau)$ is now the *frequency noise structure function*. If the frequency noise is stationary, then as $T \to 0$,

$$D_{\dot\psi}(\tau) = 2[R_{\dot\psi}(0) - R_{\dot\psi}(\tau)] \qquad (C.3.6)$$

where $R_{\dot\psi}(\tau)$ is the frequency noise correlation function. Hence frequency variation is intimately tied to the correlation process of the frequency noise. Note that for $\tau \to 0$, $D_{\dot\psi}(\tau) \to 0$, whereas for large values of τ, $D_{\dot\psi}(\tau)$ approaches twice the rms mean frequency jitter, $2\sigma(T)$. The function $D_{\dot\psi}(\tau)$ in (C.3.6) is sketched in Figure C.2. The parameter

$$\frac{(D_{\dot\psi}(\tau))^{1/2}}{\omega_c} \qquad (C.3.7)$$

is often called the *frequency stability* function [3] of the oscillator and is often listed as a system specification in oscillator design. That is, the function $(D_{\dot\psi}(\tau))^{1/2}$ is specified not to exceed a fixed value at a specified T and τ. We emphasize it is a requirement on mean frequency variation (frequency differences) rather than on mean frequency jitter. Hence frequency stability and phase stability of an oscillator are separate requirements.

Frequency stability can also be measured directly in terms of the phase noise process $\psi(t)$. In particular, we can write

$$\Delta\dot\psi = \frac{\psi(t+\tau+T) - \psi(t+\tau)}{T} - \frac{\psi(t+T) - \psi(t)}{T} \qquad (C.3.8)$$

corresponding to differences of accumulated phase over T sec time intervals separated by τ sec. Hence second differences of oscillator phase leads to its frequency stability, whereas first differences (accumulated phase) are related to its phase stability.

C.4. Phase and Frequency Noise Spectral Density

It has been shown that oscillator phase noise $\dot{\psi}(t)$ causes mean frequency jitter and frequency variations from the true oscillator frequency. When the frequency noise is considered a stationary process, many of its statistical properties can be derived from its noise power spectrum $S_\psi(\omega)$. We must caution, however, that oscillator phase noise may not necessarily be stationary, since it arises as a result of integrating a frequency noise process. For this reason, it may be more meaningful to deal with the frequency noise spectrum $S_{\dot{\psi}}(\omega)$ associated with the process $\dot{\psi}(t)$. Nevertheless, it is common practice in communication analysis to accept phase processes as being stationary, or at least stationary over the time intervals of interest.

The accumulated phase of the oscillator over a T sec interval, as defined in the numerator of (C.1.6), has a power spectrum related to the phase noise spectrum $S_\psi(\omega)$ by

$$|e^{j\omega T} - 1|^2 S_\psi(\omega) = [(\cos \omega T - 1)^2 + \sin^2 \omega T] S_\psi(\omega)$$

$$= 4 \sin^2 \left(\frac{\omega T}{2}\right) S_\psi(\omega) \tag{C.4.1}$$

The mean squared accumulated phase $D_\psi(T)$ in (C.2.3) is obtained by integrating this spectrum. Hence the mean squared frequency jitter in (C.2.4) is then

$$\sigma^2(T) = \frac{4}{\pi T^2} \int_0^\infty \sin^2 \left(\frac{\omega T}{2}\right) S_\psi(\omega) \, d\omega \tag{C.4.2}$$

Thus the phase stability of an oscillator depends directly on its phase noise spectrum. This can also be written in terms of the frequency noise spectrum $S_{\dot{\psi}}(\omega)$ by interpreting

$$S_\psi(\omega) = \frac{S_{\dot{\psi}}(\omega)}{\omega^2} \tag{C.4.3}$$

Equation (C.4.2) becomes

$$\sigma^2(T) = \frac{1}{\pi} \int_0^\infty \frac{\sin^2 (\omega T/2)}{(\omega T/2)^2} S_{\dot{\psi}}(\omega) \, d\omega \tag{C.4.4}$$

Thus frequency jitter can be determined from either the phase or frequency noise spectra. Note that (C.4.4) is the integrated power under a filtered frequency noise spectrum, where the effective filtering is via the function $[\sin^2 (\omega T/2)]/(\omega T/2)^2$.

In using (C.4.2) or (C.4.4), the existence of the integrals must be of concern. The integral in (C.4.2) exists only if $S_\psi(\omega)$ decreases to zero as $\omega \to \infty$ and increases slower than $1/\omega^3$ as $\omega \to 0$. This follows, since the integrand in (C.4.2) behaves as $\omega^2 S_\psi(\omega)$ for $\omega \to 0$. The condition for $\omega \to 0$ is of particular interest, since it basically limits the rate at which phase noise spectra can increase at the low frequencies, where inverse ω variations are typical.

The frequency structure function in (C.3.5) is also directly related to the phase noise spectrum by

$$D_{\dot\psi}(\tau) = \frac{2}{\pi} \int_{-\infty}^{\infty} [1 - \cos \omega\tau] \sin^2\left(\frac{\omega T}{2}\right) S_\psi(\omega) \, d\omega \qquad \text{(C.4.5)}$$

and to the frequency noise spectrum by

$$D_{\dot\psi}(\tau) = \frac{1}{\pi} \int_{0}^{\infty} [1 - \cos \omega\tau] \left(\frac{\sin^2 (\omega T/2)}{(\omega T/2)^2}\right) S_{\dot\psi}(\omega) \, d\omega \qquad \text{(C.4.6)}$$

The terms involving T accounts for the fact that a mean frequency over T sec is used, whereas τ accounts for the time period between frequency measurements. As $T \to 0$, (C.4.5) approaches the true frequency structure function of the oscillator noise. Note that for $\omega \to 0$ the integrand in (C.4.5) behaves as $\omega^4 S_\psi(\omega)$, and the integrability is more easily assured than in (C.4.2). Thus, in certain cases, the frequency stability function may be computable from the phase spectrum, whereas the phase stability function may not exist.

Since the spectral density of the oscillator phase noise determines its corresponding phase and frequency stability, accurate models of oscillator phase noise spectra are necessary to evaluate oscillator stability effects in a given system. Over the years, certain specific oscillator models have evolved from measurements and theoretical studies as typifying oscillator phase noise. In general, oscillator phase noise spectra tend to increase at lower frequencies and flatten out at intermediate frequencies, before rapidly falling off, as shown in Figure C.3. This behavior is modeled by the one sided spectrum

$$\hat{S}_\psi(\omega) = \frac{S_3}{\omega^2} + \frac{S_2}{\omega^2} + S_0, \qquad \omega \le \omega_u$$

$$= \frac{S_4}{\omega^4}, \qquad \omega > \omega_u \qquad \text{(C.4.7)}$$

where the S_i are a compatible set of coefficients that determine the level of

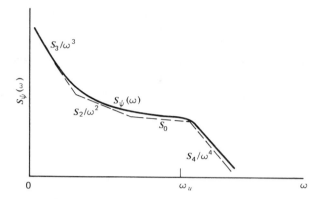

Figure C.3. Typical phase noise spectral density.

the spectrum and ω_u defines the upper edge of the spectrum. The low frequency behavior, given by the first term in (C.4.7), is the result of frequency *flicker noise* [4, 5] associated with inherent amplification within the oscillator. However, its rate of increase tends to vary among oscillators, and this rate has a predominant effect on the oscillator stability parameters, as we have seen. For this reason, accurate modeling of the actual low frequency phase spectrum is mandatory in oscillator studies. This flicker effect is predominant out to about 1–10 Hz, so that S_3 can be interpreted as the value of $S_\psi(\omega)$ at 1 Hz; that is, $S_3 = S_\psi(2\pi)(2\pi)^2$. The flattening of the spectrum due to the second and third term in (C.4.7) exhibits the effect of *white frequency noise* in the oscillator (that converts to a quadratic increase in the equivalent frequency noise spectrum). The entire phase spectrum is then filtered by the resonant filtering of the oscillator tuning cavity, which ultimately causes the spectrum to decrease rapidly at high frequencies. Although most oscillators tend to fit the basic model of Figure C.3, the various levels, slopes, and critical frequencies of each component tend to vary from one oscillator to the next.

References

1. Yaglom, A. *Theory of Stationary Random Functions*, Prentice-Hall, Englewood Cliffs, N.J., 1962.
2. Cutler, L. S. and Searle, C. L. "Some Aspects of the Theory and Measurement of Frequency Fluctuations in Frequency Standards," *Proc. IEEE*, vol. 54, No. 2, February 1966.
3. Lindsey, W. and Chie, C. "Theory of Oscillator Instability Based upon Structure Functions," *Proc. IEEE*, vol. 12, no. 64, p. 197.

4. Halford, Donald. "A General Mechanical Model for $|f|^\alpha$ Spectral Density Random Noise with Special Reference to Flicker Noise $1/|f|$," *Proc. IEEE*, vol. 56, March 1968, pp. 251–258.

5. Atkinson, W. R., Fey, L., and Newman, J. "Spectrum Analysis of Extremely Low Frequency Variations of Quartz Oscillators," *Proc. IEEE* (Corresp.), vol. 51, February 1963, p. 379.

INDEX